T0268957

NANOTECHNOLOGY COMMERCIALIZATION

FOR
MANAGERS
AND
SCIENTISTS

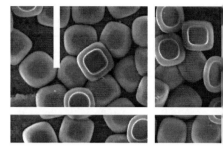

NANOTECHNOLOGY COMMERCIALIZATION FOR
MANAGERS
AND
SCIENTISTS

edited by
WIM HELWEGEN
LUCA ESCOFFIER

PAN STANFORD PUBLISHING

Published by

Pan Stanford Publishing Pte. Ltd.
Penthouse Level, Suntec Tower 3
8 Temasek Boulevard
Singapore 038988

Email: editorial@panstanford.com
Web: www.panstanford.com

British Library Cataloguing-in-Publication Data
A catalogue record for this book is available from the British Library.

Nanotechnology Commercialization for Managers and Scientists

Copyright © 2012 Pan Stanford Publishing Pte. Ltd.

Disclaimer of Liability

Although each chapter has been written with the greatest amount of expertise and care, the publisher, the editors, the authors and, if applicable, their employers cannot be deemed liable for any loss or damage caused or alleged to be caused, directly or indirectly by the information contained in this publication. For advice relating to, e.g., your product or business, always consult a qualified professional in the relevant field. The views of the authors as expressed in this publication do not necessarily represent the views of their employers.

ISBN 978-981-4316-22-4 (Hardcover)
ISBN 978-981-4364-38-6 (eBook)

Printed in the USA

Contents

Contributors

Joanna Brougher is a patent attorney whose practice focuses primarily on the preparation and prosecution of patent and trademark applications in the areas of biomedical devices, life sciences, and nanotechnology. She also assists clients on a variety of matters, including establishing patent prosecution strategy, guiding established companies in product life cycle management, conducting non-infringement analyses, and conducting due diligence for venture capital investment, mergers and acquisitions, and licensing agreements. Joanna is also an adjunct at the Harvard School of Public Health and a contributing editor for the *Biotechnology Healthcare Journal* and the *FDA Update Journal*. She has published numerous articles related to patent law and healthcare. Joanna has an undergraduate degree in microbiology, an MPH degree from the University of Rochester, and a law degree from Boston College Law School. She can be reached at joannabrougher@gmail.com.

Niklas Bruun is a professor of private law at the University of Helsinki and the director of the IPR University Center in Helsinki. He is a regular advisor to the Finnish government and EU institutions on issues of labor law and IP law. Furthermore, Prof. Bruun leads a graduate school in intellectual property law (INNOCENT) and is involved in IP activities at the Hanken School of Economics. Among his research interests are the relationship between academia, industry, and IP rights. He has been vice-chairman and chairman of the Finnish Copyright Council for about 20 years and is chairman of the Finnish Copyright Commission since 2008. He has also been chairman of Finnish Society for Industrial Property Law from 1998 to 2001 and is chairman of the Board of Good Business Practice of the Central Chamber of Commerce in Finland.

Prof. Bruun holds an honorary doctorate from the University of Stockholm.

Rachel M. Buchanan is a graduate student in the Department of Biomedical Engineering at the University of Texas at Austin. She received a BS in biomedical engineering from Rensselear Polytechnic Institute in 2009.

Bärbel Dorbeck-Jung is a professor of Regulation and Technology at the University of Twente, the Netherlands. She holds a master's degree in German law (University of München) and is a member of the ETPN Board, the EU Round Table Nanomedicine, and the Working Group on Nanotechnologies of the Dutch Standardization Institute. Prof. Dorbeck-Jung has worked and published on topics related to governance, legislation and self-regulation, good governance, and the rule of law, computer law, health care law, and technological regulation (IT and nanotechnologies). Currently she conducts empirical and theoretical studies on medical technology and nanotechnology regulation. She extensively lectures at international conferences and seminars on these issues. Prof. Dorbeck-Jung is one of the project leaders of the Dutch NanoNext Theme Risk & Technology Assessment.

Luca Escoffier graduated in law from the University of Parma, Italy, in 2001. He earned a Master of Laws in IP in 2003 (WIPO/University of Turin), interned at WIPO, and worked as an IP counsel for a nanobiotech company in Italy until 2008. He then moved to Seattle to work at the University of Washington as a visiting scholar and then as a visiting lecturer. Luca was one of the four experts selected in 2009 as Fellows at the Institute of Intellectual Property in Tokyo. He was one of the 80 students from Singularity University (in 2010) chosen from a pool of 1600 applicants to spend 10 weeks at the campus of NASA Ames in Mountain View. He is a Fellow of the Stanford-Vienna Transatlantic Technology Forum, and from May 2010 the founder and CEO of Usque Ad Sidera LLC. Luca is about to submit his PhD dissertation about nanotechnology patenting and valuation.

Kaarle Hämeri holds a professorship in Aerosol Physics at the University of Helsinki. He is an expert in studies on aerosols with focus on ultrafine and nanoparticles. He has published about 100 articles in peer-reviewed international journals and more than 250 other papers and reports. His research topics range from synthetic nanoparticles and indoor aerosols to urban air quality, aerosol measurement techniques, and aerosol-climate interaction. Prof. Hämeri has a significant role within the scientific community and holds several confidential posts in international organizations and committees. He is currently president of the International Aerosol Research Assembly and an editor of two international journals, *Atmospheric Research* and *Atmospheric Chemistry and Physics*. Prof. Hämeri has worked as an expert in various institutions and presented numerous papers in scientific conferences.

Michael Heintz specializes in environmental law, energy issues, and emerging technologies. He has frequently presented and published on issues related to nanotechnology regulation and global climate change issues. He received his BS in natural resources and environmental sciences from Purdue University, and his law degree and MS in environmental sciences from Indiana University. He currently works for the Maryland Energy Administration in Annapolis.

Wim Helwegen holds a Master of Laws degree in international and European law from Tilburg University in the Netherlands. He is specialized in the interaction of patent law and advanced technologies, such as nanotechnology and biotechnology. After having worked at a Court of Appeals in the Netherlands, Wim conducted postgraduate research at the Queen Mary Intellectual Property Research Institute at Queen Mary University of London. In 2007, he was appointed as a researcher at the IPR University Center in Helsinki. Currently, Wim is preparing a doctoral dissertation on the patenting of nanotechnology at the University of Helsinki. In addition, he is a lecturer in patent law at Hanken School of Economics.

Efrat Kasznik is a valuation expert with over 15 years of economic consulting experience. She holds an MBA from UC Berkeley and a BA in accounting and economics from the Hebrew University, Jerusalem and is the founder and president of Foresight Valuation Group, a Silicon Valley–based consulting firm providing IP valuation, litigation, and strategy services. Kasznik specializes in performing business valuations and valuations of intellectual property for a range of purposes, including mergers and acquisitions, financial reporting, technology commercialization, transfer pricing, and litigation damages. Prior to founding Foresight, she held a series of partner-level positions with leading litigation and IP consulting organizations. She has also been involved as a CFO, co-founder, and adviser to several Silicon Valley start-ups in the telecommunications, media, and cleantech fields.

Michael B. Landau is professor of law at the Georgia State University College of Law in Atlanta, Georgia. His law degree is from the University of Pennsylvania, where he won the Nathan Burkan Memorial Copyright Award. In 2005–2006 he was a Fulbright Scholar at the IPR University Center at the University of Helsinki. Prior to entering academia, Prof. Landau practiced law with the New York firms of Cravath, Swaine & Moore and Skadden, Arps, Meagher, Slate & Flom, where he represented entertainment, technology, and media clients. He has presented papers or has been an invited guest lecturer at numerous law schools in the United States and Europe, including Georgetown, NYU, Vanderbilt, Emory, Tulane, the London School of Economics, Cambridge, University of Edinburgh, University of Durham, and the Amsterdam Institute for Information Law. Before entering the legal profession, Prof. Landau was a professional musician.

Hanna R. Laurén received a Master of Science degree from the University of Turku in 2002, majoring in chemistry and minoring in biochemistry, physics, and mathematics. After graduation she worked for five years as a researcher at the University of Turku, focusing on the functionalization and solubilization of single-wall carbon nanotubes and their layer-by-layer self-assembly into polyelectrolyte multilayers with conducting polymers. Since 2007, Hanna has been working as a patent agent at the Helsinki-based patent agency Oy Jalo Ant-Wuorinen Ab, where she specializes in chemistry, chemical instrumentation, and nanotechnology.

Claes Post works at the Technology Transfer Office at Linköping University. He received his MPharm from Uppsala University and his PhD (Pharm) from Linköping University. A professor of neuropharmacology at Linköping University, Sweden, he is focusing on developing commercially viable projects from the medical faculty at the university. Dr. Post has had almost a 20-year-long career in the pharmaceutical industry as head of preclinical research at Astra Pain Control in Södertälje, Sweden, as well as at Astra Draco in Lund, Sweden. For 4 years he was also head of preclinical and clinical CNS at Pharmacia in Milan, Italy. During the last more than 10 years, Dr. Post has worked with VC-funded start-up companies in Sweden and Denmark apart from being a partner at VC funds in Sweden and Denmark. He has published more than 120 peer-reviewed scientific papers. Academically, he has been adjunct professor of neuropharmacology at Uppsala University, as well as at Lund University and Karolinska Institute.

Sarah Rouse is a registered patent attorney at Katten Muchin Rosenman LLP focused on identifying, securing, and maximizing the value of clients' intellectual property. Dr. Rouse is co-inventor on patents directed to nanomedicine. Her research led to the formation of Keystone Nano, a company providing platform technologies for nano-enabled therapeutics, and NanoSpecialties LLC, a company creating nano-based products for industrial markets. Dr. Rouse received dual undergraduate degrees from the South Dakota School of Mines and Technology and her PhD in materials science and engineering from the Pennsylvania State University. Her doctoral research focused on the synthesis, dispersion, and characterization of nanocomposite particles for bioimaging, drug delivery, and gene therapy. While at PSU, Dr. Rouse was named a National Science Foundation Fellow. She received her JD and certificate in intellectual property from DePaul University College of Law. She also interned at the World Intellectual Property Organization (WIPO) Coordination Office at the United Nations.

Christine A. Smid is a graduate student in the Department of Biomedical Engineering at the University of Texas at Austin, from where she received a BS in 2008.

Marco Spadaro has a degree in pharmaceutical chemistry and technologies. Marco has been involved in IP practice since 1990, both in private law firms and as head of the Corporate Patent Department of a primary Italian pharma company. He was a founding partner of Studio Associato Leone & Spadaro in 2006, and from 2010 a partner of Cantaluppi & Partners. Marco is an expert in drafting and prosecuting patents in chemistry, particularly pharmaceutical, biotech, nanopharma, food, polymers, and composite materials; patent strategies and patent portfolio management; and opposition and litigation proceedings. In addition, he is a lecturer in IP at the Patent Academy of the European Patent Office and at several universities and industries. He is also a tutor at the Centre d'Etudes Internationales de la Propriété Industrielle Université Robert Schumann, Strasbourg, France, since 1996.

Ennio Tasciotti, PhD, is an associate professor in the Department of Nanomedicine and Biomedical Engineering at the Methodist Hospital Research Institute. He received an MS in biological sciences from the University of Pisa in 2000 and a PhD in molecular medicine from Scuola Normale Superiore in 2005.

Pekka Valkonen is a patent manager at Fortum Corporation, a Scandinavian Euro STOXX company in the utility sector. He is responsible for intellectual property matters of major business units within the company. He deals with patents, trademarks, domain names, and research agreements. When handling the IP matters of spinoff of companies, Valkonen has made himself familiar with IP matters in small technology-based companies. Before the formation of Fortum Corporation, he worked at Neste Corporation, where he was responsible for patent matters in specialty polymers and especially electrical conducting polymers. He began his IP career in the Finnish Patent Office as an examiner, senior examiner, and consulting manager. Valkonen has lectured on utilizing patents in business, valuation of IP, and patent strategies.

Pieter de Witte obtained his PhD in supramolecular chemistry from Radboud University in Nijmegen in 2004, after which he became a postdoctoral researcher at ISIS institute of Strasbourg University, France. From 2004 to 2008 he was program officer for the Dutch national nanotechnology program NanoNed, at Technology Foundation STW, where he coordinated the utilization program and the interactions between industrial users and academic research programs. Since 2008, Dr. de Witte has been working at FOM Foundation and is responsible for collaborations with industry, in particular the coordination of the Industrial Partnership Programme (IPP).

Po Chi Wu is an adjunct professor in the School of Business and Management at the Hong Kong University of Science and Technology and a visiting professor and co-founder of the Global Innovation Research Center in the School of Software and Microelectronics at Peking University in Beijing, China. He is a co-founder and managing director of Dragon Bridge Capital, a merchant banking firm helping Chinese and US technology companies become global citizens. Dr. Wu has been a venture capitalist and entrepreneur for more than 25 years and has invested in early-stage high-tech and life science companies in Silicon Valley and Asia. He has a PhD in biochemistry and molecular biology from Princeton University and a BA in mathematics and music from the University of California at Berkeley.

Foreword

Nanotechnology holds great promise for the future of humankind, and scientists and managers should be aware of this. Public and private sector investments in nanotech research have increased exponentially in the past two decades. We are now facing a future, not too far beyond the present, in which materials and devices with astonishing properties will completely change the rules of the game. Novel products will possess features that were almost unimaginable just a few years ago.

Until recently, universities and research centers around the world had been the prime actors in this developing revolution because nanotechnology research requires the skills of interdisciplinary teams that are most readily found in academia. What we are seeing today is a paradigm shift into the entrepreneurial arena. More and more pure researchers are getting involved in spinoff ventures that spring from the academic setting, and there is a need for interdisciplinary knowledge that combines scientific and managerial skills. At the same time, managers who expect to become involved in near-term nanotechnology enterprises require basic knowledge of the wide range of current applications in this fascinating field.

This book is a valuable attempt to satisfy these objectives. Authors with diverse backgrounds offer insight and useful advice both to scientists who may be seeking to capitalize their nanotech research through the creation of a new venture and to managers who need to know how and why this unique technology domain is regulated. The book focuses strongly on the creation and monetization of the intellectual property related to nanotechnology inventions, starting from the conception of the patentable idea and progressing through the venture capital stage and also nanotechnology regulation. The full pipeline of present-day nanotechnology is examined through the expert eyes of patent attorneys, professors, regulation experts, managers, and scientists, with helpful comparisons of IP issues in the United States and Europe.

I have found this volume to be very useful in my own work. Anyone who is interested in starting a nanotechnology-based venture or who wishes to understand how to manage one should read this book to become more aware of the opportunities and challenges that nanotechnology will bring into our lives.

Robert A. Freitas Jr.

Preface

Nanotechnology will have a large impact on our future, but a lot of research and development (R&D) projects have yet to be conducted. This R&D will require extraordinary efforts from individuals and groups in universities, research institutes, and the industry. Unfortunately, scientific genius does not always equal commercial success. In order to benefit commercially from one's research, or even to prevent others from obstructing research, a myriad of factors need to be taken into account. Many of those, for example, environment, health and safety regulations, academy-industry cooperation, intellectual property, and attracting investments, come into play well before and during the research process. This book intends to provide the reader with the basics of the most relevant factors that need to be taken into account before, during, and after the R&D phase. Although some of the subjects discussed are highly complicated, the authors have written the chapters in a way that makes them understandable for professionals who are not familiar with the topic at hand.

The nanoscale brings many challenges to scientists who deal with it. Some of its unique characteristics also pose challenges in the process of commercialization. This book discusses these nano-specific challenges. While most chapters and parts of chapters are nano-specific, others are of a more general nature, either because a more general discussion is needed in order to understand the nano-specific part or because, despite not being nano-specific, they are essential in the commercialization process.

To provide our readers with the best possible information, we relied upon the expertise of a great and diverse team of authors: Joanna Brougher, Niklas Bruun, Rachel Buchanan, Bärbel Dorbeck-Jung, Michael Heintz, Kaarle Hämeri, Efrat Kasznik, Michael Landau,

Hanna Laurén, Claes Post, Sarah Rouse, Christine Smid, Marco Spadaro, Ennio Tasciotti, Pekka Valkonen, Pieter de Witte, and Po Chi Wu. We wish to extend our deepest gratitude to them for sharing their expertise and for their commitment and diligence during the entire process.

We are also very grateful to Stanford Chong, the publisher of this work, and to his editorial team for having made the realization of a book with so many authors a smooth and enjoyable experience.

Wim Helwegen and Luca Escoffier
Helsinki and Tokyo
September 2011

Chapter 1

Introduction to the World of Nanotechnology

Rachel M. Buchanan, Christine A. Smid, and Ennio Tasciotti

The Methodist Hospital Research Institute, The University of Texas at Austin, Houston, TX, USA
etasciotti@tmhs.org

1.1 History and Definition

In 1959 physicist Richard Feynman introduced the potential of atomic and molecular engineering with his revolutionary speech: "There is plenty of room at the bottom." Not but two decades later, the initiation of various research activities in nanotechnology began. Only years ago, the microscale was accepted as the smallest achievable and most desirable size and the ultimate reachable limit for designing, engineering, and building objects. Today, those limits have been largely surpassed and the new advantages of the synthesis at the nanoscale, the scale of atoms and molecules, have been made available in different areas. Presently, the spectrum of nanotechnology application has expanded to almost all areas of science and technology such as bioengineering, material science, medicine, electronics, and agriculture [1]. The word "nanotechnology"

Nanotechnology Commercialization for Managers and Scientists
Edited by Wim Helwegen and Luca Escoffier
Copyright © 2012 Pan Stanford Publishing Pte. Ltd.
ISBN 978-981-4316-22-4 (Hardcover), 978-981-4364-38-6 (eBook)
www.panstanford.com

was introduced for the first time in the late 1970s. Although today there are almost countless definitions of nanotechnology, the National Nanotechnology Initiative (NNI) defines it as (1) Research and technology development at the atomic, molecular, or macromolecular levels, in the length scale of approximately 1–10-nanometer range, (2) Creating and using structures, devices, and systems that have novel properties and functions because of their small and/or intermediate size, and (3) Ability to control or manipulate on the atomic scale [2].

Throughout the years and across the different laboratories and research institutes though, the definition of nanotechnology broadened up to the ability to detect, measure, maneuver, and manufacture things at the nanometer scale, that is, objects with dimensions or features that are less than 100 nm [3–4]. Thus, nanotechnology is not only the study of very small things, but primarily the research and development of materials, devices, and systems that exhibit unique emerging physical, chemical, and biological properties due to their size.

1.2 Fabrication and Characterization

Nanofabrication is the design and production of devices with dimensions measured in nanometers. A broad spectrum of methods have been developed and described in the literature and are summarized in Table 1.1. They can be generally grouped in two main classes: miniaturization of a process (top-down) and self-assembly (bottom-up).

Table 1.1. Comparison of the "Top-down" and "Bottom-up" approaches to nanofabrication

"Top-down" nanoscale technology	"Bottom-up" molecular nanotechnology
From big to small	From small to big
Pattern and etch	Synthesis and self-assembly
Expensive fabrication	Cheap fabrication
Less scalable material/device	More scalable material/device
Limited material selection	Vast material selection

1.2.1 *Top-Down Nanofabrication: Photolithography and Nanolithography*

The most common form of top-down nanofabrication involves lithographic patterning techniques using short-wavelength light sources [4]. The major advantage to this approach is that the parts are patterned and placed and no assembly is needed. In order to fabricate objects such as small-scale circuits and machines at the nanoscale, the use of a clean room, and of electron microscopes are required. The clean room is a necessary component, since tiny dust particles (generally a few hundreds of nanometers) are capable of interfering with the process and to cause major damage or malfunctioning of a nanoscale device or object.

Photolithography is a process used to build components using light to transfer a pattern of a part (such as from a microchip) from a photo mask to a light sensitive chemical photoresist that forms an image of the pattern on a substrate. This technique is used in various industries including the manufacturing of semiconductors, flat-panel displays, and micromachines. There are various types of lithographic techniques used in nanotechnology. Two main types include dip pin nanolithography (DPN) and thermal dip pin nanolithography (tDPN). DPN uses an AFM scanning probe tip to draw nanostructures. A probe tip is coated in liquid ink, which then flows onto the surface to make patterns. The tip makes contact leading to high-resolution patterning capabilities for a variety of molecular inks on different substrates. tDPN, an extension of DPN, uses easily melted solid inks and special AFM probes that have built-in heaters. The heaters allowed for writing to be turned on and off when desired. This technique is used to create features that are too small to be done by DPN [4].

1.2.2 *Bottom-Up Nanofabrication*

The "bottom-up" approach of nanotechnology focuses on constructing chemically and physically stable nanostructures and macromolecules based on molecular self-assembly. Self-assembly is consistently observed in nature, as seen through the formation of supramolecular structures such as DNA, ribosomes, silk fibrils, collagen, and biological membranes [3]. Although the concept of self-assembly is thoroughly recognized in nature, only recently has it gained interest in the areas of engineering fabrication and production. Molecular self-assembly is defined as the spontaneous organization of molecules under near thermodynamic equilibrium conditions into structurally well-defined and stable arrangements through noncovalent interactions [5–6]. These interactions include hydrogen bonding, electrostatic attraction, hydrophobic forces, and van der Waals interactions. Molecular self-assembly relies not only on chemical complementarity, but also structural compatibility [7]. These characteristics are essential in choosing the molecular building blocks required in the fabrication of macrostructures.

1.2.3 Electron Microscopes

Electron microscopes use electron beams instead of visual light, leading to the resolution of features down to a few nanometers. They use a beam of high energy electrons to probe the sample. Two major types of electron microscopes exist including the transmission electron microscope (TEM) and the scanning electron microscope (SEM). Using electrons that posses a much smaller wavelength than any other light source, permits imaging at magnifications that are thousand times greater than a light microscope and with resolutions of the order of 0.2 nm. This allows to finely analyze materials and to acquire information about their surface topography, geometry, chemical composition, texture, and hardness [8].

1.2.4 Scanning Probe Microscopes

Scanning probe microscopes (SPM) are used to study the surface of materials at the nanoscale. There are two primary types of SPM, scanning tunneling microscopes (STM) and atomic force microscopes (AFM). SPM allows for resolution of features down to about 1 nm in height, leading to the three-dimensional image of individual atoms at the surface. The AFM measures the interaction of forces between the probe tip and the surface of the sample [9]. Why is the study of the surface so important? In nanotechnology, the relationship between surface area and volume is important because as objects get smaller the ratio of surface area to volume increases. Considering most physical and chemical changes occur at the surface, objects with a higher surface area are prone to more physical and chemical change. The ability to manufacture at the nanoscale is well sought after for the future. In order for this to occur, tools must be created to manipulate atoms and molecules. Until this can occur, AFM and STM can be used as nanomanipulators in order to arrange atoms and order molecules in order to build objects at the nanometer scale [10].

1.3 Current and Future Applications

1.3.1 *Diagnostics*

The limitations poised by molecular diagnostics can be mitigated through the implementation of nanotechnology [11]. Petricoin *et al.* birthed the idea of using mass spectrometry of the entire set of circulating proteins expressed by an individual, referred to as the proteome, as a diagnostic tool [12]. The concept is that proteins and protein fragments produced by cancer cells or their microenvironment eventually make their way into circulation. These proteins could then be analyzed through mass spectrometry and used in combination with mathematical algorithm to diagnose disease [12–13]. There is substantial evidence suggesting that the low-molecular-weight (LMW) circulatory proteome contains information capable of detecting early stage diseases [14–16]. Mutated genes that exist in diseased patients lead to the production of altered proteins, which in turn alter normal cellular behavior, resulting in the observable disease characteristics. Current mass spectrometry technology can generate a high-resolution portrait of the proteomes. One proteomic platform often used in diagnostics is surface enhanced laser desorption/ionization-time-of-flight (SELDITOF) mass spectrometry [12]. It involves the pretreatment of a biological fluid with proteomic chips that perform extraction of proteins using hydrophobic, ion-exchange, metal binding, or other interactions. These bound proteins are then analyzed using mass spectrometry and the information can be used for diagnosis or identification of biomarkers to be analyzed using other techniques. However, there is skepticism that SELDITOF and other mass spectrometry-based proteomic technologies may be incapable of detecting such a low concentration of molecules released by only a few tumor cells or their surrounding environment. The identification of these molecules will require techniques more sensitive than current mass spectrometry that are capable of detecting concentrations at or below 10^{-12} M. This lack of sensitivity in current techniques as well as rapid enzymatic degradation of the serum proteins limits the translation of the technology into clinical practice [17]. There are emerging nanotechnology solutions that aim to amplify and harvest these biomarkers that play a key role in discovering and characterizing molecules for early disease detection, subclassification, and predictive capability of current proteomics modalities.

A high-throughput and reproducible fractionation system using nanoporous silica chips (NSC) is currently under development within the Ferrari laboratory. This chip is able to deplete most high-molecular-weight proteins to allow the enrichment and stabilization of the LMW proteins present in the circulating proteome [17–18]. The NSC can be tailored via nanopore size as well as physio-chemical properties for tight control over the molecular cut-off, harvesting, and stabilization of proteins and peptides. When combined with mass spectrometry, the NCS can perform a fast efficient and reliable fractionation for high-throughput enrichment, stabilization, and detection of LMW biomarkers.

Another method of improving the detection of biomarkers in the circulating proteome is through the use of smart hydrogel particles for the harvesting and protection of circulating LMW biomarkers [17, 19]. The hydrogel microparticles contain a certain affinity to capture and concentrate the LMW fraction of the serum molecules. The peptides and proteins are then protected from enzyme degradation by the encapsulating hydrogel. The selectivity of peptides can be altered through the structural design of the nanoporous sieve as well as the chemical modification.

Additionally, when bound to the appropriate antibody, magnetic nanoparticles can be used to label specific molecules, structures, or organisms [11]. Magnetic immunoassay techniques can then detect the magnetic field generated by the magnetically labeled targets. Genetic sequences can be detected using gold nanoparticles conjugated to short segments of DNA.

Highly sensitive biosensors that can accurately detect genetic alterations or detect low concentrations of biomarkers are vital to the early detection of disease and subsequently early prognosis and therapy response [17]. Nanowires can be used to implement field effect transistor (FET) semiconductors and provide a highly sensitive approach free of labels to detect and quantify biomolecules. This occurs through the target binding events that occur on the nanowires and result in conductance changes that can be observed to detect specific molecules. Additionally, the high surface-to-volume ratio and small diameter lends to high sensitivity due to an accumulation of carriers throughout a much larger wire cross-section [17, 20, 21]. This method has been used to detect targets such as DNA and proteins. Another category of biosensor devices is micro and nanocantilever systems [17]. These silicon cantilevers can be micro or nanofabricated with multiplexed capability and implemented with label-free biomolecular detection capabilities. The variation of the surface stress produced by specific binding events can be quantified and translated to molecular recognition. For example, a DNA microarray has been developed using cantilever technology that can simultaneously detect multiple molecules at nanomolar concentrations. Cantilever nanosensors have also been used to detect small amount of protein biomarkers at concentrations as low as ng/mL.

It is envisioned that there will once be a time where nanoparticles capable of harvesting specific serum proteins will be administered in a physician office and then sampled during a follow-up visit once given time to collect the appropriate diagnostic information. This sample containing the LMW molecules captured by the nanoparticles can be quickly analyzed using mass spectrometry. The hope is that this will result in an individual global-health profile affordable by everyone or even the ability for a small sample to reveal an image of the physiological and pathological states of every tissue in the body [13].

1.3.2 *Nanoparticles and Implantable Devices for Targeted Delivery*

The use of nanoparticles in medicine covers a large spectrum of applications. Magnetic nanoparticles guided by external fields to the tumor site and then energized by an external electromagnetic field are used in order to destroy cells that the particles are near. Nanoparticles in drug delivery are in heavy pursuit and many are in clinical trials. Targeted and controlled drug delivery relates to delivering therapeutics to a patient at the right location and time, respectively, for that individual patient's need [17] while avoiding healthy organs and tissues. Novel nanotechnology approaches involve the use of nanoparticle delivery vesicles and implantable nanodevices for tightly regulated drug delivery and to overcome the challenges of drug insolubility and biological barriers. Through the use of these nanotechnologies, it may be possible to (1) improve the delivery of poorly water-soluble drugs, (2) target delivery of drugs, (3) achieve transport of drugs through the interior of epithelial and endothelial cells and across the tight junctions between them, (4) deliver large macromolecule drugs to intracellular sites of action, (5) co-deliver for combination therapy, (6) visualize the site of treatment by incorporating imaging modalities with therapeutic agents, and (7) stabilize therapeutic agents against enzymatic degradation from nucleases and proteases [22–23].

Nanovectors are particles with nanodimensions used to delivery therapeutic or diagnostic agents through encapsulation or physical attachment of the desired moiety to the particle [17]. Some examples of nanovectors include, but are not limited to, liposomes, micelles, dendrimers drug-polymer conjugates, and nanoemulsions, which are summarized in Table 1.2 [17, 24]. Liposomes are vesicles comprised of a phospholipid membrane enclosing an aqueous volume. Liposomes can be uni- or multilamellar and vary in size, lipid composition, surface charge, and method of fabrication. Micelles are supramolecular constructs formed through the self-assembly of amphiphiles that form a core-shell structure in an aqueous environment. Typically, nanosized micelles consist of polyethylene glycol (PEG) and a low-molecular-weight hydrophobic core-forming block. Dendrimers are polymeric complexes comprised of branches

Table 1.2. Summary of nanovectors used to delivery therapeutic or diagnostic agents

Nano delivery system	Applications
Liposomes	Doxorubicin (Dox) Amphotericin B Daunorubicin
Micelles	Paclitaxel (PTX) Camptothecin (CPT) Diazepam
Nanoemulsion	Amphotericin B PTX Dexamethasone Benzathine penicillin G
Dendrimers	Indometacin 5-fluorouracel Antisense oligonucleotides
Drug-polymer conjugates	N-(2-hydroxypropyl)methacrylamide – Dox (PK1) polyglutamate-PTX (CT-2103 or Xyotax) poly(glutamic acid)-CPT (CT-2106) cyclodextrin-CPT (IT-101)

around an inner core that possesses properties similar to macromolecules. Due to high branching and a subsequently high density of functional groups, they can be functionalized with groups such as carbohydrates, peptides, and silicon. The vesicle is an attractive drug delivery system due to the presence of a central cavity and channels between the dendrons where the drugs can be trapped [17]. Drug-polymer conjugates–polymer-drug conjugates were some of the first nanotherapeutic platforms to be used for drug delivery. The mechanism of drug delivery involves the simple conjugation of drugs or proteins to water-soluble polymers. Some common polymers used are PEG, N-(2-hydroxypropyl)methacrylamide (HPMA) copolymers, and polysaccharides. Advantages of this method are increased circulation time, reduced immunogenicity, and passive targeting through what is called the enhanced permeability and retention effect (EPR). The EPR effect is the phenomenon that occurs in most solid tumors which possess extensive angiogenesis, defective vascular architecture, and compromised lymphatic systems as well as an upregulation of permeability mediators allowing for a more targeted delivery of systemic therapeutics [25]. Nanoemulsions are dispersions of oil and water in which the dispersed phase forms nanosized droplets that are stabilized with a surface active film comprised of surfactants. They are attractive for drug delivery due to their simple formation, thermodynamic stability, and optical transparency. The structure of the nanoemulsion affects the rate of drug release.

Due to their small size, nanoparticles can penetrate deep into tissue through small capillaries, cross the openings in the epithelial lining, and can be actively taken up by cells. This allows for a more efficient delivery of therapeutic agent to the desired target site, called passive targeting. Additionally, it augments the potential to further target the particles through specific ligand conjugation for various disease targets, called active targeting [23]. Passive targeting can be utilized in the case of tumors or inflamed tissue that are characterized by leaky vasculature, through which nanoparticles can escape to the diseased site [24]. In order for passive targeting to be effective, the nanoparticle must inherit the ability to circulate systemically for extended periods of time. This produces multiple opportunities for the payload to be delivered to the target site. The particles therefore must avoid being marked for destruction by the reticuloendothelial system (RES), and a common method used to circumvent this process is the chemical binding of particles to the hydrophilic and non-inflammatory polymer polyethylene glycol (PEG). Another means of passive targeting is through compromised biological barriers such as the increased permeability of the blood brain barrier in pathologies such as hypoxiaischemia, inflammatory and infectious diseases, and cancer. In the past decade, there has been significant progress made in identifying specific molecular moieties such as epitopes or receptors over expressed in certain diseases. Active targeting using these moieties can enhance the therapeutic effects of nanoparticle drug delivery in localized diseases [24]. The particles can be modified with ligands that actively target these receptors by binding to their surface. This is particularly advantageous for therapeutics that are incapable of being effectively taken up by cells and need assistance through fusion, endocytosis, or other processes to pair with their cellular target. Also, nanoparticles can be designed as "smart," meaning they are responsive to their external environment. For example, a nanoparticle can be designed to dissolve in the highly acidic environment of the stomach, thus releasing its contents in a localized manner. Nanoparticles can also be used for a diagnostic imaging approach. These particles can be coated with molecules that have an affinity for certain targets (i.e., active targeting). The particles can contain heavy metals, or they may fluoresce, enabling more localized imaging of desired sites in the body.

The biophysicochemical properties of the nanovehicle, such as size, charge, surface hydrophilicity, and the nature and density of the ligands on their surface, can all impact the circulating half-life of the particles as well as their biodistribution. Therefore, despite the fact that the presence of targeting ligands can potentially enhance cellular uptake by receptor-mediated endocytosis, drug delivery systems can be engineered to better target a particular tissue or be nonspecifically absorbed by cells merely through the optimization of their biopysicochemical properties [22]. Despite the development of targeted liposomes almost 30 years ago, the majority of currently approved nanotherapeutics lack the charac-teristics of active targeting or triggered drug delivery. For example, the currently approved products for cancer therapy accumulate in the tumor tissue through the above mentioned EPR effect and culminate in release of payload. Several factors are suggested to have contributed to this disappointment. Amongst these issues are the overall feasibility of commercial product development, the cost ineffectiveness with regard to the advantage of the system over standard delivery methods, and proving the importance and global impact of the targeted disease that would justify an advantage of the system [22].

Conventional drug delivery methods are unable to fully overcome the challenges of long-term treatment, narrow therapeutic windows, complex dosing schedules, combination therapies, individualized dosing regimens, and unstable active ingredients [26]. The ability to control the drug administration over the duration of weeks to months according to the individual needs of a patient can be addressed with the development of implantable drug delivery devices. These devices are capable of sustaining drug delivery and meet the need for multiple periodic administrations associated with conventional drug therapy. Implantable devices can provide people with relief from self-medication and/or recurrent doctor visits [17]. An ideal implant would protect the drug from the body until administration time, allow either continuous or pulsatile delivery of both liquid and solid drug formulations, and be controlled externally

[26]. One possible design is an array of small sealed reservoirs that contain one or more drug formulation and sealing and unsealing can be preferentially controlled. These systems can be tailored for specific applications by the release mechanisms, reservoir geometry, and drug formulation.

1.3.3 *Tissue Engineering and Medical Implants*

The extracellular matrix (ECM) on which cells grow in the body, is made up of various large proteins. In order to promote adequate cell differentiation and proliferation, certain features at the nanoscale are essential. To achieve optimal tissue growth, the natural extracellular environment must be mimicked for the necessary cell adhesion, mobility, and differentiation to occur [27]. Synthetic polymers meet most demands for tissue engineering (TE) scaffolds; they are capable of serving as bulk mechanical and structural platforms as well as enable the molecular interactions within the cells that are necessary to induce tissue healing. Most of the synthetic polymers used in TE are nontoxic, consistently available, inexpensive to create, and easy to alter [28]. However, they often lack the ability to create biological cues as natural polymers do to induce a desired cell response [29]. For this reason the cells rely on several topographical and physiochemical signals. These signals can be provided either by the proteins contained in the extracellular matrix (ECM) or by the growth factors that bind to the receptors present on the cell surface. As the cells move over a natural matrix or an artificial scaffold, they sense the presence of grooves and ridges through the extension and retraction of filopodia [28]. The cells determine their behavior through this interaction, adjust their response to the environment, and regulate their terminal differentiation [27]. In response to these phenomena, the relevance of chemical modifications and physical features at the nanoscale proves crucial in the development of the ideal scaffold for the repair and growth of tissue.

   The emergence of nanotechnology offers a new toolset for the discovery, engineering, and manufacturing of nanopatterned surfaces and nanostructured scaffolds for implantable devices. Moreover, nanotechnology offers novel and improved solutions for the localized release of the biomolecules and growth factors that are needed in any TE approach. Nanotechnology in TE overcomes many downfalls that microstructured implants face. This includes infection, chronic inflammation, and poor binding with the surrounding tissue. As a means to address these issues, nanoscale features have been implemented to provide enhanced biointegration [30]. Natural tissues contain various nanometer features because of the presence of collagen fibrils and other proteins that are less than 100 nm in one dimension [30]. The nanometer-scaled surface structures enhance cellular response through mimicking natural tissue. Due to the tunability and adaptability of the manufacturing processes, several different scaffold types can be obtained and ideally optimized for the particular needs and requirements of the individual patient or application. Currently, nanomaterials have been proven to assist in the restoration of several tissues and organs.

Implanted devices within the body either fulfill a structural requirement such as a bone replacement or are implanted into the bloodstream, such as stents and possibly sensors in the future. In order to promote proper integration of the implants within the body, a nanotexture on the surface of the implants is essential, so that the cell responds by excreting extracellular matrix molecules, humanizing the implant surface. Long-term implants with rubbing surfaces, such as joint replacements, generate particles as a consequence of wear. Various current materials produce microparticles, which lead to inflammation while newly developed nanomaterials generate nanoparticles, which are be less detrimental for the body and are much easier to be excreted and cleared from the system [17].

1.3.4 *Electronics*

Nanotechnology is regularly used in the electronics industry. Today, most of the computer microprocessors have features that are less than 100 nm [31]. These smaller sizes lead to an increase in processing capacities and speed. Although the computing power has significantly increased, the ultimate limit that is possible has not been reached. The demand for this power as well as improved data storage capabilities has been the driving force for research in better performance and higher component packing density chip technology over the last 3 to 4 decades [32].

Currently, the integrated circuit in these technologies is made through the top-down approach using processes such as photolithography, thin-film deposition, etching, and metallization. By this technology, a probe writes over a surface the chip circuit [33–34]. This way of building circuits in electronic chips has a limitation of around 22 nm [35]. Below 22 nm errors will occur and short circuits and silicon limitations will prevent chip manufacturing. As each technology generation passes, the tools to make these devices have improved. However, as the size of the devices reach the nanoscale, the physical effects lead to difficulties in production [36]. In the future, nanotechnology will enable the development of new circuit materials, processors, and possibly new ways of storing and transferring information. It will also likely enable greater versatility due to faster data transfer and larger data memories.

The novel field of quantum computing, the area of science aimed at developing computer technology based on the principles of quantum theory, is also emerging in electronics [37]. In this form of computing, the quantum bit, or qbit is used rather than the usual bit of information. This qbit is able to process anything between 0 and 1, leading to new calculations and higher processing speeds [38]. The first quantum computer has been built in the United States, with processing speeds of a billion times per second, the fastest on earth [39].

1.3.5 *Microelectromechanical Systems*

Microelectromechanical systems (MEMS) are devices that have a length between 1 μm and 1 mm that combine both electrical and mechanical components fabricated using integrated circuit batch-processing technologies. Some current fabrication techniques include surface silicon micromachining, bulk silicon micromachining, lithography, electrodeposition, and plastic molding and electrodischarge machining (EDM). Nanodevices also called NEMS aim to further decrease the size of these electromechanical miniature devices. Some applications include automobile airbags, keyless entry systems, dense arrays of micromirrors for high-definition optical displays, scanning electron microscope tips for imaging single atoms, micro heat exchangers for cooling electronic circuits, reactors for the separation of biological cells, and blood analyzers and pressure sensors for catheter tips [40].

The development of such devices is expected to lead to breakthroughs in information technology, computers, medicine, health, manufacturing, transportation, energy, avionics, security, and more. In particular, MEMS and NEMS have an impact on medicine and bioengineering (DNA and genetic code analysis and synthesis, drug delivery, diagnostics, and imaging), bio and information technologies avionics, and aero space (nano- and microscale actuators and sensors, smart reconfigurable geometry wings and blades, space-based flexible structures, microgyroscopes), automotive systems and transportation (transducers and accelerometers), and manufacturing and fabrication (micro- and nanoscale smart robots) [17]. Nanoscale devices and systems will allow for the manipulation of atomic scale phenomena where quantum effects are predominant. Reducing the dimenstions of MEMS systems leads to the incorporation of novel materials such as carbon nanotubes and entirely new fabrication technologies [41].

MEMS products have presented themselves for several years as accelerometer chips used to control automobile airbags, mirror chips used in projection screen TVs, inkjet nozzles for printers, and pressure sensors in medical applications. In addition to these successful products new products and improvement upon existing products is currently underway. For example, computer games that use a MEMS accelerometer [42], smaller and more heat resistant microphones [43], RF switches that reduce power loss in microwave applications [44], and blood pressure sensors with wireless data transfer implanted in patients [45].

1.3.6 *Sensors*

Silicon chips have been following Moore's law of doubling in the number of transistors they hold every 18 to 24 months for about four decades [46]. Since this trend cannot be followed indefinitely it is thought that silicon-based industries may come to a standstill around 2015. However, nanotechnology provides the tools to develop organic and molecular-based transistors which may lead to the continuation of shrinking dimensions corresponding to Moore's law.

Nanotechnology has thus far presented itself most useful in the area of sensor technology. It has enabled the creation of small, inexpensive, and efficient sensors that serve wide-ranging applications including the detection of water and air impurities, which will be touched upon later in this chapter. Some of the advantages of nanotechnology-enabled sensors include greater sensitivity and selectivity, lower fabrication costs, reduced power consumptions, and improved stability over conventional sensors. Nanoscale materials can be used to create entirely new devices whose properties can be tailored and geared toward sensing applications or they can be incorporated into existing devices as enhancements. One example is the development of nano-bio-organic elements for the measurements of intracellular events. Nanostructured features are ideal for sensing as sensitivity may increase due to tailored conduction properties, limits of detection may be lowered, small quantities can be analyzed, and specificity may be improved. Sensors that may reap the benefits of nanotechnology include but are not limited to physical sensors, electrosensors, chemical sensors, and biosensors. Highly sensitive magnetic sensors and quantum resistance useful in the electronic industry can be developed by tuning the density of states in materials through nanotechnological processes [46].

Nanotechnology can enable sensors to detect very small amounts of chemical vapors. Due to the incredibly small size of nanotubes, nanowires, or nanoparticles, a few gas molecules are adequate to change the electrical properties of the sensing elements. This allows detection of chemical vapors at minute concentrations. These detecting elements change their electrical characteristics, such as resistance or capacitance, when they absorb a gas molecule [47]. For example, zinc oxide nanowire transistors will alter their ability to conduct current in the presence of a foreign substance. Different substances increase or decrease the current by different degrees, thus specific chemicals can be identified by how much they affect the flow of current. Novel materials such as nanostructured thin films similar to polyaniline and TiO_2 thin films have been developed through nanotechnology. These materials enhance chemical sensing through the increased surface area to volume ratio they possess which allows more active sensing area to interact with the target molecules. Strong photon and phonon quenching and amplification are observed on nanosurfaces that do not exist on bulk materials currently in use. By altering on the nanoscale, optical and electronic properties can be manipulated to suit the desired application. These highly sensitive nanosensors can be useful in industrial plants with high chemical usage to monitor chemical vapor release. In the future, such sensors will be essential for hydrogen fuel cells for detecting escaped hydrogen [47].

Sensors also have significant applications in human health monitoring. The ability to detect, diagnose, and treat diseases inside the human body is a dream slowly becoming a reality with the rapid development of nanotechnology. Some examples are glucose sensors for diabetic patients, detection of high fat levels, calcium or vitamin deficiency, abnormal temperature or blood pressure, viral infection, and blood clot location.

Advances in nanotechnology will continue to improve the design of sensors incorporating sophisticated electronic signal processing with innovative transducers and actuators, sensors essential to the medical sciences, and electronic components' communication circuits. Although many sensors currently exist on the market with nanotechnology-enabled features, even smarter, cheaper, and more selective and sensitive sensors are under development.

1.3.7 *Energy*

All elements of energy conversion such as charge transfer, molecular rearrangements, and chemical reactions happen at the nanoscale. The progress of nanoscale materials with new physical and chemical properties, as well as ways to characterize, manipulate, and assemble them, has created breakthroughs in the energy sector toward sustainable energy production, storage, and use [48]. Nanomaterials are appealing because of their high surface area per unit volume, which contributes to much higher surface activity than in the bulk material. This speeds up chemical reactions, thereby improving the efficiency of various processes. Within energy

Table 1.3.   Summary of energy systems that implement nanotechnology

Energy conversion	Energy storage	Energy transmission	Energy use
Solar cells	Supercapacitors	Superconducting cables	Conservation in Manufacturing industry and construction,
Thermoelectric devices	Batteries	Hydrogen distribution	Materials for transport fuel cells
Catalysts for conversion	Hydrogen storage		
Environmental management			Catalysts for combustion
Hydrogen production			
Carbon dioxide capture and storage			

systems nanotechnology can be used in various ways, as seen in Table 1.3 [48–52].

The overall usage of batteries is a significant problem. The use of nanomaterials could lead to batteries with higher energy capacities or rechargeable batteries and accumulators, leading to a decrease in the issues of battery disposal [53–54]. For example, researchers at Stanford University used silicon-based nanowires to increase the ability of rechargeable lithium-ion batteries used to power electronics such as laptops and cell phones. This new Li-ion technology has ten times the capacity of traditional lithium batteries [55]. The use of nanotechnology can also be used to produce less flammable electrode materials for use in batteries [54].

The use of nanomaterials such as nanotubes, zeolites, and alanates are being investigated for the development of more environmentally friendly systems [56–58]. Hydrogen fuel cells with the catalyst consisting of carbon supported metal particles with diameters of 1–5 nm will improve hydrogen storage for better hydrogen energy cells [59–61]. Also, nanotechnology can help reduce of combustion engine pollution by the use of new nanoporous filters [62–63], which will remove pollutants more efficiently. The reduction of energy consumption using nanotechnology focuses on the development of more efficient lighting [64], combustion systems [65], and the use of lighter and more durable materials for transportation [62, 66]. Light-emitting diodes (LEDs) [67] and quantum-caged atoms (QCAs) [68] are also studied for energy consumption reduction.

1.3.8 *Food Production, Processing, Preservation, and Packaging*

Compared to the conventional formulations, various nanochemicals have been introduced to increase the effectiveness of treatments in agriculture. Among these, there are several pesticides containing nanosized particles or chemicals created through synthetic processes involving nanotechnology. Remainders of these products have the potential to be present in consumed products on the market, thus posing a potential risk for the health of the consumers.

Nanotechnologies are also applied in food production machinery. Examples of this type of nanotechnology are coatings of machines or the use of nanosieves (e.g., to filter out bacteria). While direct food contact with this application of nanotechnology is evident, it is expected to have negligible additional safety concerns or carryover into food in comparison with conventional techniques. The type of material (and wear-off as result of the use) of filters or coatings might require some attention, but this is not exclusively related to safety of nanotechnologies.

Nanoparticles are incorporated in packaging materials or storage containers in order to lengthen the storage time while keeping the products fresh. For instance, nanoparticles are incorporated to increase the barrier properties of packaging materials (e.g., silicate nanoparticles, nanocomposites, and nano-silver, magnesium-, and zinc-oxide). Nanoparticles can also be applied as reactive particles, or nanosensors in packaging materials. They are designed to respond to environmental changes such as temperature or moisture in storage rooms, degradation products of the food commodities, or contamination by microorganisms [69].

1.3.9 *Water Purification*

In developing countries a staggering 80% of illnesses are water related due to the consumption of contaminated water [70]. Not only is there limited availability in third World countries, but the rapid rising population and industrial demand places developed nations at risk of water shortage due to rapidly rising consumption rates. Nanotechnology provides hope for the monitoring, desalinization, and purification of water and wastewater treatment industries.

Many conventional technologies exist and are successful in the removal of bacteria, viruses, coliforms, and other contaminants. Water desalination and wastewater treatment methods exist as means to purify water sources. However, these methods face technical challenges in terms of cost and effectiveness of removing contaminants in a way that is feasible for developing countries. Another disadvantage of conventional water treatment is that it cannot remove dissolved salts and some soluble inorganic and organic substances [71]. Nanotechnology can increase the effectiveness of these existing water treatment methods and potentially reduce the costs [70]. Nanotechnology can impact water applications in three areas: treatment and remediation, sensing and detection, and pollution prevention. A range of water treatment devices that incorporate nanotechnology are already on the market and others are in advanced stages of development. These nanotechnology applications include nanofiltration membranes, nanocatalysts, magnetic nanoparticles, and nanosensors for the detection of contaminants [72].

Carbon nanotubes, nanoparticles, and dendrimers allow for more effective and cost-effective water filtration. Nanotechnology membranes include nanostructured filters made of carbon nanotubes or nanocapillary arrays and nanoreactive membranes where functionalized nanoparticles assist in the filtration process. Dendritic polymers can assist in filtration of water contaminated by organic solutes and inorganic anions. Remediation is the process of removing, reducing, or neutralizing water contaminants that threaten human health. This can be done through thermal, physiochemical, or biological methods. Solvent extraction, activated carbon adsorption, and chemical oxidation are effective yet costly and time-consuming processes. Novel nanomaterials are being developed with increased affinity, capacity, and selectivity for heavy metals and other contaminants. These improvements are due to enhanced reactivity, surface area, and sequestration characteristics [73].

The four pressure-driven membrane processes for water purification are microfiltration, ultrafiltration, nanofiltration, and reverse osmosis. Microfiltration and ultrafiltration are often used for pretreatment. Reverse osmosis removes substances that were not rejected from the pre-treatment including all soluble and minutely insoluble substances, but water [71]. Nanofiltration membranes provide the advantage of selectively rejecting substances, enabling essential nutrients, such as calcium ions present in water to be retained. Many nanofiltration membranes have been developed using nanomaterials such as carbon nanotubes and alumina fibers for use in specific filtration applications. Other materials used for nanofilters include zeolites, attapulgite clays, and nanoporous polymers. Although these materials have been used for water purification for many years, recent advances in nanotechnology has allowed for more precise control over the engineering of these filters. At Rensselaer Polytechnic Institute (USA) and Banaras Hindu University (India) carbon nanotube filters were developed capable of removing 25-nm-sized polio viruses from water as well as *E. coli* and *staphylococcus aureus* bacterial. These nanotube-based water filters are also found to be more resilient and reusable than conventional membrane filters due to their uniquely engineered design. The filters can be cleaned through heating or purging and the nanomembranes allow water to flow more rapidly due to the straighter membranes compared to conventional filters. Additionally, Argonide (USA) has developed a filter consisting of oxidized aluminum nanofibers on a glass fiber substrate capable of retaining up to 99.999% of viruses and is currently in production. The alumina fibers carry a positive charge enabling them to filter bioorganisms at a much higher flow rate than traditional filters. More importantly, this technology can be used to filter water by mere muscle force, making it ideal for rural location [71]. These are just a few examples of nanofiltration membranes that have been developed as many others exist and are continuously being improved.

A major obstacle faced by desalination of water is the high production costs mainly due to the energy required to force water through the filtration membranes. Desalination costs about USD 1 per m^3 of salt water and USD 0.60 per m^3 brackish water.

Nanotechnology provides hope that will drive down these costs and provide a more constant supply of fresh drinking water. The Long Beach Water Departments have already reduced the energy requirements by 20% to 30% through the use of a low-pressure two-staged nanofiltration process.

Nanocatalysts such as titanium dioxide and iron nanoparticles can be used to degrade organic pollutants and remove salts and heavy metals from water. By using either homogeneously dispersed catalytic particles in solution or depositing them on a membrane, pollutants can be chemically degraded instead of merely being removed and transferred elsewhere. There are several companies developing applications that are already available or will be available soon. The company EnvironmentalCare from Hong Kong has developed a nanophotocatalytic oxidation technology for the removal of bacteria and water pollutants. Nanocoated titanium dioxide filters trigger a chemical process that converts the pollutants into carbon dioxide and water. Additionally, in photocatalysis, water that passes through the nanomaterial is exposed to ultraviolet light which destroys contaminants. Recently, a scientist at the U.S. Department of Energy's Argonne National Laboratory has successfully created visible-light catalysis, using silver chloride nanowires decorated with gold nanoparticles that decompose organic molecules in polluted water. The addition of gold nanoparticles to silver chloride expands its photocatalytic properties to include wavelengths of visible light. The light excites electrons with the gold and produces a charge separation in the silver chloride nanowires allowing for decomposition of organic molecules [74].

Magnetic nanoparticles are also being developed to adsorb metals and organic compounds. The magnetic nanoparticles can be coated with different compounds that have a selective affinity for contaminants such as arsenic, thereby allowing the removal of the contaminant through a magnetic pump. The magnetic nanoparticles and the pollutant could then be separated for reuse.

Another benefit that nanotechnology provides to water purification is in the actual detection of contaminants. Nanosensors can detect single cells or even atoms, making them sensitive enough to detect biological and chemical contaminants at low concentrations

[73]. This technology can help people identify contaminants without having to send samples to laboratories for testing.

Most nanotechnology applications are still in the lab and have yet to be scaled up to the industrial level therefore hampering the reduced price claim. Additionally, those technologies that are commercially available claim a price premium [70].

1.3.10 *Air Quality*

Nanotechnology can be used to reduce air pollution in two different ways: catalysts and nanostructured membranes. Catalysts enable a chemical reaction at lower temperatures or make it more effective.

Nanotechnology improves the performance and cost of catalysts that transform the vapors produced by cars or industrial plants. The larger surface area nanoparticles possess provide for more interaction with the reacting chemicals than catalysts made from larger particles.

Toyota Central Research and Development Laboratories in Japan developed a mesoporous manganese oxide/nanogold catalyst that can remove volatile organic compounds and hazardous nitrogen and sulfur oxides from air at room temperature. Current air purifying systems use photocatalysts, adsorbent materials like activated charcoal, or ozonolysis and are incapable of eliminating organic pollutants at room temperature [71]. The system was tested on the three key components of organic air pollution (hexane, toluene, and acetaldehyde) and found that it was capable of removing all three and degrading them more efficiently than existing systems. The success is not only due to the increased surface area of the porous manganese allowing for numerous adsorption sites for pollutant molecules, but the gold incorporated onto the surfaces creates a large number of oxygen free radicals necessary for the oxidation process to degrade the adsorbed molecules [75].

Researchers at UCLA are using crystals containing nanosized pores to trap carbon dioxide (UCLA Newsroom). They have taken organic and inorganic units and combined them into a synthetic crystal which codes information in a DNA-like manner. These metal-organic frameworks (MOFs) can incorporate a large number of different functionalities on linking groups in a way that mixes the linker, rather than forming separate domains. Complex MOFs with up to eight distinct functionalities in one phase were created. This complex arrangement of several functional groups within the pores can lead to properties that are not the linear sums of those of the pure components as one would expect. The scientists in fact found that one MOF in particular exhibited up to 400% better selectivity for carbon dioxide over carbon monoxide compared with its best counterpart exhibiting homogeneous linkages.

Other nanotechnology applications currently in development for air quality include the reduction in the amount of platinum used in catalytic converters. This is achieved through a novel platinum alloy substitution and reducing emissions from power plants by converting carbon dioxide into nanotubes. Also, the conversion of carbon dioxide into methanol to power fuel-cells using air and light stable combinations of dyes and functionalized nanoscale semiconductor particles is under exploration [76]. Nanostructured membranes are being developed to separate carbon dioxide from industrial plant exhaust streams. Dr. Zhu from the University of Queensland is working toward developing a carbon nanotube membrane for gas separation that will work like a sieve to separate high volumes of methane or carbon dioxide from other gases. This technology can potentially trap moving gas up to 100 times faster than other gas separation techniques, making it ideal for large-scale plants. The membranes separate and process large volumes of gas effectively, providing an advantage over conventional membranes at the large scale required for coal-fired power plants or natural gas processing.

1.3.11 *Space*

Every research and study currently being conducted for the mapping and sensing of vast planetary terrains as well as for enhancing health management and protection systems for astronauts uses nanotechnology. This includes the development of bio-nano space-suits with robots integrated into the layers coined "All Terrain Bionano" (ATB) suits. These bionanorobots possess the properties, design, and functionalities inherent to biological materials such as individual peptides and DNA. They are "smart" in nature, capable of actuation, sensing, signaling, and information processing at the nanoscale. Robots in the outermost layer aim to detect damage to the suit, whereas in the more inner layers, the robots participate in personal interaction with the astronaut, and respond to needs such as providing drugs in the case of a medical emergency [77]. This technology employs the breach sealing method, where a breach in the suit is covered and the underlying layers are protected with bionano robots that flow through the suit layers and bind amongst themselves through self assembly. The Bionanorobot technology is also being utilized to develop a sensor capable of exploring large areas on planets for traces of elements and chemicals.

Nanotechnology may play a role in making space travel not only possible but also more practical. Advancements in nanomaterials such as carbon nanotubes can create lightweight materials for use in space. At Johnson Space center there is research committed to the development of nanotube composites designed to reduce the weight of spacecrafts [78]. The "Nanotube Project" focuses on the production, purification, and application of single-walled carbon nanotubes due to their advantageous strength-to-weight ratio. It is predicted that the nanotube composites will reduce the weight of the spacecraft by at least 50%.

By significantly reducing the weight of the rockets and space-crafts, also the amount of rocket fuel required would be reduced. These advances could lower the cost of reaching orbit and eventually make traveling in space possible and affordable in the near future.

References

1. Burke, M. (ed.) (2008) *Nanotechnology, The Business*, CRC Press, Boca Raton.
2. (n.d.). Retrieved from http://nano.gov/
3. Chen, P. (2005) Self-assembly of ionic-complementary peptides: a physicochemical viewpoint, *Colloids and Surfaces A: Physicochemical and Engineering Aspects,* 261, 3–24.
4. Mongillo, J. (ed.) (2007) *Nanotechnology 101*, Greenwood Publishing Group, Westport, CT.

5. Zhang, S. (2001) Molecular self-assembly, in *Encyclopedia of materials: science and technology* (ed. Buschow, K., Cahn, R., Hemings, M., Ilschner, B., Kramer, E., and Mahajan, S.), Elsevier, Oxford, UK, pp. 5822–5829.

6. Whitesides, G. M., Mathias, J. P., and Seto, C. T. (1991) Molecular self-assembly and nanochemistry: a chemical strategy for the synthesis of nanostructures, *Science*, 254, 1312–1319.

7. Hartgerink, J. D., Beniash, E., and Stupp, S. I. (2002) Supramolecular chemistry and self-assembly special feature: peptide-amphiphile nanofibers: a versatile scaffold for the preparation of self-assembling materials, *Proc. Natl. Acad. Sci.* 99, 5133–5138.

8. Müller, S. A., Aebi, U., and Engel, A. (2008) What transmission electron microscopes can visualize now and in the future, *Journal of Structural Biology*, **163**(3), 235–245.

9. Hansma, H. G., and Pietrasanta, L. (1998) Atomic force microscopy and other scanning probe microscopes, *Current Opinion in Chemical Biology*, **2**(5), 579–584.

10. Fahlbusch, St., Mazerolle, S., Breguet, J.-M., Steinecker, A., Agnus, J., Pérez, R., and Michler, J. (2005) Nanomanipulation in a scanning electron microscope, *Journal of Materials Processing Technology*, **167**(2–3), 371–382.

11. Jain, K.K. (2003) Nanodiagnostics: application of nanotechnology in molecular diagnostics, *Expert Review of Molecular Diagnostics*, **3**(2), 153–161.

12. Diamandis, E. P. (2004) Mass spectrometry as a diagnostic and a cancer biomarker discovery tool: opportunities and potential limitations, *Mol Cell Proteomics*, **3**(4), 367–378.

13. Liotta, L.A., Ferrari, M., and Petricoin, E. (2003) Clinical proteomics: written in blood, *Nature*, **425**(6961), 905.

14. Calvo, K. R., Liotta, L. A., and Petricoin, E. F. (2005) Clinical proteomics: from biomarker discovery and cell signaling profiles to individualized personal therapy, *Bioscience Reports*, **25**(1–2), 107–125.

15. Hu, Y., Bouamrani, A., Tasciotti, E., Li, L., Liu, X., and Ferrari, M. (2010) Tailoring of the nanotexture of mesoporous silica films and their functionalized derivatives for selectively harvesting low molecular weight protein, *ACS Nano*, **4**(1), 439–451.

16. Bouamrani, A., Hu, Y., Tasciotti, E., Li, L., Chiappini, C., Liu, X., and Ferrari, M. (2010) Mesoporous silica chips for selective enrichment and stabilization of low molecular weight proteome, *Proteomics*, **10**(3), 496–505.

17. Sakamoto, J. H., van de Ven, A. L., Godin, B., Blanco, E., Serda, R. E., Grattoni, A., *et al.* (2010) Enabling individualized therapy through nanotechnology [review], *Pharmacol Res.* **62**(2), 57–89.

18. Bouamrani, A., Hu, Y., Tasciotti, E., Li, L., Chiappini, C., Liu, X., and Ferrari, M. (2010) Mesoporous silica chips for selective enrichment and stabilization of low molecular weight proteome, *Proteomics*, **10**(3), 496–505.

19. Luchini, A., Geho, D. H., Bishop, B., Tran, D., Xia, C., Dufour, R. L., *et al.* (2008) Smart hydrogel particles: biomarker harvesting: one-step affinity purification, size exclusion, and protection against degradation, *Nano Lett.*, 8(1), 350–361.

20. Cui, Y., Wei, Q., Park, H., and Lieber, C. M. (2001) Nanowire nanosensors for highly sensitive and selective detection of biological and chemical species, *Science*, **293**(5533), 1289–1292.

21. Wang, W. U., Chen, C., Lin, K. H., Fang, Y., and Lieber, C.M. (2005) Label-free detection of small molecule-protein interactions by using nanowire nanosensors, *Proc Natl Acad Sci USA*, **102**(9), 3208–3212.

22. Farokhzad, O. C., and Robert Langer, R. (2009) Impact of nanotechnology on drug delivery, *ACS Nano.*, **3**(1), 16–20.

23. Panyam, J., and Labhasetwar, V. (2003) Biodegradable nanoparticles for drug and gene delivery to cells and tissue, *Advanced Drug Delivery Reviews*, 55, 329–347.

24. Koo, O. M., Rubinstein, I., and Onyuksel, H. (2005) Role of nanotechnology in targeted drug delivery and imaging: a concise review, *Nanomedicine: Nanotechnology, Biology, and Medicine*, **1**(3), 193–212.

25. Maeda, H., Wu, J., Sawa, T., Matsumura, Y., and Hori, K. (2000) Tumor vascular permeability and the EPR effect in macromolecular therapeutics: a review, *J. Control Release*, **65**(1–2), 271–284.

26. Staples, M., Daniel, K., Cima, M. J., and Langer, R. (2006) Application of micro- and nano-electromechanical devices to drug delivery, *Pharmaceutical Research*, **23**(5), 847–863.

27. Engel, E., Michiardi, A., Navarro, M., Lacroix, D., and Planell, J. A. (2008) Nanotechnology in regenerative medicine: the materials side, *Trends Biotechnol*, **26**(1), 39–47.

28. Place, E. S., George, J. H., Williams, C. K., and Stevens, M. M. (2009) Synthetic polymer scaffolds for tissue engineering, *Chem Soc Rev*, **38**(4), 1139–1151.

29. Harrington, D. A., Sharma, A. K., Erickson, B. A., and Cheng, E. Y. (2008) Bladder tissue engineering through nanotechnology, *World J. Urol.,* **26**(4), 315–322.

30. Chun, Y. W., and Webster, T. J. (2009) The role of nanomedicine in growing tissues. *Ann Biomed Eng.,* **37**(10), 2034–2047.

31. Komatsu, H., and Ogsawara, A. (2005) Applying nanotechnology to electronics: recent progress in Si-LSIs to extend nano-scale, *Science and Technology Trends*: *Quarterly Review*, 16, 36–45.

32. Yu, B., and Meyyappan, M. (2006) Nanotechnology: Role in emerging nanoelectronics, *Solid-State Electronics,* **50**(4), 536–544.

33. Bohr, M. T. (2002) Nanotechnology goals and challenges for electronic applications, *IEEE Transactions on Nanotechnology*, **1**(1), 56–62.

34. Lu, W., and Lieber, C. M. (2007) Nanoelectronics from the bottom up, *Nature Materials*, 6, 841–850.

35. Gasser, B. (2003). *Nanotechnology in electron devices, Intel Corporation electronic publications*, Retrieved from http://leitl.org/

36. Cui, Y., and Lieber, C. M. (2001) Functional nanoscale electronic devices assembled using silicon nanowire building blocks, *Science,* **291**(5505), 851–853.

37. Nielsen, M. A., and Chuang, I. L. (eds) (2000) *Quantum Computation and Quantum Information*, Cambridge University Press, Cambridge, UK.

38. Joachim, C. (2002) Bonding more atoms together for a single molecule computer, *Nanotechnology*, **13**(2), R1–R7 (7).

39. Knill, E. (2010) Physics: Quantum computing, *Nature*, 463, 441–443.

40. Gad-el-Hak, M. (ed.) (2005) *MEMS: Introduction and Fundamentals*, Taylor and Francis Group, Boca Raton.

41. Lyshevski, S. E. (ed.) (2002) *MEMS and NEMS*, CRC Press LLC, Boca Raton.

42. Retrieved from http://www.analog.com/en/press-release/May_09_2006_ADI_Nintendo_Collaboration/press.html

43. Retrieved from http://www.memsinvestorjournal.com/2006/04/mems_microphone.html

44. *Understanding nano: MEMS.* (2006). Retrieved from http://www.understandingnano.com/water.html

45. Grayson, A. C. R., Shawgo, R. S., Johnson, A. M., Flynn, N. T., Yawen, L. I., Cima, M. J., and Langer, R. (2004) A bioMEMS review: MEMS technology for physiologically integrated devices, *Proceedings of the IEEE*, **92**(1), 6–21.

46. Fry, B., and Kourosh Kalantar-zadeh, K. (eds) (2008) *Nanotechnology-Enabled Sensors*, Springer Science & Business Media, LLC, New York.

47. *Understanding nano: Chemical Sensors and Nanotechnology.* (2006). Retrieved from http://www.understandingnano.com/water.html

48. Tegart, G. (2009) Energy and nanotechnologies: priority areas for Australia's future, *Technological Forecasting & Social Change,* 76, 1240–1246.

49. *Nanomaterials for next generation energy sources, technology tracking.* (2005). Retrieved from http://technology-tracking.com/

50. *Road maps for nanotechnology in energy, The Institute for Nanotechnology,* Edinburgh. (2006, September). Retrieved from http://www.nanoroadmap.it/roadmaps/NRM_Energy.pdf

51. Walsh, B. (2007). *Environmentally beneficial nanotechnologies barriers and opportunities.* Retrieved from http://www.defra.gov.uk/environment/quality/nanotech/documents/envbeneficial-report.pdf

52. *Nanotechnologies and Energy White Paper*, Cientica, London, (2007) Retrieved from www.cientica.com.

53. Armand, M., and Tarascon, J. M. (2008) Building better batteries, *Nature,* 451, 652–657.

54. Panero, S., Scrosati, B., Wachtler, M., and Croce, F. (2004) Nanotechnology for the progress of lithium batteries R&D, *Journal of Power Sources,* **129**(1), 90–95.

55. Chan, C. K., Peng, H., Liu, G., McIlwrath, K., Zhang, X. F., Huggins, R. A., and Cui, Y. (2008) High-performance lithium battery anodes using silicon nanowires, *Nanotechnology*, **3**(1), 31–35.

56. Kumar, M., and Ando, Y. (2007) Carbon nanotubes from camphor: An environment-friendly nanotechnology, *Journal of Physics: Conference Series*, 61, 643–646.

57. Yoo, E., Gao, L., Komatsu, T., Yagai, N., Arai, K., Matsumoto, T., and Nakamura, J. (2004) Atomic hydrogen storage in carbon nanotubes promoted by metal catalysts, *J. Phys. Chem.,* **108**(49), 18903–18907.

58. Frackowiak, E., and Beguin, F. (2002) Electrochemical storage of energy in carbon nanotubes and nanostructured carbons, *Carbon*, **40**(10), 1775–1787.

59. Li, W., Zhou, W., Li, H., Zhou, Z., Zhou, B., Sun, G., and Xin, Q. (2004) Nano-stuctured Pt-Fe/C as cathode catalyst in direct methanol fuel cell, *Electrochimica Acta*, **49**(7), 1045–1055.

60. Choi, J. S., Chung, W. S., Ha, H. Y., Lim, T. H., Oh, I. H., Hong, S. A., and Lee, H. I. (2006) Nano-structured Pt-Cr anode catalyst over carbon support, for direct methanol fuel cell, *Journal of Power*, 156(2), 466–471.

61. Liu, H., Song, C., Zhang, L., Zhang, J., Wang, H., (2006) A review of anode catalysis in the direct methanol fuel cell, *Journal of Power*, 155(2), 95–110.

62. Presting, H., König, U., (2003) Future nanotechnology developments for automotive applications, *Materials Science and* Engineering: C, 23(6–8), 737–741.

63. Dutta, J., (2005) Nanotechnology in environmental protection and pollution, *Science and Technology of Advanced Materials*, 6, 219–220.

64. Hood, E., (2004) Nanotechnology: looking as we leap, *Environmental Health Perspectives*, 112(13), A740–A749.

65. Yetter, R. A., Risha, G. A., and Son, S. F. (2009) Metal particle combustion and nanotechnology, *Proceedings of the Combustion Institute*, 32(2), 1819–1838.

66. Bing-qiu, A.O. (2002) Latest development of lightweighting automotive materials, *Automobile Technology & Material.*

67. Bao, J., Zimmler, M. A., Capasso, F., Wang, X., and Ren, Z. F. (2006) Broadband ZnO single-nanowire light-emitting diode, *Nano Lett*, 6(8), 1719–1722.

68. Dutta, P., and Gupta, S. (eds) (2008) *Understanding of Nano Science and Technology*, Global Vision Publishing House, Daryaganj, New Delhi.

69. Bouwmeester, H., Dekkers, S., Noordam, M. Y., Hagens, W. I., Bulder, A. S., de Heer, C., *et al.* (2009) Review of health safety aspects of nanotechnologies in food production, *Regulatory Toxicology and Pharmacology*, 53(1), 52–62.

70. Berger, M. (2007). *Water, nanotechnology's promises, and economic reality.* Retrieved from http://www.nanowerk.com/spotlight/spotid=2372.php

71. *Nanotechnology, Water and Development, Meridian Institute: connecting people to solve problems.* (2007). Retrieved from http://www.merid.org

72. *Understanding nano: water pollution and nanotechnology.* (2006). Retrieved from http://www.understandingnano.com/water.html

73. Theron, J., Walker, J. A., and Cloete, T. E. (2008) Nanotechnology and water treatment: applications and emerging opportunities, *Critical Reviews in Microbiology*, 34, 43–69.

74. DOE/Argonne National Laboratory (2010, June 21). Gold nanoparticles create visible-light catalysis in nanowires. *ScienceDaily*. Retrieved August 12, 2010, from http://www.sciencedaily.com /releases/2010/ 06/100615122534.htm

75. Sinha, A. K., Suzuki, K., Takahara, M., Azuma, H., and Nonaka, T. (2007) Mesostructured Manganese Oxide/Gold Composites for Extensive Air Purifications, *Angew. Chem. Int. Ed.*, **46**(16), 2891–2894.

76. *Understanding nano: Air Pollution and nanotechnology*. (2006). Retrieved from http://www.understandingnano.com/water.html

77. Mavroidis, C. (2006, July). *Bio-nano-machines for space applications*. Retrieved from http://www.niac.usra.edu/files/studies/ final_report/914Mavroidis.pdf

78. Retrieved from http://mmptdpublic.jsc.nasa.gov/jscnano/default.asp

Chapter 2

Overview of Intellectual Property Rights

Wim Helwegen[a] and Luca Escoffier[b]

[a] P.O. Box 4 (Yliopistonkatu 3, 4th floor), 00014, University of Helsinki, Finland
[b] Waseda University, 1-104 Totsukamachi, Shinjuku-ku, Tokyo,
169-8050, Japan, and Usque ad Sidera LLC
wim.helwegen@helsinki.fi; ceo@usqueadsidera.com

The most common instruments to secure investments in high tech R&D are patents and trade secrets. Patents can be used to create a temporary monopoly on a certain invention, while trade secrets can be used to keep certain technologies or sensitive information concealed. In addition, database rights can be used to protect certain collections of research results from undue use. Other intellectual property rights (IPRs) such as trademarks are only relevant once the end-product nears its market entry. This chapter will give an overview of the IPRs that are most relevant to the nanotechnology sector and other high tech sectors. Due to their importance for nanotechnology, patents and trade secrets will be discussed more details than utility models, trademarks, design protection, and copyrights.

2.1 Patents

Patents are the most commonly used method of IP protection in the high tech sector. Under the European Patent Convention only one

Nanotechnology Commercialization for Managers and Scientists
Edited by Wim Helwegen and Luca Escoffier
Copyright © 2012 Pan Stanford Publishing Pte. Ltd.
ISBN 978-981-4316-22-4 (Hardcover), 978-981-4364-38-6 (eBook)
www.panstanford.com

kind of patents exists, whereas the US system encompasses several kinds of patents: utility patents, plant patents, and design patents. The patents discussed in this chapter are referred to as "utility patents" in the United States.

A patent provides the proprietor with a negative right: the right to exclude. Depending on the jurisdiction in which the patent is valid, a patentee can exclude others from for example, making, using, selling, leasing, importing, and keeping in stock the patented invention. In addition, the mere offer of any of the acts mentioned above could be forbidden by the patentee, again depending on the jurisdiction.

In no way does a patent confer a positive right on the patentee. If the authorities require regulatory permission to make, use or market the invention, this permission should be sought separately from a patent.

Patents are normally valid for 20 years, although exceptions exist for patents on pharmaceutical products that have to be subjected to clinical trials. In the EU such pharmaceutical patents can qualify for a Supplementary Protection Certificate, which provides additional protection after the expiry of the 20 year term of the patent. In the United States, a similar provision, known as Patent Term Extension, exists. These extensions are meant to compensate for the time during which the 20 year term of the patent has started but the patented invention could not be marketed due to the absence of regulatory permission.

2.1.1 *Requirements*

The right to exclude all others from using the patented invention can lead to a *de facto* monopoly for the patentee during the term of the patent. Because monopolies are undesirable from an economic perspective, patents are only granted if certain requirements are met. These requirements will be discussed briefly in this chapter and in more detail in Chapters 3 and 4.

Under the European Patent Convention (EPC), which is directly valid in 38 countries, the requirements for patentability are that the subject matter for which the patent is sought is an invention, susceptible of industrial application, which is novel and involves an inventive step (Art. 51 (1) EPC). In the United States, patents can be granted to "whoever invents or discovers any new and useful process, machine, manufacture, or composition of matter, or any new and useful improvement thereof" (35 USC 101). The articles following Section 101 specify the requirement. Europe and the United States share the following criteria for patentability: (1) patentable subject matter, (2) novelty, (3) industrial applicability/utility, and (4) inventiveness/non-obviousness.

2.1.1.1 Patentable subject matter

Under the EPC, the first requirement is that the subject matter for which the patent is sought is an invention. The EPC does not provide a definition of "invention" but does name a number of categories which certainly do not qualify as inventions: discoveries, scientific theories, mathematical methods, aesthetic creations, schemes, rules, and methods for performing mental acts, playing games or doing business, and programs for computers and presentations of information (Art. 52 (2) EPC). For the purpose of this chapter, the most important excluded category is "discoveries." As a general rule, one can say that a discovery is the finding of something that existed before. If, however, a practical use is found for a discovery, it might qualify as an invention.

In the United States, the invention or discovery should be a "process, machine, manufacture, or composition of matter." Courts have interpreted these four categories in such a broad sense that little seems to be excluded. It has been argued that "if you can name it, you can claim it" [1]. However, abstract ideas, laws of nature, and physical phenomena, such as the discovery of a plant, are not patentable subject matter [2].

Because one of the functions of patent law is to stimulate innovation and provoke new and better products, mere statements of facts on existing matter or methods are not patentable, whether in Europe or the United States. In order to qualify for patentability, a practical use has to be added to the discovery.

2.1.1.2 Novelty

The next requirement is that the invention is new. The novelty requirement of the EPC is absolute. This means that "everything made available to the public by means of a written or oral description, by use, or in any other way, before the date of filing of the European patent application" is not regarded as novel (Art. 54 EPC). The effect of this requirement is that every disclosure made to third parties not governed by a non-disclosure agreement, destroys novelty and thus destroys patentability if the patent application has not been filed at the European Patent Office (EPO) before the disclosure. After the date of filing, you are free to disclose as much information as you wish. Even if you do not disclose anything yourself, the entire patent application will be published by the EPO 18 months after the date of filing.

The strict novelty requirement can be particularly burdensome in technologically advanced and research intensive fields like nanotechnology. Submitting a paper that describes the invention to a magazine destroys novelty regardless of whether the paper is published or not — the magazine editor who read the paper is also a third party. The same goes for presenting inventions at a conference or in a business meeting where persons from outside the inventors own organization are present. Even disclosing your invention to a friend or complete stranger can destroy novelty; although this risk can be theoretical because the burden to prove that such disclosure took place lies with the patent office or the person opposing or litigating against the patent. Nevertheless these risks should be avoided at any time.

US Patent law can be a lot more lenient in case of the situations described above. In the United States, there is a so-called grace period of one year within which you can apply for a patent after having disclosed your own invention (35 USC 102 (b)). This leaves some time for evaluation after a publication or market entry. Another difference is that in the United States the patent is awarded to the first to invent, while in Europe it is the first to file the application at the patent office who will be awarded the patent. However, in September 2011, the "Patent Reform Act of 2011" was enacted in the US. The Act replaced the first-to-file principle for the first-to-file principle. As far as the first-to-file principle is concerned, the Act takes effect 18 months after the enactment in September 2011.

The advantage of the US system is that it leaves more room for trials and exercises less pressure on the inventor and the patent attorney. In Europe, the pressure to file the patent application as soon as possible is a lot higher. However, under the pre-patent reform US system a lot of resources have to be dedicated toward proving that you are the first to invent. This requires a precise administration because it can easily become the subject of a legal procedure. In Europe no such discussion exists because the patent will be awarded to the first person to submit the application at the patent office, which can be proven with ease.

2.1.1.3 Industrial application/utility

If an invention can be "made or used in any kind of industry, including agriculture," it is deemed to have industrial applicability (Art 57 EPC). The term "industry" is interpreted broadly and includes "any physical activity of technical character" [3]. The requirement mainly serves to exclude matter which is purely aesthetic or which cannot work because it operates in contradiction with the laws of nature, for example, a perpetual motion machine [4]. In the US, the so-called utility requirement of 35 USC 101 serves a similar function as the industrial applicability requirement, but approaches the subject matter differently, by requiring that an invention must be useful.

2.1.1.4 Inventive step/non-obviousness

The inventive step requirement stipulates that an invention must not be "obvious to a person skilled in the art" (Art. 56 EPC). In the United States, where it is referred to as the non-obviousness requirement, no patent can be obtained if the invention "would be obvious at the time the invention was made to a person having ordinary skill in the art" (35 USC 103 (a)). The requirement's *raison d'être* is that society has no interest in granting exclusivity for inventions that do not advance a technology far enough to merit exclusivity.

The skilled person adds a subjective dimension to the inventive step requirement. The advantage is that it makes the requirement able to evolve with a technology. The knowledge of skilled persons increases over time, and so does the inventiveness threshold. The more prior art exists, whether patents, scientific papers, newspaper articles or any other information, the more knowledge is attributed to the skilled person, and the more difficult it becomes to meet the requirement. The downside of the wide margin of appreciation that exists when determining the knowledge of the skilled person is that it creates a lot of space for disputes. Hence, the inventive step requirement plays an important role in most patent lawsuits.

2.1.1.5 Disclosure

Once the examiner of the EPO has established that the invention meets the aforementioned patentability requirements, there are several other requirements that the patent application has to meet. One of those requirements is the disclosure requirement. The EPC stipulates that the patent application shall disclose the invention "in a manner sufficiently clear and complete for it to be carried out by a person skilled in the art" (Art. 83 EPC). In the United States, a similar requirement exists that requires the patent applicant to describe the invention "in such full, clear, concise, and exact terms as to enable any person skilled in the art to which it pertains [...] to make and use the same" (35 USC 112). In addition, an applicant for a US patent also has to describe the best mode in which the invention can be made.

This requirement serves two main functions: an information and a demarcation function. European and US patent applications become publicly available documents 18 months after the filing date. As a result, the inventor contributes the technical information that led to the invention to the public knowledge. The information should be clear enough to enable a skilled person to replicate the invention without undue burden. In this way, competitors who monitor the EPO or USPTO database can get an insight into your R&D process, but if this requirement would not exist, it would be possible to obtain a patent that is burdensome for society without giving society the benefit of the information contained in the patent. In return for this contribution to the common knowledge, the inventor receives the temporary right to exclude. In addition, patents have a demarcation function: you show in the patent what you claim as your property. If you own a plot of land and want to prevent people from entering it, you have to make clear somehow that the land is your private property, the same goes for patents. If the invention is not described in sufficiently clear terms, courts can invalidate the patent. Nevertheless, keeping the claims of a patent slightly ambiguous is attractive, and sometimes necessary, for a patent applicant. If a court has to decide whether the patent is infringed, it will look at the claims of the patent. The broader the claims, the more competing inventions can be held to infringe the patent. In addition, if the language of the patent is broad and ambiguous, competitors will get less insight in the applicants R&D process. Narrow and precise claims, however, may be too easy to circumvent and render a patent useless. It is understandable that most patentees aim for as broad as possible claims, but it is very important to keep in mind that the line between a valid broad claim and an insufficient disclosure can be extremely thin. Because an insufficient disclosure can lead to the invalidity of the entire patent, it is necessary to seek the assistance of highly specialized patent attorneys, as they are aware of all the nuances and pitfalls in your particular field of technology.

2.1.2 *Post Grant*

Once a patent is granted, it is entirely up to the patentee to decide how to use — or not to use — it.

Many patents end up in patent a company's patent portfolio, never to be looked at again. This is of course not the route to successful commercialization. In order to benefit from the patent, a myriad of possibilities exists. The patent can, for example, be sold (assignment), the right to make or use the invention can be "rented" out (license), the patent can be used as a collateral for debts, as a bargaining chip when being accused of infringement of another patent and, of course, the patent can be used to exclude all others from making, using, selling, etc., the invention. In order to utilize the patent in the best way possible, it is essential to implement a patent — or IP — strategy well before the first patent is granted.

In case of litigation, a big difference exists between Europe and the United States. Whereas a single US patent is valid for the entire United States, European patents are often described as bundles of national patents. Accordingly, a US court has jurisdiction for the entire United States and a court within Europe only has jurisdiction for the country in which it is based. If a European patent is infringed in multiple member states, a separate infringement suit has to be filed in each member state where the patentee wants to forbid the making, using, selling, etc. of the patented invention. These courts will apply their national patent act and legal precedents formed by case law within that country. A lot of effort is put in harmonizing the legal rules that apply to patents in Europe, but in the current situation, rules and case law still differ from country to country. It is possible that a court in the United Kingdom rules under UK law that a certain product infringes on a patent, while a court in Germany rules under German law that the same product does not infringe the patent.

2.1.3 Exempted Uses

Once the patent has been granted, not every possible use can be prohibited by the patent holder. In Europe, most countries have limited the reach of patents. In the United States, exemptions from patent infringement do exist but their scope is very narrow.

In the various national laws of Europe it is commonly accepted that patents only extend to commercial use. Commercial must be understood as meaning "non private." As a result, the non-private use of a patented invention causes liability for infringement even if the entity that uses the invention is a non-commercial entity such as a public hospital, a university, or a charitable organization. "Private" is interpreted narrowly and usually only applies to use within ones personal environment. In the United States, private — or personal — use is not exempted from infringement. "[W]hoever without authority makes, uses, offers to sell, or sells any patented invention, within the United States or imports into the United States" infringes that patent (35 USC 271). Whether the use is private or non-private is not an issue in terms of infringement.

In addition to the aforementioned private use exemptions, most European jurisdictions contain a research exemption. Research exemptions provide that acts that are conducted for experimental purposes "relating to the subject matter of the invention" cannot be held infringing. The exact definition of subject matter of the invention differs between member states, but in general one can say that it is allowed to verify the claims of a patent or conduct research into, for example, improving the patented product as long as the patented product is not used outside the experimental context without the permission of the patent holder. Whether or not the experiment is conducted by a commercial or non-commercial entity is not relevant, neither is the intent of the research. If a company conducts experiments on a patented invention in order to develop a new product based on that invention, it can still rely on the research exemption. If the experiment is successful and a marketable product is developed in which the patented invention in included, the product can only be released with the permission of the patent holder, for example, via a license agreement.

US patent law also contains a research exemption. However, the exemption is that narrow that the practical use for industries and even universities is very limited. If an act is part of the legitimate business of an entity and it is not conducted "solely for amusement, to satisfy idle curiosity, or for strictly philosophical inquiry," the research exemption does not apply [5]. As a result of the legitimate business requirement, universities can also infringe a patent in the course of experimentation, because experimentation is part of their legitimate business.

2.2 Other Forms of Intellectual Property

Beside patents, other different types of Intellectual Property Rights (IPRs) might be envisioned for a more or less effective protection of nanotechnology innovations. Among them we can mention: copyright, trademarks, integrated circuit layouts, designs, and a variant of patents, the utility model. Utility models are not foreseen in all jurisdictions and it consists of a right protecting something less than an invention, usually an improvement or betterment of an existing product or process, so it covers some sort of incremental invention. This kind of intellectual property right is also commonly referred to as "petty patent" or "small patent."

2.2.1 *Utility Models*

If a product has a lesser inventive character, a utility model can be a good method of obtaining protection. Utility model applications are generally not examined as thoroughly as patents and the requirements are usually lower, especially the inventiveness requirement. A utility model provides a shorter period of protection than the usual 20 years granted to ordinary patents. For instance, German utility models are valid for 10 years [6]. As to the possibility of protecting nanotech-related inventions through utility models, this could, for example, be useful in instances where the "minor" invention involves the employment of nanotech-related materials instead of a known traditional material. If the invention meets the less stringent requirements of a utility model and not those of a patent, the inventor will be able to protect the claimed invention through this residual kind of IP protection.

2.2.2 *Copyright*

Copyright can be used to protect the expression of an idea. Such expression can be a production in the literary, scientific and artistic domain, whatever may be the mode or form of its expression, such as books, pamphlets and other writings; lectures, addresses, sermons and other works of the same nature; dramatic or dramatico-musical works; choreographic works and entertainments in dumb show; musical compositions with or without words; cinematographic works to which are assimilated works expressed by a process analogous to cinematography; works of drawing, painting, architecture, sculpture, engraving and lithography; photographic works to which are assimilated works expressed by a process analogous to photography; works of applied art; illustrations, maps, plans, sketches and three-dimensional works relative to geography, topography, architecture or science.

(Article 2 Berne Convention for the Protection of Literary and Artistic Works)

So, as is easily inferable from the above definition, copyright covers most of human ingenuity's expressions. Because copyright can only be used to protect the expression of ideas in, for example, journal publications or conference presentations, it cannot be used to protect the idea itself. For protecting the ideas from being used by others, you can only rely on patents, and to a lesser degree on utility models. For copyrights, no formal registration procedure exists. The protection arises as soon as the work is created and meets a certain standard of originality, which can differ between different jurisdictions. However, in the US, it might be wise to register the copyright with the Copyright Office because such registration is necessary in case of a lawsuit. Copyright terms and the ways in which they are calculated differ between jurisdictions, but are very long compared to patents. For example, in the EU, a copyright lasts for the life of the author plus 70 years. Contrary to popular belief, it is not necessary to use the © symbol in combination with the year of creation and name of the creator — or any other statement — in order to obtain a copyright. It can be useful, however, to use this statement in order to prove when the work was created.

2.2.2.1 Economic rights, moral rights, and other features

For sake of completeness, it is worth mentioning that there are economic rights and moral rights attached to copyrighted works. When we talk about the economic rights, we refer to all kinds of monetary benefits that can be related to the exploitation of copyrighted materials. Some examples are the licensing or assignment of the works. As to the moral rights, some jurisdictions recognize the right of the authors to be mentioned as the creators of the works and "to object to any distortion, mutilation or other modification of, or other derogatory action in relation to, the said work, which would be prejudicial to his honor or reputation" (Art. 6b is Berne Convention). Moral rights are usually not alienable.

2.2.3 *Trademarks*

Trademarks are distinctive "signs" that are used to distinguish the products or services of an undertaking from those offered by others. Their main goal is to avoid confusion about the products origin among consumers. The word "sign" can be tricky sometimes though. In fact, by "sign" usually trademark laws cover things like music, smells, tridimensional shapes, and alike. To be more precise, in the

European Union, for example, Article 4 of the Community Trademark Regulation provides that a

> A Community trademark may consist of any signs capable of being represented graphically, particularly words, including personal names, designs, letters, numerals, the shape of goods or of their packaging, provided that such signs are capable of distinguishing the goods or services of one undertaking from those of other undertakings.

So, for example, a trademark can easily protect a distinctive word or the peculiar shape of a bottle of water. For nanotechnology, the important thing to understand, though, is that trademarks do not protect the functional features of a product. These cannot be trademarked but should be patented instead. The reason is simple: trademarks have a potentially unlimited lifespan. As long as the trademark holder extends the mark in compliance with the applicable procedures, the trademark will continue to exist. If you want to protect the technical features of a product you can get a patent that, in exchange of the temporary legal monopoly, will allow the use of the technology by the public at large after 20 years. In fact, the development of technology may not be hindered by a potentially infinite legal monopoly of the patentee.

2.2.3.1 Requirements and characteristics

To get a trademark, the sign in question should be distinctive. By distinctive we mean that it must not be descriptive with regard to the product or service that you wish to trademark. So, for example, "Apple" is a distinctive sign for computers, but not for a fruit seller. If a fruit seller would obtain a trademark to the word "Apple," this fruit seller could forbid all others from using the word "Apple" in relation to the sale of fruits. This, of course, would be highly undesirable. To prevent that established words are taken from the public domain and are claimed as private property, the distinctiveness requirement exists. When filing your trademark application, you have to designate for which class of goods and/or services you wish to register the trademark, for example, antibiotics, carbon, skin care, or transistors. These classes have been established in the Nice Agreement. The trademark office in Europe, either the national trademark office of the relevant member state, or the Office for Harmonization of the Internal Market (OHIM) and in the United States the USPTO, will determine whether the trademark is distinctive for the relevant product classes.

If the mark applied for is a "reproduction, an imitation, or a translation, liable to create confusion," of a well-known mark, both the European trademark offices and the USPTO will reject the application (Art. 6b is Paris Convention for the Protection of Industrial Property). In case the mark applied for is identical to an existing but not well-known mark in the applicable product classes, the USPTO will reject the application whereas in Europe it is the holder of the existing trade mark who must file an opposition to the application trademark office where the application is filed. Trademark applications that are not identical, but similar to existing trademarks have to be opposed by the right holder both in Europe and the United States.

2.2.4 *Industrial Designs*

Industrial designs can be used to protect the appearance of a product, but not its functional aspects. Because copyrights and trademarks can also be used to protect the appearance of products, overlaps of rights may occur. In fact, it is very common that products are protected by several independent IPRs at the same time. To be granted protection in Europe, the design must be novel and possess a distinctive character (Art. 3 (2) Directive 98/71/EC). In the United States, design patents are available for products that are "new, original and ornamental" (35 USC 171). The ornamental, or decorative, character is important because the design does not protect functional features. Furthermore, US design patents can only protect products that have a utility, in order to exclude purely aesthetic works from protection through design patents.

Like in copyrights and trademarks, the protection attached to nanotechnology can be solely indirect. In fact, let us assume that the unique appearance of a product is determined by the use of nanomaterials or nanotech-related process, in this case the product, where it meets the registrability requirements of a design, might well be protected but what the protection will cover is exclusively the appearance itself. So, let us say that a piece of furniture is made using carbon nanofibers and as a result its appearance can be protected as a design, in this case what the owner will be able to do is to enforce her rights against infringers who copy the appearance of the product but not the way the product is made.

2.3 Trade Secrets

Because nanotechnology can be hard to reverse engineer, trade secrecy may at present be an attractive form of protecting knowledge in the nanotechnology industry [7]. Trade secrets can be used to prevent that any kind of commercially valuable information is being made available to, or used by, others, but once the information has become public via an independent discovery, there is nothing that can be done against the use of that information. The way in which trade secrets are regulated varies between different jurisdictions, but the Agreement on Trade Related Intellectual Property Rights (TRIPs) sets minimum levels of protection that are applicable in every member state (153 [8]). In the TRIPs context, a minimum level of protection means that member states should provide at least the protection required by the TRIPs agreement, but are allowed to impose further reaching protection for, say, trade secrets. Every TRIPs member state must, at least, provide legal measures that enable the lawful possessor of information to prevent that information from being disclosed to, acquired by, or used by others if that information:

(a) is secret in the sense that it is not, as a body or in the precise configuration and assembly of its components, generally known among or readily accessible to persons within the circles that normally deal with the kind of information in question;
(b) has commercial value because it is secret; and
(c) has been subject to reasonable steps under the circumstances, by the person lawfully in control of the information, to keep it secret. (Art. 39 section 2 TRIPs).

Treating certain information as a trade secret has some obvious benefits over applying for patent protection for the possessor of that information. Unlike patents, that are subjected to a disclosure requirement, a trade secret, by its very nature, does not inform competitors about past or ongoing R&D activities. Furthermore, there is no formal procedure that one has to follow to obtain a trade secret and the subject matter to which it can be applied is much broader than patents. Patents are only available for restricted categories of inventions, but trade secrecy can be applied to any information that meets the requirements of the TRIPs Agreement, although the exact way in which these requirements are applied differs between different jurisdictions. If it is questionable whether an invention will withstand the examination proceeding at the patent office, trade secrecy might be an option. A major advantage of trade secrets over patents is that the duration of a trade secret is potentially unlimited, unlike patents, which are restricted to a 20 year term. The downside of the undefined term is that there is no certainty on how long the trade secret can be maintained. If a competitor independently obtains the same information after, say, one year, a patent would have been a better way of protecting the invention. A patent could in such case be used to forbid the competitor to use the invention, whereas a trade secret loses its entire value as soon as it publicly available. If the invention is of such a complexity that it would take competitors more than 20 years to develop it independently, trade secrecy might be more attractive.

The aforementioned lack of a formal procedure does certainly not mean that trade secrets are easier to obtain and maintain than patents. To make sure that the information remains a secret, a strict secrecy policy has to be applied and detailed records have to be kept. To a certain extent this overlaps with the secrecy and record keeping that is required to obtain patent rights.

References

1. Adelman, Rader, and Thomas (2009) *Cases and Materials on Patent Law*, Thomson Reuters, p. 58.

2. Ibid.

3. Guidelines for Examination in the European Patent Office, Part C, Chapter IV, 5.1

4. Ibid.

5. 307 F.3d 1351 (Fed Cir. 2002)

6. § 23 section 1 Gebrauchsmustergesetz.

7. Lemley, and Mark A. (November 2005) Patenting nanotechnology, *Stanford Law Review*, 58, p. 19. Available at SSRN: http://ssrn.com/abstract=741326 or doi:10.2139/ssrn.741326

8. http://www.wto.org/english/thewto_e/whatis_e/tif_e/org6_e.htm

Chapter 3

Nanotechnology Patent Procurement and Litigation in Europe

Hanna R. Laurén
Oy Jalo Ant-Wuorinen Ab, Iso Roobertinkatu 4-6 A, FI-00120 Helsinki, Finland
hanna.lauren@jalopat.fi

3.1 Obtaining Patent Protection in Europe

In order to obtain patent protection in Europe, there are three possible alternatives for an inventor: the national, European, and international routes. The national and European procedures lead to the grant of a patent, while an international patent application must be continued as national or regional patent applications after the international stage of prosecution has ended in order to obtain patents in designated countries or regions. The most important characteristics of the different routes are summarized in the following.

3.1.1 *National Route*

If the applicant is only interested in protecting his invention in a very limited amount of European countries, for example, only one or

Nanotechnology Commercialization for Managers and Scientists
Edited by Wim Helwegen and Luca Escoffier
Copyright © 2012 Pan Stanford Publishing Pte. Ltd.
ISBN 978-981-4316-22-4 (Hardcover), 978-981-4364-38-6 (eBook)
www.panstanford.com

two, the most cost-effective choice is to use the national route and to file separate national applications before each country's national patent office. In order to utilize the priority right arising from the first patent application concerning the invention, the applicant must remember to file all subsequent applications concerning the same invention within the 12 months' priority period.

When the national route is used, the patent applications shall conform to the formal and substantive requirements set by national legislation. The substantive patent law has been extensively harmonized in the contracting states of the European Patent Convention. However, since the application procedures are differently structured and the applicant is obliged to prosecute multiple applications in parallel, the scope of protection obtained will usually not be uniform.

As to the formal requirements, the application text must usually be translated in the official language of the country already in an early stage of prosecution to enable the examination of the application. Normally it will be necessary to employ a local patent attorney in each country to represent the applicant before the national patent office and to facilitate communication between the applicant and the office, as the language of the proceedings is usually an official language of the country.

3.1.2 *European Route*

When a wider patent protection in Europe is desired, it will be more reasonable to file a single European patent application before the European Patent Office, instead of filing multiple national applications. It has been estimated that a European patent costs about as much as three or four national patents. This estimate takes into account the official and attorney fees [1].

The grant procedure of European patents is governed by the European Patent Convention (EPC) [2, 3]. There are currently 38 contracting states of the EPC, and additionally European patents can be extended to the so-called extension states, which at present are Bosnia and Herzegovina, and Montenegro [4]. Russia, Belarus, Ukraine, and Moldova are not parties to the EPC.

A European patent confers to the patent proprietor the same rights as would be conferred by national patents granted by the respective contracting states [5]. In essence, the European route is a way to obtain a set of national patents in a single procedure. It enables the applicant to streamline the prosecution in the application stage as there will be no separate correspondence with each national office via local attorneys. The language of proceedings before the EPO and the language of the European patent application is one of the three official languages of the EPO: English, French, or German.

European patent applications are searched and examined centrally by the European Patent Office (EPO). The procedure is divided in two principal stages: the first stage comprises preparation of a search report and a preliminary opinion on patentability. The search report lists the most pertinent prior art publications. In the second stage the substantive examination follows, which means that an examining division and the applicant (or his representative) communicate in order to agree on the appropriate scope of protection, that is, the wording of the claims. The applicant may need to amend the claims to make them conform to the patentability requirements of the EPC. If the application does not contain any patentable subject matter that could form a basis for acceptable claims, the application has to be refused. Decisions of the first instance divisions of the EPO can be appealed to the Boards of Appeal of the EPO [6].

Upon grant, the claims of the European patent must be translated to the other two official languages of the EPO. Additionally, most contracting states require that the claims and the description be translated to their official language in order for the patent to take effect in that state [7].

As the search and examination of nanotechnology inventions requires special competence from patent examiners, it may be preferable to use the European route instead of prosecuting direct national applications. The EPO has taken measures to ensure that nanotechnology applications are directed to examiners with expertise in the correct field, and it has established a special nanotechnology classification system that facilitates prior art searches. The EPO also collaborates with the Japan Patent Office and the United States Patent and Trademark Office to tackle shared issues with regard to nanotechnology [8]. Therefore the applicants can expect to obtain high-quality searches and concomitantly strong and enforceable patent rights by using the European route.

3.1.3 *International Route*

Instead of filing a European patent application directly at the European Patent Office, it is possible to achieve the same result by using the international route, which is based on the Patent Cooperation Treaty (PCT) [9–11]. An international patent application (a PCT application) enables the applicant to get a patent application pending with the future option of continuing the application nationally to desired PCT contracting states (144, currently [12]). The European Patent Organisation is a party to the PCT, which means that a PCT application can be continued as a European patent application and thereby it has the effect of a regular European patent application [13].

The PCT system does not provide any global or multinational patent; it is a procedure to get an application pending and to obtain more time for decision making. In the so-called international phase, the international application is searched by one of the international search authorities. The search produces an international search report citing the most relevant prior art documents and a preliminary opinion on patentability (a written opinion). If the preliminary opinion does not seem promising with regard to obtaining a patent, the applicant has the further choice of filing a request for substantive examination (a demand) and amending the claims in that connection. The international examination produces a report on patentability. Alternatively, the applicant may only file comments to the preliminary opinion without requesting examination if he considers that the opinion is ungrounded.

The national phase follows after the international phase. Usually the national phase begins 30 months after filing the PCT application, but in the case of continuing to the EPO the relevant time limit is 31 months. The opinions issued during the international phase as well as the responses of the applicant are forwarded to the designated national and regional offices. Although the designated offices are not obliged to follow the result of the international phase, it is usually worthwhile to aim at obtaining a favorable opinion on patentability already during the international phase. In this way the applicant will save time and money during the prosecution of the national/regional applications as the claims have already been refined centrally before the international patent authorities.

As noted above with regard to the European route, also the international route offers the advantage of obtaining a high-quality search together with a patentability opinion before the ultimate decision on the designated states is to be made. The international route is clearly the most effective way to obtain more time, and it also enables the applicant to postpone the paying of fees to the national offices. However, if only European countries are of interest, the European (or national) route should be chosen as more cost effective.

3.2 Post-Grant Proceedings for a European Patent

Once a European patent has been granted, a nine-month opposition period follows. During the opposition period anyone may give the EPO a notice of opposition on the grounds that the invention is not patentable under the EPC, the disclosure of the invention is insufficient, or the subject matter of the patent extends beyond the content of the application as filed. The notice of opposition is examined by an opposition division of the EPO. As a result of the examination, the opposition division may maintain the patent as granted, maintain it in amended form, or revoke it [14].

As nanotechnology is an emerging field, patent offices, including the EPO, are in the process of learning how to search and examine applications concerning nanotechnology inventions. Therefore it is advisable that companies working in the field conduct patent searches and monitor granted patents so that they can immediately react to overly broad patents granted to competitors and file a notice of opposition. Decisions of the opposition divisions take effect in all designated states, which means that filing a notice of opposition is a very effective way to revoke or limit a competitor's weak patent *ab initio*. Even patent offices in other countries may pay attention to the result of opposition proceedings before the EPO when examining applications related to the same invention.

If the patent proprietor wants to amend the claims of a granted European patent or revoke it entirely, he may use the central limitation/revocation procedure provided by the EPC. The EPO examines the limitation or revocation request and issues a decision. The limitation or revocation of the patent will apply *ab initio* in all the designated states [15].

From a legal point of view a European patent equals a set of national patents, and therefore the enforcement of the patent right shall take place before a competent national court [16]. In the case of infringement the patent proprietor may raise an infringement action before a national court, while the accused infringer may resort to raising a revocation counteraction challenging the validity of the patent. If the accused infringer acts in multiple European countries, the patent proprietor is obliged to raise an infringement action separately in each country. Equally, the accused infringer needs to prosecute multiple revocation actions before national courts, as the European Patent Convention does not provide any centralized revocation procedure for third parties after the nine-month opposition period has lapsed.

Among the contracting states of the European Patent Convention, there are substantial differences in the procedures and practices of patent litigation. Some countries, such as Austria, Finland, Italy, and Sweden, have established specialized patent courts. In others, the plaintiff may choose in which court to raise the action. The costs of litigation vary a great deal, being generally higher in common law countries (UK) when compared to civil law countries (e.g., Germany, France, and the Netherlands). Other differences stem from the speed of the procedure, the availability of preliminary injunctions, the technical competence of the courts, the use of external experts and their statements as evidence, and the level of damages. In some countries, such as Germany, infringement actions and revocation actions are dealt with by different courts.

Litigation before national courts of the contracting states presents many challenges and uncertainties in the case of nanotechnology patents. Particularly, the technical competence of the national courts and judges, and the possibility to use technical experts and statements as evidence is crucial in order to successfully litigate a nanotechnology patent. Although the national courts in principle apply the same substantive law, divergent interpretations are possible. Decisions of national courts cannot be appealed to the Boards of Appeal of the EPO; only national appeal routes are available.

There have been plans to establish an integrated European patent court that would have jurisdiction to hear all infringement and revocation actions concerning European patents [17]. Establishment of an integrated litigation system would reduce the costs of parties by removing the need for parallel actions. It would also speed up and streamline the litigation procedures, and reduce case law divergence [18, 19].

3.3 Patentability of Nanotechnology in Europe

3.3.1 *Patentable Inventions*

According to the European Patent Convention, European patents shall be granted for any inventions, in all fields of technology, provided that they are new, involve an inventive step, and are susceptible of industrial application [20]. However, the EPC also defines certain exclusions to this basic principle. Subject matter categories that are excluded from patentability are, for example, discoveries and scientific theories; business methods and computer programs as such; inventions the commercial exploitation of which would be contrary to "ordre public" or morality; plant or animal varieties or essentially biological processes for the production of plants or animals; and surgical, therapeutic, and diagnostic methods [21].

The most relevant of the exclusions is perhaps the exclusion of discoveries, as certain nanostructures occur also naturally. The invention may still be patentable if the claims are directed suitably to the new subject matter (*vide infra*). The concept of "ordre public" has been interpreted by the Boards of Appeal as encompassing also the protection of the environment [22]. If the exploitation of a nanotechnology invention is likely to seriously prejudice the environment, it will probably be excluded from patentability. The exclusion of surgical, therapeutic, and diagnostic methods does not extend to products for use in these methods. Therefore, for example, instruments, devices, compositions, and substances for medical purposes are patentable under the EPC.

None of the exclusions mentioned in the EPC directly addresses nanotechnology inventions, meaning that there are no categorical restrictions for patenting nanotechnology. If the invention satisfies the requirements set for novelty and inventive step, a broad scope of protection may be available for pioneering nanotechnology inventions.

3.3.2 *Novelty*

In order to be patentable, the invention must be new. Article 54(1) of the European Patent Convention defines the novelty requirement in the following way:

> An invention shall be considered to be new if it does not form part of the state of the art.

The state of the art shall be held to comprise everything made available to the public by means of a written or oral description, by use, or in any other way, before the date of filing of the European patent application.

The meaning of this provision is that the invention must not be disclosed in any prior publications or used publicly before the filing date, or if priority is claimed, before the priority date. European patent law does not recognize any novelty grace period, and the novelty requirement is absolute: the invention must not be published in any way or by anybody before filing the patent application. This applies similarly to the inventor or applicant himself as well as to any other people disclosing the invention to the public. It is irrelevant in which language or in which country the prior disclosure takes place, or whether it is a question of a patent publication or a scientific article: all kinds of disclosures may function as a novelty-bar.

During the examination of the patent application, novelty is judged for each independent and dependent claim separately. A claim is not new if all features of it are disclosed in a single prior art publication. Publications are usually not combined, except if there is an explicit reference to another publication with regard to certain features or for obtaining a more detailed description of them.

Nanotechnology is a multidisciplinary field that is characterized by the small scale: current definitions of nanotechnology restrict it to structures exhibiting dimensions below 100 nm [23]. The essential feature of nanotechnology is that the small dimensions provide unexpected and previously unknown properties to the materials and devices, which immediately raises many questions with regard to patentability. Is an invention new if the only difference to the prior art is the smaller scale? Is a selection of a sub-range from a larger known range patentable? Does the discovery of new size-related properties render the nanoscale material new?

3.3.2.1 Selection inventions

In patent law the term "selection invention" refers to an invention that concerns a sub-range within a larger previously described range. In claims features such as the size of particles or the width of flow channels may be defined by means of a numerical range of values. Most often selection inventions relate to numerical ranges, but also non-numerical selection inventions are possible, such as the selection of one or more individual chemical compounds from a previously known generic formula covering a larger set of chemical compounds.

For macroscale inventions the dimensions of the material are usually not defined very specifically in patent publications. It may be that only an approximate upper limit is given, or alternatively the dimensions are not defined at all, as the skilled person is assumed to be able to arrive at appropriate dimensions even without any instructions. This leads to a situation where a later nanoscale invention falls under the macroscale dimensions disclosed in a prior publication. Is such a sub-range new and patentable?

According to the practice of the EPO, numerical selection inventions are patentable if the following criteria are met:

- The selected sub-range is narrow compared to the known range.
- The selected sub-range is sufficiently far removed from any specific examples disclosed in the prior art and from the end-points of the known range.

- The selected sub-range is not an arbitrary specimen of the prior art, that is, not a mere embodiment of the prior art, but another invention (purposive selection, new technical teaching) [24, 25].

The meaning of "narrow" and "sufficiently far removed" should be decided on a case-by-case basis. In general, we can say that nanotechnology inventions relate to a very narrow range of dimensions (1 to 100 nm or even narrower) in comparison to macroscale products and materials, which usually are measured in micrometers or millimetres. The nanoscale is also rather far removed from the macroscale. Therefore it is usually not difficult to meet the first and the second criteria.

As for the third criterion, the applicant should be prepared to show that the selection is purposive and not arbitrary. Particularly, the special nanoscale effects should exist in the whole of the selected sub-range and not anywhere else. At the least, there should be a clear enhancement or improvement of particular properties in the selected sub-range which cannot be obtained by making the selection of dimensions in some other way. Comparative experimental data is an effective way to demonstrate that the selection is purposive. The data should preferably be presented in the patent application in a way that supports the claims and emphasizes the correlation between the selected dimensions and the technical effects.

Although the EPO recognizes the novelty of numerical selection inventions that meet the above criteria and grants European patents for them, German courts are likely to invalidate such patents on the basis of the interpretation adopted by the German Federal Supreme Court (BGH) [26]. The standpoint of the BGH has been that the disclosure of a numerical interval is a simplified notation of all the values lying between the minimum and maximum values. Thus the disclosure of a numerical range takes away novelty from all individual values and sub-ranges falling within the known range. According to this "arithmetic doctrine" neither the location of the sub-range within the known range nor the special effects that can be attained play any role when novelty is judged.

While a large proportion of all European patent litigation takes place before German courts, the divergent interpretations of the EPO and the BGH potentially create a significant barrier for enforcing nanotechnology patents in Germany and may also affect out-of-court settlements elsewhere. The applicants should prepare for such enforceability problems by including claims designed particularly for German courts. One effective solution could be to direct claims not to the nanomaterial itself but to the use of the nanomaterial for a certain technical purpose or for achieving a certain effect. Use claims render novelty to the selection invention and traditionally offer a relatively broad protection scope in Germany [27].

3.3.2.2 Size-related properties

In macroscale, discovering a new property for a known substance does not make the substance new, because the property is considered to be inherently included in the substance. In nanoscale this general principle does not apply, as the occurrence of new properties relates to the dimensions of the material: the properties can only be observed if the dimensions of the material are below a certain threshold or within a limited range in the nanoscale. It will not be possible to foresee the new properties by examining the macroscale counterpart.

However, the situation is usually not as simple as it sounds. Sometimes the nanoscale version exhibits similar properties as the macroscale version but they may be present as enhanced or improved. Such a phenomenon may be due to the fact that, for example, only part of the molecule population has been modified. Nanomaterials showing enhanced properties with regard to the macroscale counterpart are within the grey zone when novelty is judged.

In the history, there are examples of situations where nanotechnology has been used without knowing that it is actually question of nanotechnology. Damascus steel was developed in the Middle East for over a thousand years ago, and swords made of it were particularly strong and sharp. In 2006, it was found out that the strength of the steel material was caused by the carbon nanotubes contained in it. However, with regard to the patentability of Damascus steel, the novelty has already been lost, even though the true cause of the strength was discovered only in 2006 [28].

In many cases the knowledge of a detailed reaction mechanism in a process or the exact molecular structure of a material may remain unknown for a long period of time even though the reactions or materials themselves are well-known and in use. Obtaining such missing information does not make the processes or materials new and patentable, although the information is valuable in a scientific sense.

3.3.2.3 Naturally occurring products

Nanotechnology inventions may relate to substances or compositions that occur also naturally to some extent. For example, fullerenes and carbon nanotubes have been reported to be found in ash-type materials even without any human interference. Similarly products of gene technology — genes — did exist in nature already before any innovations for using or modifying them was made.

The question that arises is whether such naturally occurring products can be patented. The European Patent Convention excludes mere discoveries from patentability [29]. However, if the applicant is able to show that he has provided a way of preparing or isolating the substance in question, then the exclusion does not apply. In the case of nanomaterials, the problem of finding a working synthesis or isolation method often forms a big obstacle, even though the materials themselves have perhaps been known already for a while from nature or described or predicted in prior publications. The essential point is that if the prior publication fails to enable a skilled person to isolate the substance from its natural environment and to prepare it in pure form, then the substance is not considered to be known and the publication is not novelty-destroying. Inventing a first isolation method for a naturally occurring nanomaterial may well be worth a broad patent with claims directed both to the process, which certainly is new, but also to the product, which now can be prepared by following the teaching given in the patent application [30].

3.3.2.4 Case law of the boards of appeal

The case law of the Boards of Appeal of the EPO provides some guidelines on when a nanotechnology invention may be considered new. The Boards have established that a substance that exhibits a new crystal structure is considered to be new with regard to other crystal structures of the same substance or with regard to an amorphous crystal structure. Therefore a nanocrystalline structure should be able to bring novelty to a substance.

In the decision T 0915/00 (2002), Integran Technologies had appealed to the Board of Appeal on a decision of an opposition division, according to which their European patent had been revoked. The contested patent related to a method for preparing nanocrystalline nickel material with a crystal size of less than 100 nm. Corresponding macroscale methods were known previously, but they were unable to produce the material in a nanocrystalline form. Therefore the method according to the invention was considered to be new.

According to the decisions T 0552/00 (2003) and T 0006/02 (2005) miniaturization of known structures, for example, decreasing the particle size, is sufficient for obtaining novelty.

In T 0552/00 (2003), the patent holder, Smithkline Beecham Biologicals, and the opponent, Wyeth Holdings, had appealed to the Board of Appeal on a decision of an opposition division, according to which the patent was maintained in amended form. The invention related to a vaccine composition comprising particles of an antigen and a 3-O deacylated monophosphoryl lipid A. Corresponding particles with a size in the range of 80 to 500 nm were known from a prior publication. In the contested patent the particle size was 60 to 120 nm, whereby the overlapping region was in the range of 80 to 120 nm. The Board considered that the prior publication did not provide the skilled person with any guidance for preparing particles with any specific size within the broad range of 80 to 500 nm. Neither did it indicate any particularly preferred sub-range within this range. Therefore the publication did not seriously contemplate applying the technical teaching described therein in the range of overlap and particles of this size had not been made available to the public in the sense of Article 54 EPC. The invention was thus new.

In T 0006/02 (2005), the opponent, Rhodia Acetow, had appealed on a decision of an opposition division according to which the European patent of Celanese Acetate had been maintained in amended form. The patent concerned a cigarette filter the fibres of which contained a cellulose ester and TiO_2 particles with an average particle size of less than 100 nm. A corresponding product with a particle size distribution of 10 to 1000 nm, preferably 50 to 500 nm, was known from a previous publication. However, on the basis of this piece of information it was not possible to estimate what the average particle size was. Particularly the Board noted that the lowest value of the particle size distribution (10 nm) in the prior publication did not amount to an average particle size of 10 nm and thus could not be used as starting point for any calculation. Further, the only specific average particle size disclosed in the examples of the publication (300 nm) was outside the claimed range.

On the basis of the foregoing discussion, an important factor during the judgement of the novelty of a nanotechnology invention is whether the prior art actually enables the skilled person to prepare the nanoscale material. Ambiguous, unclear or otherwise insufficient disclosures cannot function as a novelty bar for later inventions [31].

3.3.3 *Inventive Step*

A patentable invention not only needs to be new, but there should also be a so-called inventive step with regard to the prior art. This means that the invention shall not be obvious: modifications and improvements that every skilled person would immediately suggest are not patentable. In the European Patent Convention this requirement is defined in Article 56:

> An invention shall be considered as involving an inventive step if, having regard to the state of the art, it is not obvious to a person skilled in the art.

Inventiveness is not an objective concept in the same way as novelty is. To assess inventiveness, it is necessary to judge how an average skilled person would act on the basis of the prior art and the common general knowledge. Would he arrive at the invention or not? The skilled person should be thought of as a hypothetical person who is aware of all prior publications and disclosures (which is clearly impossible for any real person working in any field of technology) and who possesses average practical skills and is aware of the common general knowledge in the field. The common general knowledge means the information contained in basic handbooks, monographs, and textbooks. If the field is so new and fast-developing that textbooks do not cover it yet, the common general knowledge may also include the information contained in patent publications and scientific articles [32].

Nanotechnology is a unique field in the sense that inventions often involve elements from several different fields of technology, such as chemistry, physics, medicine, and biology. Therefore it is appropriate to consider that the concept of a skilled person in this case means a group of skilled people, each of which has average skills and knowledge of his own field. For biotechnological inventions, a similar concept of a group of skilled people has been used in the decisions of the Boards of Appeal. Such an interpretation inevitably leads to a higher bar for inventiveness: what is non-obvious to one person may be obvious to a group of people and thus non-patentable.

3.3.3.1 Argumentation

The Boards of Appeal of the EPO have established a "problem and solution approach" for assessing inventive step [33, 34]. During the prosecution of a European patent application, it is highly preferable to follow this approach when responding to an objection under Article 56. In brief, the problem and solution approach consist of the following steps:

(a) Identifying "the closest prior art," that is, the prior art publication having the most in common with the invention
(b) Defining the objective technical problem to be solved by studying the features distinguishing the invention from the closest prior art
(c) Considering whether or not the claimed invention, starting from the closest prior art and the objective technical problem, would have been obvious to the skilled person

In step (c) the question is not whether the skilled person *could* have arrived at the invention by modifying the known technology but rather whether he *would* have done so on the basis of the teachings of the prior art.

The Guidelines for Examination in the European Patent Office ("Guidelines") provide examples of circumstances where an invention may be regarded as involving an inventive step and where not. If the invention provides an unexpected technical effect, such as the special properties occurring in the nanoscale, it may be an indication of an inventive step. Also if the invention fulfils a long-felt need or involves overcoming a technical prejudice, an inventive step probably exists [35].

3.3.3.2 Obviousness of miniaturization

As concluded above, reducing the scale makes an invention novel, since different dimensions constitute a distinguishing feature. A more difficult question is whether reducing the scale is non-obvious. Miniaturization is certainly a common goal in many fields as it provides many advantages in the form of material savings, improved speed, and smaller and lighter devices, etc. According to the EPO practice, if the invention can be arrived at merely by a simple extrapolation in a straightforward way from the known art, it might lack an inventive step [36]. Therefore the applicant should be prepared to demonstrate that it is non-obvious for a skilled person to miniaturize known devices and processes to the nanoscale.

Preparation and characterization of nanostructures often involves issues that are completely different from the known macroscale phenomena. In order to manufacture devices and materials with dimensions below 100 nm, it may not be possible to proceed by utilizing the so-called top-down approach, that is, by merely making smaller and smaller versions of the known devices by suitable machining methods. As an alternative to the traditional top-down methods, bottom-up processes have been developed. The bottom-up approach is based on building nanoscale structures by assembling small substructures to larger entities until the desired final structure has been attained. The fact that different or entirely new methods are needed for manufacturing nanoscale versions of known macrostructures makes the nanostructures non-obvious and thus patentable.

3.3.3.3 Case law of the boards of appeal

In T 0268/96 (2001) the invention related to manufacturing a semiconductor device with so small dimensions that conventional patterning techniques were no longer applicable. The Board considered that the new method according to the invention was inventive as it could not be directly derived from the prior art.

In T 0070/99 (2003) the contested patent related to a method of separating a subpopulation of cells from a cell-containing liquid sample. In the method the cell sample was passed in microscale (100–500 µm) flow passages comprising immobilized proteins for separating target cells from the sample. The target cells that had been bound to the proteins were released by changing the conditions in the passages. Treating cells in the above manner was known in the bench-scale. Thus the invention related to miniaturizing the method and performing it with smaller sample volumes. According to the Board the only advantage from the miniaturization was the reduction in the size of the device and the amounts of the reagents. As miniaturization of analysis and processing devices is a common goal in the field, the Board was of the opinion that the skilled person would have aimed at narrowing the flow passages of the known bench-scale method. Therefore the claimed method was not inventive.

3.3.4 *Industrial Applicability*

A patentable invention shall be industrially applicable. This means that it should be possible to manufacture or use the invention in some field of industry [37]. The industrial applicability of the invention has to be described in the application explicitly by giving credible examples, unless the applicability is obvious on the basis of the nature of the description or the nature of the invention [38].

On traditional fields of technology, such as mechanics, the industrial applications of new inventions are usually evident. Nanotechnology, on the other hand, is by its nature a non-predictable and fast-developing field. Therefore the industrial applications of nanotechnology inventions may not be obvious to an average skilled person, even though the structural details of the invention had been disclosed carefully in the patent application.

Preferably the practical applications of a nanotechnology invention should be exemplified in the patent application by giving at least one credible application. If the invention relates to a very specialized or newly emerged subfield of nanotechnology, then more time and space should be devoted for describing different applications of the invention. If at the time of filing the patent application the practical uses of the invention are not clear and further testing is still required, it is probably better to postpone the filing until enough data has been collected and more justified conclusions can be drawn on the industrial applicability.

3.4 Drafting a European Patent Application

3.4.1 *General Considerations*

A European patent application comprises a description part (possibly including figures), and one or more claims. While the description discloses the invention to the public, the claims briefly state where the borders of the exclusive patent right lie. Therefore the claims form the most important part of any European patent. In the European Patent Convention this is defined in Article 69:

The extent of the protection conferred by a European patent or a European patent application shall be determined by the claims. Nevertheless, the description and drawings shall be used to interpret the claims.

Thus the claims of a European patent are not to be read in isolation of the other parts of the patent publication. The EPC includes a Protocol on the interpretation of Article 69. In the Protocol it is clarified that a strictly literal interpretation of claims is not an appropriate approach, nor should the claims serve only as a guideline. A correct position is somewhere between these two extremes, combining a fair protection for a patent proprietor with a reasonable degree of legal certainty for third parties. The Protocol also states that due account shall be taken of any element which is equivalent to an element specified in the claims. In patent law, this principle is called the doctrine of equivalence.

When writing a patent application, a good starting point is to draft claims first, as they are the primary means for communicating the invention briefly to the examining division. Despite the existence of the doctrine of equivalence, it is better to draft rather broad claims than to use overly detailed claim language. During the prosecution of a European application, it is easier to limit the claims than to broaden them. According to the EPC, the European patent application or European patent may not be amended in such a way that it contains subject matter which extends beyond the content of the application as filed. After grant, the European patent may not be amended in such a way as to extend the protection it confers [39]. However, focused claim drafting may also be a reasonable strategy if the applicant aims at attracting investors who might be vary of invalid and unenforceable patent rights [40].

After drafting the claims, a description should be produced in which the invention is explained in sufficient detail with a set of workable examples that cover the whole of the claimed range.

In scientific articles the results are often presented by drawing conclusions from authors' own experiments and other researchers' prior work in a logical manner and by emphasizing the parallels between the results obtained by different groups. However, such an approach is not optimal for writing a good patent application. In a patent application the invention should be presented in a way that emphasizes the technical problem to be solved and the non-obvious solution provided by the invention. To deserve a patent, the invention should not be derivable from the prior art in an obvious manner. Instead, there should be aspects that are surprising to a skilled person.

As nanotechnology inventions are often created in multidisciplinary environments, also the writing of patent applications should preferably take place through the involvement of multidisciplinary teams. If the inventors have different technical backgrounds, they should all participate in the application drafting work together with one or more patent attorneys having expertise in the most relevant of these technical fields.

3.4.2 *Sufficient Disclosure of the Invention*

According to Article 83 EPC, the European patent application shall disclose the invention in a manner sufficiently clear and complete for it to be carried out by a person skilled in the art. The logic behind this disclosure requirement is that as a reward for disclosing his invention to the public, the applicant is granted a patent right. To meet the requirement the application shall contain all information that is necessary for using and reproducing the invention.

To assess whether an invention has been properly disclosed in the application, it is again necessary to use the concept of the person skilled in the art. With regard to the disclosure requirement this concept is somewhat different from how it is perceived in the case of judging inventiveness (*vide supra*). The skilled person is now assumed to be aware of the teaching of the patent application and to possess the common general knowledge in the field. It is however not assumed that the skilled person is aware of *all* prior art.

When writing the description of the invention, the capabilities of the average skilled person should be taken into account and a sufficiently — but not overly — detailed description of the structure and function of the invention provided. Often a description of the structural parts of the invention will suffice, but sometimes a description of function may be more appropriate, such as in the case of inventions related to computers or gene technology.

Preparing and processing materials in the nanoscale often involves the use of special methods and tailor-made equipment. To the extent that such methods and equipment are not easily available to the average skilled person working in the field, they have to be described in the patent application to meet the disclosure requirement. Otherwise it would be impossible for the skilled person to reproduce the invention reliably [41].

As the patent application should enable the skilled person to practice the invention over the whole of the claimed range, the description part should contain enough examples. With regard to nanotechnology inventions, it is particularly important to keep the breadth of the claims in line with the breadth of the examples presented in the description.

While sufficient disclosure of the invention is a necessary condition for obtaining a patent, overdisclosure may also create problems. It can be very tempting to describe potential future applications or uses of the invention, even though they have not been fully developed yet. By including such prospective subject matter in the application, the applicant creates a trap for himself: when the development work actually matures to a second patent application, the earlier speculations presented in the first application form prior art and may function as a bar to patentability. The later invention may seem obvious, as the applicant himself has provided hints at that direction in the earlier application. Therefore any ungrounded visions of future applications should be avoided. Preferably each patent application should only contain subject matter that is necessary to support the claims [42].

3.4.3 *Claims*

The claims of a European patent application should be drafted in terms of the technical features of the invention [43]. The features are normally written as structural limitations, but also functional limitations can be used. The form of a claim in a European patent is a one-sentence statement divided in two parts: the first part describes the previously known features of the invention and the second part states the new features. The claims can be formulated either as a product claim, which lists the structural elements of the invention and how they are connected to each other, or as a process claim, which lists the method steps that are to be performed in the invention. A product claim confers a wider protection scope as it protects the product independent of the method used for producing it. A process claim only protects the activity of performing the process and products that are directly obtainable by the process.

In the case of patent applications directed to nanotechnology inventions, usually one or more of the following four claim types are used: composition claims (e.g., nanomaterials); device, system, or apparatus claims (e.g., devices in which nanomaterials have been used or tools for processing nanomaterials); process or method claims (e.g., methods for preparing nanomaterials or new uses for nanomaterials or nanodevices); and product-by-process claims [44].

In nanoscale it may turn out to be difficult or even impossible to properly define an invention concerning a nanostructure in a claim by using structural features only. The precise characterization of the nanostructure might be too challenging or the literal description of the structure would lead to an unnecessarily long and complicated claim. An alternative to a structural definition is a product-by-process claim. In such a claim a new and inventive product is claimed by listing the process steps leading to the product. The protection scope obtained in this way equals the scope of an ordinary product claim; it is only the format of claiming the new product that differs. If the product were prepared by some other process, it would still fall under the scope of the product-by-process claim.

Another way of defining a nanotechnology invention is to use a functional claim, that is, to define the invention by means of its functional properties. It is generally much easier and more reproducible to characterize the new properties that are due to the nanoscale dimensions than to characterize the actual structural features causing these new properties. However, claiming the invention by merely stating the underlying technical problem is not allowable [45].

If the core of the invention lies in solving problems caused by the small size, claims should be formulated in such a way that they are directed to the solution of this particular problem. Claims that only emphasize the size difference in comparison to the prior art may be refused by the EPO due to lack of an inventive step. For example, the development of a nanodevice may have required that new manufacturing methods and new materials were developed. By directing the claims not only to the device but also to the new methods and materials, obtaining a patent will be on a more secure basis and the protection scope becomes wider [46].

According to the EPO practice, use of relative terms, such as "thin" or "small," should be avoided in claims, unless the term has a well-recognized meaning in the art [47]. This is an important point to remember when drafting claims, as it may turn out to be impossible to remove an unclear term after grant, for example, in opposition proceedings. According to Article 123(3) EPC, a European patent may not be amended in such a way as to extend the protection it confers.

3.4.4 *Terminology*

In the field of nanotechnology, the terminology develops continuously, and often multiple synonymous terms are in use for depicting similar structures. Therefore the terms used in the claims of a patent application should always be defined carefully and unambiguously in the description part of the application. An unclear or erroneous term in a claim may lead to a protection scope that is too narrow or difficult to enforce before a court [48].

Whenever possible, appropriate dictionaries and terminology lists should be consulted when selecting terms for defining a nanotechnology invention. While it is in principle possible for an applicant to define entirely new terms in the description, and in this way to act as his own lexicographer, it is generally more reasonable to rely on established terminology. There are currently many private-sector and public-sector projects which aim at developing and standardizing the terminology and nomenclature used in the field of nanotechnology. For example, the Institute of Nanotechnology has published a glossary of terms [49], and the American National Standards Institute has established a nanotechnology standards panel [50].

References

1. European Patent Office (2010) How to get a European patent, *Guide for Applicants, Part 1*, 13th edition, May 2010.
2. European Patent Office (2010). *European Patent Convention*, 14th edition, August 2010.
3. Visser, D. (2009). *The Annotated European Patent Convention*, H. Tel, Publisher, Veldhoven.
4. Status on 26 July 2011, see http://www.epo.org/about-us/epo/member-states.html.
5. Articles 2(2) and 64(1) EPC.
6. Decisions of the Boards of Appeal are available at http://www.epo.org/patents/appeals/search-decisions.html.
7. See http://www.epo.org/law-practice/legal-texts/london-agreement.html for additional information on the translation requirements and the London Agreement.

8. See http://www.epo.org/topics/issues/nanotechnology.html.

9. Mulder, C. (2009) *The Cross-Referenced Patent Cooperation Treaty*, 2009 edition, Helze, Geldrop.

10. European Patent Office (2008) How to get a European patent EURO-PCT, *Guide for Applicants, Part 2*, 4th edition, April 2008.

11. See also the PCT Applicant's Guide. Available at http://www. wipo.int/pct/en/appguide/index.jsp.

12. Status on 26 July 2011, see http://www.wipo.int/pct/en/ for further information.

13. Article 11(3) PCT and Article 153(2) EPC.

14. Articles 99-101 EPC.

15. Articles 105a, 105b, and 68 EPC.

16. Articles 2(2), 64(1) and 64(3) EPC.

17. Document 11533/11 of the Council of the European Union, Draft agreement on a Unified Patent Court and draft Statute - Presidency text, 14 June 2011. Available at http://register.consilium.europa.eu/pdf/en/11/st11/st11533.en11.pdf.

18. Harhoff, D. (2009) *Economic Cost–Benefit Analysis of a Unified and Integrated European Patent Litigation System*, 29 February 2009. Available at http://ec.europa.eu/internal_market/indprop/docs/patent/studies/litigation_system_en.pdf.

19. European Patent Office acting as secretariat of the Working Party on Litigation (2006) *Assessment of the Impact of the European Patent Litigation Agreement (EPLA) on Litigation of European Patents*, February 2006. Available via http://www.epo.org/patents/law/legislative-initiatives/epla.html.

20. Article 52(1) EPC.

21. Articles 52(2), 52(3), and 53 EPC.

22. Decision T 356/93 of the Boards of Appeal, OJ, 1995, 545.

23. See http://www.epo.org/topics/issues/nanotechnology.html for the EPO definition of nanotechnology.

24. European Patent Office (2010) *Guidelines for Examination in the European Patent Office ("Guidelines")*, C-IV 9.8. Available at http://www.epo.org/patents/law/legal-texts/guidelines.html. See also the decisions T 198/84, OJ 1990, 59, and T 279/89.

25. Kallinger, Ch., Veefkind, V., Michalitsch, R., Verbandt, Y., Neumann, A., Scheu, M., and Forster, W. (2008) Patenting nanotechnology: a European Patent Office perspective, *Nanotechnol. L. & Bus.*, **5**, 95–105.

26. *Crackkatalysator I*, 1990, 150, *GRUR*; *Chrom-Nickel-Legierung*, 1992, 842, *GRUR*; *Inkrustierungsinhibitoren*, 2000, 591, *GRUR*.

27. Huebner, S. R. (2008) The validity of European nanotechnology patents in Germany, *Nanotechnol. L. & Bus.*, **5**, 353–357.

28. Reibold, M., Paufler, P., Levin, A. A., Kochmann, W., Pätzke, N., and Meyer, D. C. (2006) Carbon nanotubes in an ancient Damascus sabre, *Nature*, 444, 282.

29. Article 52(2) EPC.

30. Zech, H. (2009) Nanotechnology – new challenges for patent law? *SCRIPTed*, 6, 147–154.

31. Schellekens, M. H. M. (2008) Patenting nanotechnology in Europe: making a good start? *An Analysis of Issues in Law and Regulation*. Tilburg University Legal Studies Working Paper No. 008/2008. Available at http://ssrn.com/abstract=1139080.

32. Guidelines, C-II 4.1. See also the decisions T 171/84, OJ 4/1986, 95, and T 51/87, OJ 3/1991, 177.

33. European Patent Office (2010) *Case Law of the Boards of Appeal of the European Patent Office*, 6th edition, July 2010, p. 162.

34. Guidelines C-IV 11.5.

35. Guidelines, C-IV 11.10.

36. Guidelines, C-IV – Annex.

37. Article 57 EPC.

38. Rule 42(1)(f) EPC.

39. Articles 123(2) and 123(3) EPC.

40. O'Neill, S., Hermann, K., Klein, M., Landes, J., and Bawa, R. (2007) Broad claiming in nanotechnology patents: is litigation inevitable? *Nanotechnol. L. & Bus.*, **4**, 29–40.

41. Schellekens, M. H. M. (2008) Patenting nanotechnology in Europe: making a good start? *An Analysis of Issues in Law and Regulation*. Tilburg University Legal Studies Working Paper No. 008/2008. Available at http://ssrn.com/abstract=1139080.

42. Axford, L. A. (2006) Patent drafting considerations for nanotechnology inventions, *Nanotechnol. L. & Bus.*, **3**, 305–308.

43. Rule 43(1) EPC; Guidelines, C-III 2.1.

44. Miller, J. C., Serrato, R. M., Represas-Cardenas, J. M., and Kundahl, G. A. (2005) Intellectual property, in *The Handbook of Nanotechnology: Business, Policy, and Intellectual Property Law*, John Wiley & Sons, Hoboken, chap. 13.

45. Guidelines, C-III 4.10.

46. Bleeker, R. A., Troilo, L. M., and Ciminello, D. P. (2004) Patenting nanotechnology, *Materials Today*, February 2004, 44–48.

47. Guidelines C-III 4.6.

48. Germinario, C. (2006) Biotech to Nanotech Inventions — Effective Patenting Strategies, *Nanobiotechnology*, 2006, 1–3.

49. See http://www.nano.org.uk/nano/glossary.htm.

50. See "Standards Activities" at http://www.ansi.org.

Chapter 4

Nanotechnology Patent Procurement and Litigation in the United States

Sarah M. Rouse

Katten Muchin Rosenman LLP, 2900 K Street NW, North Tower — Suite 200, Washington, DC 20007-5118, USA

sarah.rouse@kattenlaw.com

4.1 Patent Procurement for Nanotechnology-Based Inventions: US Perspective

As the United States is a major market for nanotechnology-based products, the patenting of nanotechnology inventions in the United States is often times a critical component of product commercialization and industrial success.

4.2 The US Patent System

The United States Constitution grants Congress the power to enact patent laws, as follows: "Congress shall have the power ... to promote the progress of science and useful arts, by securing for limited times to authors and inventors the exclusive right to their respective writings and discoveries" [US Constitution, Art. I, § 8, cl. 8].

Nanotechnology Commercialization for Managers and Scientists
Edited by Wim Helwegen and Luca Escoffier
Copyright © 2012 Pan Stanford Publishing Pte. Ltd.
ISBN 978-981-4316-22-4 (Hardcover), 978-981-4364-38-6 (eBook)
www.panstanford.com

Congress exercises its authority by creating statutory schemes granting certain rights to patentees. These rights include

> a grant to the patentee, his heirs or assigns, of the right to exclude others from making, using, offering for sale, or selling the invention throughout the United States or importing the invention into the United States, and, if the invention is a process, of the right to exclude others from using, offering for sale or selling throughout the United States, or importing into the United States, products made by that process, referring to the specification for the particulars thereof [35 U.S.C. § 154 (a)(1)].

As articulated above, the rights granted to a US patentee are proscriptive and intended to protect the holder of a patent from infringement, that is, the right is not the right to make, use, offer for sale, sell, or import, but rather the right to exclude others from making, using, offering for sale, selling, or importing the invention. US patent grants are effective only within the United States, its territories, and possessions.

The US patent laws are codified in Title 35 of the US Code and the corresponding rules are organized in Title 37 of the Code of Federal Regulations [35 U.S.C. §§ 101 et seq. and 37 C.F.R. §§ 1.1 et seq.].

The United States Patent and Trademark Office (USPTO) is the Federal agency for granting US patents. The USPTO Manual of Patent Examining Procedure or "MPEP" interprets the patent laws and rules and provides guidance on the standards for examination of patent applications.

4.2.1 *Patent Prosecution Overview*

4.2.1.1 Patent application submission and examination

To obtain a patent, an inventor must first submit a patent application to the USPTO. The process of submitting the patent application and the subsequent interactions with the USPTO is called "patent prosecution."

The rules of the USPTO mandate that inventors apply for a US patent only in his or her own name. An inventor, however, may transfer all or part of his or her interest in the patent application or patent to a person or company by an assignment.

Inventors may also license these rights exclusively or non-exclusively [http://www.uspto.gov/inventors/patents.jsp]. The key components of a US patent application include the specification, abstract, drawings, and claims. The specification is a narrative that describes and distinguishes the invention. The particular parts of the specification include title of the invention, background of the invention, brief summary of the invention, and detailed description of the invention. One or more drawings are to be submitted if they are necessary for showing how the invention works. Some applications (e.g., a pure chemical nanotechnology application) need not include a drawing, unless a process can be diagrammed by a flowchart.

The most important part of the patent application is the claims. The claims are detailed statements of exactly what the invention covers, that is, what the patent holder has the right to exclude others from doing. The claims should be written broadly enough to adequately protect the invention, but narrow enough to avoid encompassing potential prior art references or risk invalidity.

Once the patent application has met the USPTO filing requirements, it is assigned to an Art Unit for examination. Each Art Unit handles a specific type of invention, as defined by the USPTO's classification system. The USPTO has created an Art Unit for nanotechnology inventions: Class 977, Nanotechnology, as well as over 250 cross-reference subclasses. The creation of Class 977 has enhanced the quality of USPTO Examination and contributed to an improved capability to track nanotechnology-related patents in the United States [http://www.uspto.gov/web/patents/classification/uspc977/defs977.htm].

"Examination" is the process by which the USPTO determines whether a patent application meets the requirements for granting a patent. The USPTO Examiner may "allow" one or more claims, finding that the claims comply with the requirements for patentability. In situations where the Examiner finds that the application does not comply with the requirements or the Examiner has identified one or more references that allegedly describes the claims of the patent application, an Office action is issued detailing the Examiner's reasons for rejection. The applicant may respond to the Office action by arguing in support of the application or by making amendments to the claims. This process is to be repeated until the patent is in a form suitable for grant or an appeal is arranged to resolve the matter. If the Examiner's rejections cannot be overcome, the application may be abandoned.

4.2.2 Types of Patent Applications

4.2.2.1 Non-provisional patent application

A non-provisional patent application is a utility patent application and must include a specification, including one or more claims; drawings, when necessary; an oath or declaration; and the prescribed filing, search, and examination fees. The application for a non-provisional patent begins the examination process by the USPTO. A non-provisional patent application may or may not result in the grant of a patent, depending on the outcome of the USPTO examination [http://www.uspto.gov/patents/resources/types/utility.jsp].

4.2.2.2 Provisional patent application

The filing of a provisional patent application allows the applicant to obtain a filing date, thereby securing a priority date, without the initial expense and formalities accompanying the filing of a non-provisional application. A provisional application expires one year after the initial filing date. As such, the applicant has up to 12 months to file a non-provisional application based on the provisional application. The provisional application's one-year period of protection can be used to evaluate the commercial potential of a nanotechnology invention before committing to a full non-provisional patent. The claimed subject matter in the later filed non-provisional application is entitled to the benefit of the filing date of the provisional application if the subject matter has support in the provisional application. The 12-month pendency for the provisional application is not counted toward the 20-year patent term [http://www.uspto.gov/patents/resources/types/provapp.jsp].

4.2.2.3 Continuation application

A continuation application is a patent application filed by an applicant pursuing additional claims to an invention disclosed in one of applicant's earlier applications (the "parent application") that has not yet been issued or abandoned. The continuation application utilizes the specification of the pending parent application and claims priority based on the parent application filing date. The continuation must have at least one inventor in common with the parent application [MPEP 201.07 "Continuation Application" [R-3] — 200 Types, Cross-Noting and Status of Application]. A continuation application is useful where the USPTO Examiner has allowed some, but not all of the claims in a pending nanotechnology application or where an applicant has identified other meaningful ways of claiming different embodiments of the invention disclosed in the parent application.

4.2.2.4 Continuation-in-part application

A continuation-in-part (CIP) application permits the applicant to submit claims to subject matter not disclosed in the parent application, but still utilizing a substantial portion of the specification of the parent application. The CIP application claims priority based on the filing date of the parent application and must name at least one inventor in common with the parent application [MPEP 201.08 "Continuation-in-Part Application" [R-3] — 200 Types, Cross-Noting and Status of Application]. The CIP application is a convenient way to claim enhancements to an invention developed after the filing of the parent application.

4.2.2.5 Divisional application

A divisional application is an application that has been "divided" from an existing patent application (the parent application). A divisional application retains the filing and priority date of the parent and as such can only contain subject matter present in the parent application. This type of application need not name any of the inventors named in the parent application. A divisional application is typically filed in response to a restriction requirement by the USTPO (i.e., requirement by the USTPO that the patent be limited to a single claimed invention) [MPEP 201.06 "Divisional Application" [R-2] — 200 Types, Cross-Noting and Status of Application; MPEP 802.02 "Definition of Restriction" [R-3] — 800 Restriction in Applications Filed Under 35 U.S.C. 111; Double Patenting].

4.2.3 *Pursuing a US Patent via the Patent Cooperation Treaty (PCT)*

The Patent Cooperation Treaty (PCT) allows an applicant to seek patent protection for an invention simultaneously in multiple countries by filing a single application (i.e., "international application") instead of several separate national or regional patent applications [World Intellectual Property Organization (WIPO), Applicant's Guide — International Phase — Annex A, http://www.wipo.int/pct/guide/en/gdvol1/annexes/annexa/ax_a.pdf].

The applicant of an international patent application under the PCT designating the United States must comply with the requirements for entry into the US national phase within 30 months of the PCT priority date. In order for the application to enter the US national phase, the applicant must supply the USPTO with a variety of items, including a copy of the international application, an English translation of the international application (if not originally filed in English), an oath or declaration of inventorship, a filing fee, and an English translation of any annexes to the international preliminary examination report [35 U.S.C. § 371; MPEP 1893.01 "Commencement and Entry" [R-3] — 1800 Patent Cooperation Treaty]. Once the applicant has met all of the requirements, the US national phase application will receive the benefit of the filing date of the international application and the application may constitute a prior art reference in the United States against another patent application [35 U.S.C. § 102(e)].

4.2.4 *Patent Term*

The term of a new US patent is 20 years from the date on which the application for the patent was filed or, if the application contains a specific reference to an earlier filed application under 35 U.S.C. § 120, 121, or 365(c), from the date the earliest application was filed. The term is subject to the payment of maintenance fees. A maintenance fee is due 3.5, 7.5, and 11.5 years after the original grant of the patent. The terms may be extended for certain pharmaceuticals and for certain circumstances as provided by law [MPEP 2701 "Patent

Term" [R-2] — 2700 Patent Terms and Extensions; MPEP 2506 "Times for Submitting Maintenance Fee Payments" [R-2] — 2500 Maintenance Fees]. After the patent has expired, anyone may make, use, offer for sale, or sell or import the invention without permission of the patentee, provided that such activities do not infringe the claims of another unexpired patent.

4.2.5 Conditions for Obtaining a US Patent

The Patent Act provides that in order to be patentable, an invention must be new, useful, non-obvious, and adequately described for one of ordinary skill in the art to make and use the invention [35 U.S.C. §§§§ 101, 102, 103, and 112]. These conditions are described in detail below.

4.2.5.1 Novelty

In order for an invention to be patentable in the United States, the invention must be novel, as defined in the Patent Act, and cannot be patented if

> "(a) the invention was known or used by others in this country, or patented or described in a printed publication in this or a foreign country, before the invention thereof by the applicant for patent," or "(b) the invention was patented or described in a printed publication in this or a foreign country or in public use or on sale in this country more than one year prior to the application for patent in the United States …." [35 U.S.C. § 102].

In other words, if an invention has previously been described in a printed publication anywhere in the world, or if the invention has been in public use or "on sale" in the United States before the date that the applicant made the invention, a patent cannot be obtained [35 U.S.C. 102(b)]. Moreover, if the invention has been described in a printed publication anywhere, or has been in public use or on sale in the United States more than one year before the date on which the US patent application was filed, a patent cannot be obtained. In this situation, it is immaterial when the invention was made, or whether the printed publication, or public use was by the inventor himself/herself or by a third party. If the inventor describes the invention in a printed publication or uses the invention publicly or places it on sale, the inventor must apply for a patent within one year, otherwise any right to a patent will be permanently lost.

As many nanotechnology inventions are simply compounds or devices that already exist at the macroscale, some of these inventions could be deemed anticipated by their larger-sized counterparts. Further, a nanotechnology invention is not novel if the elements and features perform the same function as those in the prior art without giving an unobvious and unexpected result [*In re Rose*, 220 F.2d 459 (1984).] However, it is important to note that an invention need not be entirely new to qualify for patent protection. The invention can instead be an improvement on an existing article or method. Even a small functional improvement may be patentable as well as extremely valuable from a commercial perspective.

When drafting the claims of a nanotechnology patent application, the improvement offered by the invention should be emphasized. For instance, claims could be directed to a modification of one or more existing materials or new processing steps. The modification or processing can transform a non-patentable invention into an object of manufacture that is patentable. Note further that abstract ideas, natural phenomena, and laws of nature are not eligible for patenting. For example, novel semiconductor nanocrystals may be patentable, but the quantum effects underlying the crystals are not.

4.2.5.2 Utility

The Patent Act provides that patents may be granted only for "useful" inventions [35 U.S.C. § 101]. An invention is "useful" if it is capable of providing some identifiable benefit.[1] This utility require-ment for patentability excludes from protection inventions that are inoperable, immoral, or contrary to public policy [MPEP 706.03(a) "Rejections Under 35 U.S.C. 101" [R-5] — 700 Examination of Applications]. The utility requirement also excludes composition

---

[1] *Juicy Whip, Inc. v. Orange Bang, Inc.*, 185 F.3d 1364 (Fed. Cir. 1999). *See also Brenner v. Manson*, 383 U.S. 519, 534 (1996); *Brooktree Corp. v. Advanced Micro Devices, Inc.*, 977 F.2d 1555, 1571 (Fed. Cir. 1992).

inventions that have no practical use except for further research or experimentation.[2]

In order to meet the utility requirement, an invention that is claimed in a patent application must be shown to be "operable," or "capable of being used to effect the object proposed."[3] The invention does not lack utility simply because one or more embodiments disclosed in the application is not perfect or performs crudely.[4] A commercially successful product is also not essential.[5]

It is advisable when drafting a US patent application to include a list of various objects, or assertions of utility, of the nanotechnology invention in the Summary of the Invention section of the patent application.[6] It is important that these assertions are practical applications and include more than experimental endeavors. It is also advisable that information (e.g., experimental data) be included in the application that demonstrates that the nanotechnology invention is operable, or works, for the intended practical objective.

4.2.5.3 Non-obviousness

A patent claim is obvious under the Patent Act if every element or limitation of the claim is disclosed in a combination of references, which collectively would have suggested the invention to one of ordinary skill in the art at the time the invention was made [35 U.S.C. § 103(a)]. Section 103(a) provides in pertinent part as follows:

> (a) A patent may not be obtained though the invention is not identically disclosed or described as set forth in section 102 of this title, if the differences between the subject matter sought to be patented and the prior art are such that the subject matter as a whole would have been obvious at the time the invention was made to a person having ordinary skill in the art to which said subject matter pertains. Patentability shall not be negatived by the manner in which the invention was made... [35 U.S.C. § 103(a)].

---

[2] *Brenner*, 383 U.S. at 535.
[3] *Mitchell v. Tilghman*, 86 U.S. 287, 396 (1873).
[4] *Hildreth v. Mastoras*, 257 U.S. 27, 34 (1921).
[5] *In re Anthony*, 414 F.3d 1383, 1396 (C.C.P.A. 1969).
[6] *Cross v. Iizuka*, 753 F.2d 1040 (Fed. Cir. 1985).

An obviousness inquiry, therefore, assesses the differences between the invention sought to be patented and the prior art to ascertain whether the invention as a whole would have been obvious at the time the invention was made to a person having ordinary skill in the technology to which the invention pertains [Manual of Patent Examining Procedure (MPEP) 2141 "Examination Guidelines for Determining Obviousness" [R-6] — 2100 Patentability].

A mere decrease in size to the nanoscale may be viewed as "obvious." However, in many nanotechnology inventions, the decrease in size requires the solution of a new problem that did not exist at the macroscale. Where the invention is not merely a reduction in size, but rather is the solution of a new problem, the claims should focus on the solution.

In addition, where the nanotechnology invention yields previously unrecognized and unexpected results, there are strong arguments that the invention is novel and non-obvious to a person of ordinary in skill in the art. As such, it is important to clearly document (e.g., via laboratory notebook or memorandum of invention) and emphasize any unexpected results.

4.2.5.4 Written description and enablement

The Patent Act also provides that a patent be specifically described in writing. Patent protection is granted in return only for a complete disclosure of an invention, not for vague descriptions of general ideas that may or not be workable [35 U.S.C. § 112]. Section 112 provides

> The specification shall contain a written description of the invention and of the manner and process of making and using it, in such full, clear, concise, and exact terms as to enable any person skilled in the art to which it pertains, or with which it is most closely connected, to make and use the same, and shall set forth the best mode contemplated by the inventor of carrying out his invention [35 U.S.C. § 112].

Thus, the specification of the patent application must (i) provide a written explanation or description of the invention (the "written description requirement") and (ii) describe how to make and use the invention in such full, clear, concise and exact terms to enable a person skilled in the technology to carry out the invention without undue experimentation (the "enablement requirement").

The use of overly broad or unspecific terminology in nanotechnology applications may result in a rejection under 35 U.S.C. § 112. Whenever possible, general, well-known terms of art should be used to describe the invention. Any ambiguous or unclear terms should be clearly defined and consistently used in the specification. Moreover, although efforts have been made to develop a common terminology for nanotechnology inventions (see, e.g., ASTM's Terminology for Nanotechnology Standard,[7] it is still recommended to include these agreed-upon definitions in the application. For example, the terms "nanoparticle" or "nanosize" should be clearly defined in the application. Nanoparticle characterization techniques (e.g., atomic force microscopy (AFM), scanning electron microscopy (SEM), particle size analyzers, zeta potential measurements), and sample preparation should also be clearly referenced in the specification.

Enablement is a substantial issue for numerous nanotechnology patent applications. As commercialization significantly lags development, inventions often exist on "paper" and have not been fully realized in a laboratory setting. Enablement is determined as of the filing date of the patent, and patent-owners cannot use experiments conducted post-application to establish the validity of their patents. As such, the timing for filing a patent application should be carefully considered.

4.2.5.5 Best mode

The Patent Act also requires that patents disclose the "best mode" of the invention:

> The specification . . . shall set forth the best mode contemplated by the inventor of carrying out his invention [35 U.S.C. § 103(a)].

---

[7]ASTM E 2456 – Terminology for Nanotechnology (2006).

The best mode requirement establishes a statutory bargained-for-exchange under which a patentee obtains the right to exclude others from practicing the claimed invention for a set period of time, and the public receives knowledge of the preferred embodiments for practicing the patented invention [35 U.S.C. § 112]. This statutory safeguard protects against the desire of some applicants to seek patent protection without making a full disclosure of the invention. Under the best mode requirement, applicants are not permitted to disclose their second-best embodiments, while retaining the best embodiments for themselves.

When preparing a patent application, complying with the best mode requirement may be facilitated by directly asking each inventor whether the best mode of practicing the nanotechnology invention has been described in the patent application. As the best mode requirement is determined as of the filing date, it is important to confirm with the inventors as close to the filing date as possible that the application contains the best mode.

4.3 US Patent Post-Grant Proceedings

4.3.1 *Interference*

Occasionally, two or more applications are filed by different inventors claiming substantially the same patentable invention. The patent can only be granted to one of the parties, and the USPTO commences a proceeding known as an "interference" to determine who is the "first inventor" and who is subsequently entitled to the patent. Interference proceedings may also be instituted between a patent application and an issued patent, provided the patent has not been issued for more than one year prior to the filing of the conflicting application, and provided that the conflicting application is not barred from being patentable for some other reason [MPEP 2138.01 "Interference Practice" [R-3] — 2100 Patentability].

The US patent system is a "first-to-invent" system, and is distinguishable from other countries that use a "first-to-file" system. Invention in the United States is generally defined to comprise two components: (1) conception of the invention and (2) reduction to practice of the invention. When an inventor conceives of an invention and then proceeds to diligently reduce the invention to practice (e.g., files a patent application, practices the invention), the date of invention is the date of conception. Therefore, provided that the inventor was diligent in reducing the invention to practice, the inventor will be the "first inventor" and will be entitled to patent protection. This is the case even if another party files for a patent application, thereby constructively reducing the invention to practice, before the first inventor [MPEP 2138.05 "Reduction to Practice" [R-5] — 2100 Patentability].

The interference proceeding is conducted by a panel of administrative patent judges sitting on the Board of Patent Appeals and Interferences of the USPTO. Appeals from the tribunal may be heard before either the United States Court of Appeals for the Federal Circuit or the United States District Court for the District of Columbia [35 U.S.C. § 144 and 146].

4.3.2 *Reissue*

A patentee may seek to reissue a previously issued patent in order to correct an inadvertent error in the patent that occurred without any deceptive intention on the part of the patentee. Typical errors correctable by a reissue include a defective specification or drawing, claim scope or failure to properly reference priority documents. In such instances, the patentee must admit that the patent is wholly or partly inoperative or invalid and surrender the original patent when the reissue patent grants. The reissue process cannot be used to introduce new matter into the patent. Rather, the claims sought in the reissue must fall within the invention, as described in the original patent. A reissue application will be examined in the same manner as a non-provisional patent application and the reissue patent has the same life as the patent from which is has been reissued [35 U.S.C. § 251; MPEP 2138.05 "Reduction to Practice" [R-5] — 2100 Patentability; MPEP 1405 "Reissue and Patent Term" [R-2] — 1400 Correction of Patents; MPEP 1412.01 "Reissue Claims Must be for the Same General Invention" [R-7] — 1400 Correction of Patents].

4.3.2.1 Broadening reissue

A reissue may be utilized to broaden certain claims of the original patent. For example, a reissue may be sought where the claims of the patent as issued are unnecessarily narrow and therefore fail to provide the patentee with the scope of protection to which the patentee is entitled. Proposed broadening claims must be presented in a broadening reissue application filed by the second anniversary of the issuance of the original patent, else the opportunity to broaden the claims will be lost. In addition, the patentee must unequivocally indicate this intent in the reissue application [35 U.S.C. § 251; MPEP 1412.03 "Broadening Reissue Claims" [R-7] – 1400 Correction of Patents].

The "recapture rule," which prohibits subject matter from being claimed if it was surrendered during prosecution of the original patent, also restricts the patentee's ability to broaden claims via reissue [MPEP 1412.02 "Recapture of Canceled Subject Matter" [R-7] — 1400 Correction of Patents].

4.3.2.2 Narrowing reissue

Under certain circumstances, the claims of an issued patent may be overly broad and narrowing the claims through a reissue may preserve validity. In cases where the reissue seeks to narrow the claims, the application can be filed at any time prior to the expiration of the patent [35 U.S.C. § 251].

4.3.2.3 Doctrine of intervening rights

Under the doctrine of intervening rights, a patentee's ability to recover for infringing activity based on reissued claims may be limited:

> A reissued patent shall not abridge or affect the right of any person or that person's successors in business who, prior to the grant of a reissue, made, purchased, offered to sell, or used within the United States, anything patented by the reissued patent, to continue the use of, to offer to sell, or to sell to others to be used, offered for sale, or sold, the specific thing so made, purchased, offered for sale, used, or imported unless the making, using, offering for sale, or

selling of such thing infringes a valid claims of the reissued patent which was in the original patent [35 U.S.C. § 252].

4.3.3 *Reexamination*

4.3.3.1 *Ex parte* reexamination

In an *ex parte* reexamination, either a third party challenger or the patent owner may seek reexamination of a patent based on prior art consisting of one or more patents or printed publications [35 U.S.C. § 301 and 302; 37 CFR § 1.510; MPEP 2202 "Citation of Prior Art" [R-2] — 2200 Citation of Prior Art and *Ex Parte* Reexamination of Patents]. The party seeking reexamination files a request for reexamination with the USTPO, upon which the USPTO will determine whether a substantial new question of patentability exists. If a substantial new question of patentability is found to exist, the USPTO will order a reexamination of the patent. A determination that there is no substantial new question is final and non-appealable [35 U.S.C. § 303 and 304; MPEP 2216 "Substantial New Question of Patentability" [R-2] — 2200 Citation of Prior Art and *Ex Parte* Reexamination of Patents].

If the USPTO orders reexamination, the patent holder is provided an opportunity to file a statement concerning the new question of patentability, including amendments or new claims [35 U.S.C. § 304 and 305; 37 CFR § 1.530(b)]. If the patent holder files such a statement, the statement must be served on the person requesting reexamination and the requester is provided with two months to file a reply [35 U.S.C. § 304; 37 CFR § 1.535]. The USPTO then examines the claims without additional participation by the requester. Following the *ex parte* reexamination, the USPTO issues a certificate canceling any claim determined to be unpatentable, confirming any claims determined to be patentable and incorporating in the patent any new claim or amended claim determined to be patentable [35 U.S.C. § 307; 37 CFR § 1.570].

Owners of nanotechnology patents that lack clear definitions of claim terms or whose claims are overly broad may face uncertainty with regard to third party challenges. Careful patent drafting will minimize these downstream problems.

4.3.3.2 *Inter partes* reexamination

Inter partes reexaminations are similar to *ex parte* reexaminations. Both *ex parte* and *inter partes* reexaminations are initiated by a request for reexamination, the ground for seeking reexamination is prior art consisting of one or more patents or printed publications and the USPTO determines whether the request raises a substantial new question of patentability [35 U.S.C. § 302, 303, 311(a), 312(a); 37 CFR § 1.510, 1.513, 1.931; MPEP 2609 "Inter Partes Reexamination" [R-7] — 2600 Optional *Inter Partes* Reexamination].

If the USPTO determines that there is a substantial new question of patentability, an *inter partes* reexamination is ordered. During this process, the third-party requester may interact with the USPTO by replying to the Office Actions and the patent applicant's Responses [35 U.S.C. § 313 and 314 (b); 37 CFR § 1.903, 1.931, and 1.947]. After the Examiner reaches a final decision in the *inter partes* proceeding, either the patent owner or the requester may appeal an adverse decision to the USPTO Board of Patent Appeals and Interferences. After the Board reaches a decision, either party may appeal to the U.S. Court of Appeals for the Federal Circuit [35 U.S.C. § 315; 37 CFR § 1.959].

A critical difference between *inter partes* and *ex parte* proceedings is estoppel. Unlike the requester of an *ex parte* reexamination, an *inter partes* reexamination requester will be estopped in subsequent litigation in a district court, or in a later *inter partes* proceedings, from asserting the invalidity of any claim finally determined to be valid on any ground that is raised or could have been raised in the *inter partes* reexamination [35 U.S.C. § 315(c) and 317(b)].

4.3.4 *Statutory and Terminal Disclaimers*

The scope and duration of a granted patent can be altered by statutory disclaimers, in which one or more invalid patent claims are disclaimed, and terminal disclaimers, which are used to disclaim some or all of the term of a patent. A statutory disclaimer is particularly useful after discovery of prior art that invalidates a patent claim where *ex parte* reexamination or reissue is not otherwise needed. During such proceedings, however, it is acceptable to simply cancel the claim [35 U.S.C. § 253; MPEP 1490 "Disclaimers" [R-7] — 1400 Correction of Patents].

4.3.5 *Certificate of Correction*

Corrections made to a patent using a certificate of correction filed with the USPTO are considered to have the same effect as if the patent had been originally issued in the corrected form [35 U.S.C. § 255]. A change requested via a certificate of correction cannot be so substantive that examination of patentability is necessitated. Such substantive corrections are more properly made via reissue. Errors that can provide a basis for a reissue application, which can also be corrected via a certificate of correction, such as some changes to priority or inventorship, should be corrected by a certificate of correction whenever possible. Note, however, that when a substantive error is corrected in a reissue application, minor errors that are typically corrected via a certificate of correction can also be corrected in the reissue application [35 U.S.C. § 255].

In addition, the USPTO may issue a certificate of correction for its own mistakes, at no charge to the patentee, if those mistakes are clear from the record [35 U.S.C. § 254].

4.4 US Patent Reform

On September 16, 2011, the Leahy-Smith America Invents Act ("Patent Reform Act of 2011" Pub.L. 112-29, H.R. 1249) was enacted, thereby culminating a six-year debate on patent reform. The passage of the act brings with it sweeping changes to the patent system in the United States.

4.4.1 *First-Inventor-to-File*

The Patent Reform Act of 2011 shifts the US from a first-to-invent system to a first-to-file system for establishing priority of patent applications. This first-to-file system is used by virtually every other patent system in the world.

The Act amends 35 U.S.C. §§ 102 and 103 to preclude the patenting of claimed inventions, and obvious variations thereof, that were "patented, described in a printed publication, or in public use, on sale, or otherwise available to public before the effective filing date of the claimed invention." However, a one-year grace period is established for disclosures made by an inventor, by a joint inventor, or "by another who obtained the subject matter disclosed directly or indirectly from the inventor or a joint inventor."

In addition to the amendments to § 102 that establish the first-to-file system, the Act subtly changes the scope of available prior art for novelty and obviousness determinations to include patents, patent applications, evidence of public use, sales, or information *otherwise available to the public* before the filing date of the patent application. This catch-all provision appears to eliminate the previous distinctions between domestic and foreign disclosures in defining prior art. Other amendments appear to extend safe harbors for prior art authored by an inventor or joint inventor and prior art commonly owned at the time of disclosure and filing of the claimed invention.

4.4.2 *Post-Grant Review*

The Patent Reform Act of 2011 establishes a new post-grant review process, similar to the current European Patent practice, in which a patent may be challenged under any ground of invalidity within one year after the patent is issued (or re-issued). The petitioner can assert "any ground that could be raised under 35 U.S.C. §§ 282(b)(2) or (3). The petition may be granted if the information presented in the petition, if not rebutted, "would demonstrate that it is more likely than not that at least one of the claims challenged in the petition is unpatentable."

Any petition initiating a post-grant review must identify all real parties in interest, identify with particularity the grounds on which the challenge to each claim is based, and must be filed before the petitioner files civil action challenging the patent. Once instituted, post-grant review must be completed within one year, with a possible six-month extension.

The initial authority for hearing claims rests with the newly created Patent Trial and Appeal Board. All final decisions may be appealed directly to the Court of Appeals for the Federal Circuit. Furthermore, any final decision of validity in a post-grant review proceeding precludes the petitioner or real party in interest from (i) requesting any proceeding before the Office on the basis of any ground that the petitioner raised, or reasonably could have raised, during the post-grant review; and (ii) asserting invalidity in a civil action or International Trade Commission proceeding on the basis of any grounds raised by the petitioner in the post-grant review proceeding. Settlement prior to a final decision eliminates all estoppel provisions.

4.4.3 Inter Partes *Review*

The Patent Reform Act of 2011 creates a new *inter partes* review procedure. The USPTO standard for initiation of *inter partes* review is whether there is a "reasonable likelihood that the petitioner would prevail with respect to at least one of the claims challenged in the petition." This would appear to be a higher standard than the "substantial new question of patentability" for current *inter partes* reexaminations. *Inter partes* review may be commenced only within one year after that grant of a patent or the termination of post-grant review, if one is filed, whichever is later.

The *inter partes* review process may not be instituted if (i) the petitioner has previously challenged the patent in a civil action (excluding counterclaims in defense of infringement) or (ii) the petition is filed more than one year after the date on which the petitioner/real party in interest is served with a complaint.

A significant departure from current *inter partes* reexamination is the imposition in § 316 of a one year time limit for completing the *inter partes* review, which may be extended by up to six months. Another significant departure is that *inter partes* review will be conducted by the Patent Trial and Appeal Board, with an appeal directly to the Federal Circuit. This is in sharp contrast to the current system, in which decisions by the examiner in *inter partes* reexaminations are first appealed to the Board of Appeals, and only after that appeal is completed, to the Federal Circuit.

4.4.4 *Patent Trial and Appeal Board*

The Patent Reform Act of 2011 establishes a new appeals division at the USPTO entitled the Patent Trial and Appeal Board. This board will have the authority to (i) review adverse decisions of examiners under § 35 U.S.C. § 134(a); (ii) review appeals of reexaminations pursuant to 35 U.S.C. § 134 (b); (iii) conduct derivation (inventorship) proceedings; and (iv) conduct *inter partes* reviews and post-grant reviews.

4.4.5 *Prior User Rights Defense*

Under current US law, prior user rights may be offered as a defense to infringement under 35 U.S.C. § 273 only in the limited context of business method patents. The Patent Reform Act of 2011 expands the scope of the prior user defense to include all inventions, including processes producing commercial products that were practiced in secret, which is a departure from existing case law. However, in order to assert a prior use defense, the entity asserting the defense must have reduced the subject matter of the patent to practice and commercially used the subject matter at least one year before the effective filing date of the patent.

The Patent Reform Act of 2011 does not make the defense available if (i) the subject matter of which the defense is based was derived from the patentee or (ii) the claimed invention was disclosed to the public and the date of commercialization by the prior user is less than one year before the date of such disclosure.

4.4.6 *False Patent Marking Suits*

In response to the surge in false marking *qui tam* filings, the Patent Reform Act of 2011 effectively eliminates these actions by non-competing third parties, by revising the remedy to allow only a party that has suffered a competitive injury as a result of false marking to seek compensatory damages in an amount related to the actual injuries suffered.

4.4.7 *Elimination of Best Mode Invalidity Defense*

Under the Patent Reform Act of 2011, 35 U.S.C. § 282(b) is amended to eliminate as a defense to patent infringement the patentee's failure to comply with the best mode requirement of 35 U.S.C. § 112 for both invalidity and unenforceability claims.

4.5 Patentability of Nanotechnology in the United States

4.5.1 *Strategic Prosecution of Nanotechnology Inventions*

As the field of nanotechnology progresses and more nanotechnology patent applications are filed, it becomes increasingly difficult to both meet patentability requirements and achieve a commercially valuable claim set because the state of the art is harder to distinguish. This trend illustrates the importance of articulating a strategic patent prosecution strategy.

The overall goal when preparing a patent application should be to build a patent position, not just achieve allowed claims. Building a patent portfolio should be viewed from both an offensive and defensive standpoint. In other words, a patent portfolio should ideally create enforcement opportunities as well as protect against allegations of patent infringement.

The specification of the patent application should be expansive and clearly address all of the embodiments and alternative uses of the nanotechnology invention. It is also recommended that the specification describe how specific embodiments differ from each other. Examples and representative data should be included to support the best mode of the invention and the independent claims. As stated above, it is important to use general, well-known terms of art to describe the invention and any ambiguous or unclear terms should be clearly defined. Using poorly defined or ambiguous terms as a means to introduce breadth into the application is not recommended. When listing and discussing prior art references, confirm that the art uses technical terms in the same manner as in the specification. Moreover, any prior art discussed in the specification should be described using terminology consistent with that used to describe the invention.

Before writing any claim, consider how a competitor would infringe the invention and write claims that could be potentially infringed by a single entity. Consider the types of claims available: composition or device claims, the method of making the invention, the methods of using the invention, kits, and so on. Each claim should also be complete, so that it covers the inventive feature and enough elements to put the invention in context. Include a few narrow claims directed to specific embodiments that are provided as examples in the application and that are supported with data.

4.5.2 *US Nanotechnology "Patent Thicket"*

In the United States, broad patent rights are traditionally granted to "pioneering inventions," while claims to inventions that are not "pioneering" are narrowed during patent prosecution or post-grant proceeding or are limited in scope in the initial patent application [Shapiro, 2001]. Nevertheless, broad claims have been unnecessarily extended by the USPTO to many "non-pioneering" nanotechnology-based inventions, which may have resulted in a so-called "patent thicket," or circumstance where the unreasonable breadth of patent claims of relevant issued patents makes commercialization difficult or impossible for a new entrant in a particular business sector [Shapiro and Miller *et al.*, 2005]. Numerous practitioners fear that the patent thicket will hamper research and innovation in nanotechnology, thereby stifling the goal of replicating the rapid economic growth of the biotechnology industry [Maebius, 2005]. In addition, the grant of broad, overlapping, and conflicting patent rights likely creates a scenario ripe for future patent litigation [Van Lente, 2006].

The current nanotechnology landscape of broad, overlapping patent rights may be a result of a number of factors, including the lack of prior nanotechnology publications upon which to base a prosecution rejection, deficiency of technical expertise of patent Examiners assigned to examine nanotechnology patent applications (especially prior to the creation of USPTO Class 977: Nanotechnology), and the resultant ease of obtaining patent protection [Koppikar, 2004]. Other factors, such as the lack of government-mandated licensing programs, the absence of university policies against patenting, and defensive patenting may also contribute to the formation of the US nanotechnology patent thicket [Lemley, 2005].

4.5.3 *Nanotechnology Commercialization*

Despite the foregoing, the successful patenting and commercialization of nanotechnology and nanomedical products can be achieved and, in certain circumstances, may even be facilitated as a result of the grant of the broad overlapping patents [Lemley, at 623]. One factor driving this trend is the general need for nanotechnology start-up companies to present an expansive patent portfolio in order to attract investment dollars.

Venture capital groups and angel investors are a source of significant financial activity in the nanotechnology business arena. The evaluation of an investment candidate will include a detailed analysis of the patent portfolio. This usually includes an inventory of all pending patent applications and issued patents, and an examination of claim scope, patent filing strategy, and how that strategy fits with the company's overall business plan. The evaluation will also include an assessment as to whether the nanotechnology company is free to practice its technology.

As such, it is important to posture a patent portfolio to reflect that the nanotechnology company is free to make and sell its unique protected product. It is also beneficial that the portfolio allow for monetization (e.g., licensing opportunities) so that some royalty revenue may be realized, or provide for enforcement opportunities.

4.5.4 *Alternative Mechanisms for Commercialization of Nanoproducts*

Patent protection is often the best means to protect nanotechnology. However, alternative legal theories are available to combat the nanotechnology patent thicket and facilitate the successful commercialization of nanotechnology-based products. Such alternatives may include receipt of invalidity or right-to-practice opinions, trade secret protection, litigation, open licensing, and the creation of nanotechnology patent pools.

In certain instances, a nanotechnology company may optimize its competitive advantage by maintaining the technology as a trade secret. Although a trade secret need not be novel, there must be an inherent economic value stemming from the secrecy of the information. To maintain a trade secret, the technology must be protected and kept secret by reasonable efforts, such as through the use of non-disclosure agreements. If the information becomes publicly available, legal protection under trade secret law is no longer available [18 U.S.C. § 1839].

Another solution to the nanotechnology patent thicket is "open" or widespread licensing [Lemley, at 623]. In particular, cross-licensing can assist in the commercialization of nanotechnology-based products. Another example of open licensing is patent pooling [Van Lente, at 203]. In a patent pool, parties assemble overlapping patent rights into a single agreement, with each party taking exclusive or non-exclusive rights to a particular field of use covered by the combined patents. The patent pool may clear commercialization channels blocked to individual parties. Patent pools, however, may promote the risk that parties will overvalue their own contributions to future inventions and set prices too high, thereby foreclosing participation by new parties [Van Lente, at 209].

4.5.5 *US Patent Infringement and Litigation*

Patent infringement is the unauthorized making, using, offering for sale, or selling a patented invention within the United States, or importing a patented invention into the United States. When patent infringement occurs, the patentee may sue for relief in the appropriate Federal court. The patentee may ask the court for an injunction to prevent the continuation of the patent infringement and may also ask the court for an award of damages. The USPTO has no jurisdiction over questions relating to infringement of patents. In examining patent applications, no determination is made as to whether the invention sought to be patented infringes any prior patent. Nanotechnology patents may be difficult to enforce because of difficulties associated with the detection of infringement [Lemley, at 623]. The inability to ascertain potential infringers may undermine the value of such patents to the patent holder.

The first step in any patent infringement analysis is to determine the meaning and scope of the patent claims. A Markman hearing is a pretrial hearing during which a judge examines evidence from the parties on the appropriate meanings of relevant keywords used in the patent claims. This analysis, called "claim construction," begins with an examination of "intrinsic evidence," that is, the claim language, the patent specification and the prosecution history.[8] The words in a claim generally are given their ordinary meaning, unless a special meaning other than the ordinary meaning is clearly stated in the specification or file history.[9]

The infringing activity must fall within the claims of an issued US patent, whether literally, or under the doctrine of equivalents. The term "literal infringement" means that each and every element recited in a claim has identical correspondence in the allegedly infringing composition, device, or process. That is, for a device or product literally to infringe, every limitation set forth in the patent claim must be found in the accused product or process.[10] Any deviation from the claim, even as to only one limitation of the claim, precludes a finding of literal infringement.[11] The patent owner bears the burden of proving infringement by preponderance of the evidence.[12]

Even in the absence of literal infringement, an accused product or process may infringe under the judicially created "doctrine of equivalents." The doctrine of equivalents allows the patentee to claim those insubstantial alterations that were not captured in drafting the original patent claim, but which could be created through trivial changes.[13] Equivalence may be found if the claimed

[8] *Vitronics Corp. v. Conceptronic, Inc.*, 90 F.3d 1576, 1582 (Fed. Cir. 1996); *Markman v. Westview Instruments, Inc.*, 52 F.3d 967, 979 (Fed. Cir. 1995) (en banc), *aff'd*, 517 U.S. 370 (1996).
[9] *Vitronics*, 90 F.3d at 1582; *see also Ekchian v. Home Depot, Inc.*, 104 F.3d 1299, 1303 (Fed. Cir. 1997) ("Because the specification does not use the term 'conductive' in a special or unique way, its ordinary meaning to one skilled in the art controls.").
[10] *Laitram Corp. v. Rexnord, Inc.*, 939 F.2d 1533, 1535 (Fed. Cir. 1991).
[11] *Litton Sys. Inc. v. Honeywell, Inc.*, 140 F.3d 1449, 1454 (Fed. Cir. 1998); *Texas Instruments Inc. v. Cypress Semiconductor Corp.*, 90 F.3d 1558, 1563–65 (Fed. Cir. 1996).
[12] *Id.*
[13] *Festo Corp. v. Shokatsu Kinzoku Kogyo Kabushiki Co.*, 535 U.S. 722, 733 (2002).

and accused elements differ only insubstantially from one another from the perspective of one skilled in the art.[14] Whether the accused product or process produces substantially the same result may also be a relevant test for the determination of infringement under the doctrine of equivalents.[15] "[T]he proper time for evaluating equivalency — and thus knowledge of interchangeability between element — is at the time of infringement, not at the time the patent was issued."[16]

The scope of claim coverage permitted by the doctrine of equivalents is limited by prior art and the doctrine of "prosecution history estoppel." It is well established that limitations in a claim cannot be given a range of equivalents so wide as to cause the claim to encompass the prior art.[17] If a claim encompasses the prior art, it may be properly deemed as invalid. Prosecution history estoppel precludes a patentee from obtaining protection for subject matter which it relinquished during prosecution in order to obtain allowance of the claims."[18] To determine what subject matter has been relinquished, an objective test is applied, inquiring "whether a competitor would reasonably believe that the applicant has surrendered the relevant subject matter."[19]

---

[14] *Hilton Davis Chem. Co. v. Warner-Jenkinson Co.*, 62 F.3d 1512, 1518 (Fed. Cir. 1995), *rev'd and remanded on other grounds*, 520 U.S. 17, *reaff'd and remanded*, 114 F.3d 1161 (Fed. Cir. 1997).

[15] *See Id.; Ethicon Endo-Surgery, Inc. v. U.S. Surgical Corp.*, 149 F.3d 1309, 1315–16 (Fed. Cir. 1998).

[16] *Warner-Jenkinson Co. v. Hilton Davis Chem. Co.*, 520 U.S. 17, 37 (1997).

[17] *Wilson Sporting Goods Co. v. David Geoffrey & Assocs.*, 904 F.2d 677, 684 (Fed. Cir. 1990), *overruled in part on other grounds, Cardinal Chem. Co. v. Morton Int'l*, 508 U.S. 83 (1993); *Stewart-Warner Crop., v. City of Pontiac*, 767 F.2d 1563, 1572 (Fed. Cir. 1985).

[18] *Lockwood v. Am. Airlines, Inc.*, 107 F.3d 1565 (Fed. Cir. 1997); *Wang Labs., Inc. v. Mitsubishi Elecs. Am., Inc.*, 103 F.3d 1571, 1577-1578 (Fed. Cir. 1997) ("Prosecution history estoppel . . . preclud[es] a patentee from regaining, through litigation, coverage of subject matter relinquished during prosecution of the application for the patent. Were it otherwise, the inventor might avoid the PTO's gatekeeping role and seek to recapture in an infringement action the very subject matter surrendered as a condition of receiving the patent.").

[19] *Pharmacia v. Mylan Pharm.*, 170 F.3d 1373, 1376-78 (Fed. Cir. 1999).

References

Koppikar, Vivek Stephen B. Maebius, and J. Steven Rutt (2004) Current trends in nanotech patents: a view from inside the Patent Office, *Nanotechnology Law & Business*, **1**(1), article 4.

Lemley, Mark (November 2005) Patenting nanotechnology, *Stanford Law Review*, **58**. Available at http://ssrn.com/abstract=741326 or doi:10.2139/ssrn.741326.

Maebius, Stephen B. and Leon Radomsky (2005) The Nanotech IP Landscape: Increasing Patent Thickets Will Drive Cross-Licensing. Available at http://www.foley.com/files/tbl_s31Publications/File Upload137/2955/Document1.pdf.

Miller, John C. Ruben Serrato, Jose Miguel Represas-Cardenas, and Griffith Kundahl (2005) *The Handbook of Nanotechnology Business, Policy, and Intellectual Property Laws*, John Wiley & Sons, Inc., Hoboken, New Jersey.

Michael A. Van Lente (2006) Building the new world of nanotechnology, *Case Western Reserve Journal of International Law*, **38**(1), 173–215.

Shapiro, Carl (2001) *Navigating the Patent Thicket: Cross Licenses, Patent Pools and Standard Setting, Innovation Policy and the Economy* (ed. A. B. Jaffe *et al.*), MIT Press, Cambridge.

Chapter 5

How to Set Up an Effective IP Strategy and Manage a Nanotechnology-Based Patent Portfolio

Different IP Strategies, Including the Way to Integrate and Manage a Nanotech Patent Portfolios in Companies and Universities Fortum

Pekka Valkonen

Fortum Corporation, Keilaniementie 1, Espoo, Finland
pekka.valkonen@fortum.com

> *A good deal of the corporate planning I have observed is like a ritual rain dance: it has no effect on the weather that follows, but those who engage in it think it does.*
> Russell L. Ackoff, *Creating the Corporate Future*, 1981, p. ix [1]

5.1 Strategy: Why It's Ambiguous

When talking about IPR or patent strategy, there are two things that make it a difficult question: the complex nature of the concept of strategy and the right to ban associated with intellectual properties. Due to the inaccuracy of the concept of strategy, often it is necessary to begin presentations of strategy by describing its military origin, and at times a presentation is begun by describing the strategy used

Nanotechnology Commercialization for Managers and Scientists
Edited by Wim Helwegen and Luca Escoffier
Copyright © 2012 Pan Stanford Publishing Pte. Ltd.
ISBN 978-981-4316-22-4 (Hardcover), 978-981-4364-38-6 (eBook)
www.panstanford.com

in the Battle of Thermopylae. One basic reason for the vagueness is the multiplicity of the concept of strategy.

This was presented extremely well by Mintzberg [2] in his proposal that several different concepts of strategy can be merged under five headings, Mintzberg's 5 Ps for Strategy. In brief, these perspectives are as follows:

1. Strategy as plan: a direction, guide, course of action — intention rather than actual
2. Strategy as ploy: a maneuver intended to outwit a competitor
3. Strategy as pattern: a consistent pattern of past behavior — realized rather than intended
4. Strategy as position: locating of brands, products, or companies within the conceptual framework of consumers or other stakeholders — strategy determined primarily by factors outside the firm
5. Strategy as perspective: strategy determined primarily by a master strategist

Later on, Mintzberg *et al.* [3] proposed a slightly more extensive collection of strategies by classifying various strategies under ten schools.

Another factor that makes the concept more difficult, especially in terms of IPR, is that a patent is not a positive right but rather a right to exclude. As for patents, this has been stated by Earl Halsbury in the House of Lords on February 20, 1985 [4]: "A patent does not give you the right to do make something or to do anything except to appear in court as the plaintiff in an action for infringement." On the other hand, as a tool for business IPR is a most strategic tool due to its long duration of effect and its financial significance. The validity of a patent, 20 years, exceeds many other spans of planning in financial administration, while a trademark may remain valid for an unlimited period.

This presentation mainly concerns patents, even though IPR also includes

- Utility patents and applications
- Design patents
- Trademarks

- Semiconductor rights or chip design
- Plant variety rights

However, the significance of rights other than patents is smaller in most fields of technology and also nanotechnology, and above all, these other IPRs do not seem to have the special features associated with nanotechnology.

5.2 Pros and Cons of Patents

In general, one can say that the reason for a company's patents is mostly to obtain a competitive advantage through exclusive rights to the patented invention. There may also be other reasons for patenting, such as raising the company's image, and sometimes recognized scientists can also be rewarded by patenting inventions with less financial significance.

Patenting is advantageous to a company mostly in the following cases:

- Preventing others from utilizing the results of one's own product development
- Preventing others from getting a patent that would be harmful to oneself
- Pricing one's own products as desired by taking advantage of exclusive rights
- Increasing one's own patent base, which provides for increased freedom of action in case of patent infringement disputes
- Obtaining concrete, clearly determined material for technological cooperation
- Possibly obtaining license revenue without one's own production input
- Promoting marketing by creating a favorable image

On the other hand, patenting may be harmful because of the costs and inventions becoming available to the public. Even though the costs of a single patenting phase of an individual invention are modest, comprehensive international protection of a product or method will cost a lot. In patenting, inventions generally become public within 18 months after the application filing. The right of prior use makes it possible to continue existing operations even if the subject matter is subsequently patented by someone.

In addition to these benefits and disadvantages, there are some alternatives to patenting. The main alternatives are publishing new innovations or keeping them business secrets. The use of these alternatives is supported mainly by the following factors:

Early publication to create prior art is favorable especially in cases where

- there is no need to prohibit third party use
- the freedom to operate is already sufficient
- the invention does not differ much from prior art and/or use of the innovation cannot be foreseen
- the economical benefit is small

Trade secrets are favorable if

- there is a real chance to keep the invention a secret
- the product/technology has a short life cycle
- the innovation reveals too much know-how, which cannot be discovered from the product itself
- patent infringement cannot be discovered easily

In addition to this, companies aim to create competitive advantage and maintain it through utilization of their leading position, complexity of the product or process, and fast product development; at times, open innovation is proposed as an alternative to IPR.

The utilization of patents in business requires creating a patent strategy adapted to the company's overall strategy, a functional patent policy, and good familiarity with the industry.

5.3 IPR Strategies in the Literature

Especially in the United States, a considerable growth in patenting began in the 1980s, and to explain this phenomenon, Kortum and Lerner [5] proposed that the jump in patenting reflects an increase in US innovation spurred by changes in the management of research.

Parallel to increased patenting, a rather comprehensive literature on IPR management emerged, with Petr Hanel's [6] extensive literature review on it. Of the many writings on intellectual property rights business management practices presented by Hanel, it is possible to introduce only a few presentations that shed light on patent strategy as a whole.

Of Mintzberg's schools of strategy presented in the beginning, Knight [7] mainly represents the Planning School, and he has clearly grounded definitions for three different organizational levels in patent strategy. The first one applies to the broadest level of the organization, the product line: *Patent strategy for a product line is the science and art of employing the business, technical, and legal resources of a company to afford the maximum support to adopted policies with and without competition.* The definition of the next level, on the other hand, is *Patent strategy for a technology area is the science and art of managing research to meet competitors in the marketplace under advantageous conditions.* The definition for narrowest entity, the product, is *Patent strategy for an invention is a careful plan for gaining an end, including clever schemes for outwitting a competitor.*

On the other hand, the IPR strategy cannot be developed separately from other management of the company, as patents are only one business tool alongside others. The development of intellectual property management has introduced a new addition to the concepts of IPR strategy, and Davis and Harrison [8] have developed a value hierarchy for an organization's IPR activity.

According to the book *Edison in the Boardroom*, IPRs in different companies form a value hierarchy as a pyramid with five levels (Fig. 5.1). Each level represents a different expectation that the company has about the contribution that its intellectual property or intellectual asset function should be making to the corporate goals.

Level 1 of the value hierarchy is the defense level. At this level, the IP function provides a patent shield to protect the company from litigation.

Level 2 is the cost control level, in which companies focus on how to reduce the costs of filing and maintaining their IP portfolios. Intellectual property is still viewed primary as a legal asset.

Figure 5.1. Value hierarchy of IPR. Redrawn from Ref. 8.

Level 3 of the value hierarchy is the profit center level. Companies at this level turn their attention to more proactive strategies that can generate remarkable additional revenues while further continuing to trim costs. Passing from previous level of activity to this one requires a major change in a company attitude and event its organization.

Level 4 is the integrated level. In this level the IP organization reaches outwardly beyond its own department to serve a greater purpose within the organization as a whole. In essence, its activities are integrated with those of other functions and embedded in the company's day-to-day operations, procedures, and strategies.

Level 5, the final level, is the visionary level. Few companies have reached this level of looking outside the company and into the future. In this level, the IP function, having already become deeply ingrained in the company, takes on the challenge of identifying future trends in the industry and consumer preferences.

In addition to adaptation to the overall business process, business always involves a competitive situation. One is not alone in the market with patents, either; patent strategy should take other parties' patents into consideration as well.

Often, a single individual patent does not offer much protection against competitors, as a single patent is reasonably easy to circumvent. This is why companies aim to patent their products and methods through several patents, and end up with patent portfolios.

Granstrand [9] presents a group of patent strategies as a two-dimensional patent landscape in which gradually advancing R&D activity is filling up the landscape, describing a field of technology with circles representing other parties' patents. These circles are located according to the claim of each patent with regard to similar solutions, and the size of the circles can be used to describe the extent of the scope of protection. The arrows in landscape presents ways to new routes or technical solutions. Figure 5.2 illustrates a variety of patent strategies using these symbols.

In the figure Granstrand presents six different patent strategies in this technology landscape.

Ad hoc blocking and "inventing around." In this strategy small resources are used, and one or a few patents are used to protect an innovation in specific technology. This strategy is quite easy to invent around, it means to find out other technical solutions to the problem.

Strategic patent searching. This strategy consists of a single patent with significant patent claims, and this patent is called strategic patent. It is expensive to invent around or find another technical solution.

Blanketing and flooding. In this case the whole technical area is filled with many patents and patent applications. Applications in the area are filed in a less systematic way and the inventions are often minor. This strategy may be suitable for an emerging technology where uncertainty is high regarding the direction of technology.

Fencing. In this case a block of patents is directed to a certain technological area, and application is drafted more systematically. This strategy is suitable for a range of possible different technical solutions for achieving a similar functional result.

Surrounding. In this strategy an important patent, for example, aforementioned strategic patent, is surrounded by a lot of other patents with less importance. Surrounding is used for ensuring area for applications or for providing possibilities to license out.

Combination into patent networks. In this strategy a patent portfolio is built with patents of various kinds and configurations to build together strength position in technology area.

Competitors' patenting has been added to the patent landscape above. Porter [10], a significant developer of business competitive analysis, introduced other competitive factors with an effect on the industry to the competitive situation, Porter's Five Forces: including rivalry, the threat of substitutes, buyer power, supplier power, and barriers to entry.

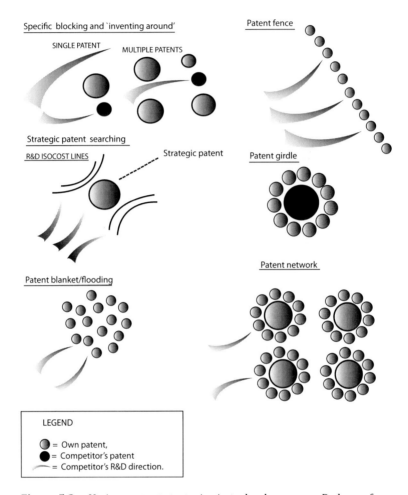

Figure 5.2. Various patent strategies in technology space. Redrawn from Ref. 9.

In Mintzberg's strategy classification, Porter belongs to the Positioning School. Porter's model of five competitive forces model is usable in many respects, both in patent strategy work and also practical work with patents when composing an individual patent application. This is presented in more detail in connection with the practical operating models at the end of the chapter.

5.4 Appropriability of Patents

Appropriability is a term in economics area, which is used describing the environmental factors that govern an innovator's ability to capture profits generated by an innovation.

The term has often been used to describe the significance of patent protection in different fields of industry. A seminal empirical study on the utilization of patents was the so-called Yale study made by Levin *et al.* [11], in which extensive questionnaire data were used to find out different ways of protecting innovation and utilizing patents within the manufacturing industry in the United States. According to the study, patenting is not the most important mechanism of utilization in most fields of industry; the most significant ones were nondisclosure, learning curve, and the sales and servicing network. Only in a few fields, such as the pharmaceutical industry and chemistry, were patents the most significant means of utilization.

Another significant study of US manufacturing, the study by Cohen *et al.* [12] Carnegie Mellon survey sometimes called "Yale II" survey, confirmed that patents were the most commonly used means of protection only in certain chemistry-related fields of industry. It also emerged that companies use various combined means of utilization. The study showed that in many fields, the purpose of patents is not to protect the invention but rather for strategic reasons, such as obtaining negotiating power, cross-licensing, and gaining reputation.

The same characteristic of using patents in a more effective and versatile manner than for only protecting innovations is evident in several comparisons between Japanese and Western companies. Such comparisons between Japan and other countries in the patenting practices of companies have been made with regard to British companies [13], American companies [14], and Swedish companies [9].

At the same time as the increase in patenting in the 1980s, especially in the United States, research in the field of economics also began to get more interested in the utilization of patents, and this has resulted in a number of empirical studies. While earlier studies were made using US data, the focus of the studies has more recently been on Europe. This is partly due to the CIS (Community Innovation Survey) studies by European statistics agencies.

Blind *et al.* [15] present a summary table of six significant empirical studies on the motives of patenting: the study data is from 1993 to 2003 and from several countries. Some results from these studies are listed here

- Arundel *et al.* [16]: data from 1993, Germany, Italy, Great Britain
- Duguet and Kabla [17]: data from 1998, France
- Cohen *et al.* [18]: data from 1994, United States and Japan
- Pitkethly [13]: data from 1994, Japan and Great Britain
- Schalk *et al.* [19]: data from 1997, Germany
- OECD [20]: data from 2003, many countries

In all of these studies, the traditional motive — protection of innovation — emerged as the main motive for patenting. The order of significance of the strategic motives in these six studies was as follows:

1. Offensive blockade
2. Defensive blockade
3. Exchange potential/negotiating power
4. International market extension
5. Make own invention the standard
6. Licensing revenues
7. Reputation/technical image
8. Internal performance indicator/motivation

(order calculated by the author from the means of the order in the table by Blind *et al.* [15])

The conclusion of Blind *et al.* [15] on the basis of extensive German data from 2002 is that the study confirms the significance of strategic patenting. One conclusion is the ambition to increase the company's reputation and negotiating position with other companies, and encouraging the R&D personnel to declare inventions. Another conclusion is the clear difference between patenting motives between "discrete" and "complex" fields. A third conclusion is that large companies particularly strongly prefer new, strategic patenting motives, such as the potential for using patents in exchange, and utilizing them in measuring the R&D performance and rewarding.

5.5 Appropriability of Patents in Nanotechnology

Nanotechnology is a field of technology that has experienced a strong growth in patenting. Also, with the field considered an important future industry, the copious amount of patenting in the field has given rise to significant discussion on the effect of IPRs on the development of the industry.

Lemley [21] proposes three significant reasons why nanotechnology differs from other fields of technology when it comes to patenting. First, this is almost the first new field in a century in which the basic ideas are being patented at the outset. A second factor distinguishing nanotechnology is its unique cross-industry structure. Third, a large number of the basic nanotechnology patents have been issued to universities, which have become far more active in patenting in the last 25 years.

The strong growth in patenting has resulted in several bibliometric surveys of the area, and their results are often called patent landscapes. They present companies with the most patents, the relationships between these companies with the help of reference publications, the countries in which the most applications emerge, geographical areas with proliferative patents in the field, technology surveys, and division into different fields of technology. A typical such study is by Li *et al.* [22] on three major patent offices (USPTO, EPO, and JPO) between 1976 and 2004. It uses the patent databases as indicators of nanotechnology trends via bibliographic analysis, content map analysis, and citation network analysis on nanotechnology patents per country, institution, and technology field.

The results show that there are differences between the countries in terms of the numbers of patents, fields of technology, and quotations. It also includes lists of the biggest applicants in different offices and the most popular classes of patents. Preschitschek and Bresser [23] go further in their presentation of their technology landscape in China and Germany, which itemizes the patenting strategies of companies in either country. According to the results, China has become a significant country in the field, also when it comes to applied nanotechnology. In China, the patent applicant is by far more often a university or a research institute, and the results are more often applied research. In Germany, on the other hand, universities often focus more on basic research. The problem with these studies on technology landscapes and, in general, bibliographic studies, is that the results do not often contain much new. Examples of this are the results of the OECD's [24] extremely extensive study whose abstract's main headings are as follows:

Conclusions and future work

This analysis aims at capturing current inventive activities in nanotechnologies based on the analysis of patent applications to the EPO. The current status of nanotechnology, the recent acceleration in inventive activities, and the strong linkages between scientific and inventive activities have been documented.

1. Nanotechnology — a set of technologies on the nanometer scale, not a single technological field
2. The recent rise of inventive activities in nanotechnology
3. Science fuels technological development in diverse ways

Studies of this kind can act as the foundation when creating the competitor analysis required in creating a patent strategy, or even more extensively an analysis of companies in the field. One can use the publications as a model for shaping information searches, which companies operate in the field, which countries or regions are significant, and what other activity is there in the nanotechnology field.

The special nanotechnology classifications made by key patent offices facilitate patenting in nanotechnology and also making patent surveys concerning patent landscapes. According to Huang *et al.* [25] the USPTO (United States Patent and Trademark Office), the EPO (European Patent Office), and the JPO (Japan Patent Office), have made intense efforts to improve their own classi-fication systems and combine all nanotechnology-related patents into one single patent class. The USPTO established an informal nanotechnology classification Class 977 in 2004, the EPO created the Y01N tag specifically for nanoscience and technology in 2004, and the JPO created the ZNM class.

In addition to patent landscapes concerning nanotechnology, quite much has been written on patenting in this technological field in economics and legal science. These have much focused on the problems and functioning of the actual patent system in the field, such as Lemley [21]. However, few empirical studies have been made from the point of view of companies on how patenting is used in business in nanotechnology. There are even fewer empirical studies on the patenting behavior of companies. One of these very few is Andrea Fernández-Ribas' study [26] on the patenting of US-based nanotechnology companies of different sizes utilizing WO patenting (WO stands for World Intellectual Property Organization). The study data consisted of WO patent applications submitted in 1996–2006 included in the CNS-ASU (Center for Nanotechnology in Society at Arizona State University) database. According to the results, even though both large companies and SMEs patented internationally, there were also differences.

Large companies usually apply for patents in developing countries, such as China or Korea, while SMEs tend to patent in more ordinary locations, already-industrialized countries with strong IPR systems. SME companies begin foreign patenting at a very early phase; in the field of nanotechnology, they could be called "born global." For large companies, globalization was seen in the form of an invention developed abroad.

In another empirical study, Munari and Toschi [27] review the effects of technology companies' patenting activities on investments by venture capitalists. The study data included 332 nanotechnology companies from various countries that had received venture capital financing in 1985–2006. Factors that explained patenting were the total number of patent applications in EPO, the number of patents belonging to the core of nanotechnology, and the extent of the patent claims measured by IPC classes. According to the results of the study, the number of patent applications did not, per se, have an effect on the financing received by the companies, but the number of patents in the core of nanotechnology increased the value of the companies by 15% in the first VC round. The extent of patent protection was not observed to have statistical significance for the financing received by the companies.

5.6 IPR Policy: Implementing Strategy from the Bottom Up

While strategy shows the long-term direction for measures to reach the objectives, policy, on the other hand, simplifies decision making and actions. Certainly, policy can be considered as belonging to the Pattern group in Mintzberg's strategy classification — strategy is a pattern — specifically, a pattern in a stream of actions.

The bottom-up tactic/manner is a practical, easy, and useful way of creating an IPR policy. It is created by reviewing all IPR-related matters, especially the main matters to be decided on, and by specifying procedures for them. The creation of an IPR policy is not only about creating a decision-making process, but also training those taking part in it and committing them to courses of action. Therefore, it is important that management at a sufficiently high level, expertise in R&D, key business functions and, if necessary, support functions — such as HR or quality assurance — can be involved in the creation of the policy. Similarly, it is necessary to ensure that a sufficient understanding of practical IPR matters and R&D activity is involved.

Depending on the size of the organization, the policy may involve several levels; in a large company as many as three levels (group, business area, business unit), and a single level in a small company. Specific responsibilities should be defined for each level so that all IPR matters are effectively dealt with.

Here is a list of things that should be organized, and the long list is separated in three quite woolly areas: organizational matters, decision content, and other matters:

- Who owns the IPRs, and whether there are differences between patents, trademarks, or other rights
- Whom the IPR reports to — the most common options are legal, R&D, QA, marketing department, or a single business unit (usually an exceptionally patent-intensive unit in the company)

- Budgeting and allocation of resources at a general level and, if decentralized, it should be specified separately for each level
- What the level of decision making is, whether decisions are made on invention reports, application for patent, patent application abroad, maintenance of patents, and termination of patents in the same way
- Decision-process procedure — whether decisions are made in the linear organizations or in R&D matrices, whether evaluation teams are used and how IT facilities can support this.
- Any obligation to inform other business functions of the expiry of patents
- How and by whom should patents and other IPRs be taken into account in strategies and operating plans

Here is a list of things related to decision content:

- Overall attitude toward patents owned by third parties and investigations of patent situation early in the product development process and more specific surveys prior to the adoption of new processes or launch of products
- In which areas are patents desired: product, process, applications, or industry standards
- Patenting in which geographical area, what about the others (trademarks, designs)
- Taking the patent landscape into account, that is, the adaptation of patenting to other parties' IPR strategies — is the aim an individual, extensive patent, or a group of patents covering a more extensive scope
- Timing — is the aim to submit the patent applications as early or as late as possible?
- The routes of patenting — whether to start in the company's or inventor's home country, in which languages to compose the applications, whether the PCT route is continued, whether EPO is used, what is the importance of design rights, utility patents, or trademarks

A list of things related to other matters to take into account is as follows:

- Make sure whose responsibility it is and at which stage the IPR surveys are needed in business acquisitions and divestments, and at which stage the IPR unit is involved.
- Pay attention to how (in and out) licensed patents are taken care of, including surveys associated with their preparation, decision processes, assurances, and possession of routines.
- Take care of IPR-related agreements and research cooperation NDAs, their documentation.
- Adapt patenting in the research community to the company's decision process; additionally, take into account the impact of any public financing on the functions.
- Determine responsibilities of documentation of decisions, matters concerning the storage and destruction of decisions, and other correspondence.
- Get maximum information on employee inventions and what different courses of action they result in, as regards different countries and research cooperation.
- Alongside formal reporting, inform the parties involved of the processing of patent-related matters and decide how to take care of it.
- Increase and provide training in competence related to IPR. Different organizations and different levels need different kinds of knowledge of IPR matters. Senior management needs information on risks and financial importance, the finance department has its own needs for costs, R&D on the needs for development, business functions on the competitive benefits of IPR, etc.
- Determined what the attitude to inventions offered by outsiders is and also how to avoid related contamination of technical information.

In order to complement this list, some views have been presented below with regard to the centralization and decentralization of the most important matters to be specified in the policy.

The benefit of a centralized organization is that the decisions to be made are more easily unidirectional. It also offers the possibility of closer relationships with senior management and the required support of the management. Centralized management of IPR matters is more cost-efficient, as the economy of scale applies.

Also, it makes the organization of the harmonization, control, and follow-up of operations easier.

The problem in a centralized IPR system is that the activity is too detached from the reality of the business functions. At the same time, it is more difficult to organize the connections with inventors, and the required support for decisions in the business functions is more remote. Centralized financing of IPR activity, especially in multinational companies, requires carefully going into matters of accounting and taxation in several countries.

The benefits of a decentralized IPR organization are several of the downsides of a centralized system: close relationships with the business and inventors, more detailed allocation of operating costs, the possibility of customizing taxation, and accounting on a country-specific basis. Special problems in decentralized IPR activity include the question of whether the organization has the critical mass needed for professional activity, its continuity and development. Also, the use of subcontractors involves special additional needs for coordination, which also results in additional costs. Decentralized activity also involves situations where different business units have conflicting needs and wishes. Furthermore, in a decentralized model the reporting and responsibilities of IPR activity may be in different units. In particular, the company's overall strategy, relative strength with regard to IPRs, and how competitors use their IPRs as competitive advantages have an effect on these decisions made in the IPR policy and choices regarding centralization.

Most of the company's strategic basic choices are naturally aligned with the IPR choices. Such basic choices include whether the product strategy is broad or narrow, if the market strategy is global or regionally limited, what the technology strategy is — whether it is based on in-house R&D or if it is of the assembly-type — and whether the strategy emphasizes market share or seeks profitability.

Finally, one should also remember that IPR activity involves a variety of routines whose management requires sufficient resources: transfer deeds and letters of authority, invoices for work by external agents, preparation of decision-making, archiving, annual fees, due date control, monitoring of competitors, and training in the field. All of these must be taken into account when dimensioning the resources of IPR activity.

5.7 IPR Policy: Some Practical Measures

5.7.1 *Implementing IPR Policy by Learning*

The IPR function has two ways of proving the value of patents to the company's senior management:

1. *"Quick and dirty."* This means a case in court.
2. *"The hard way."* This means (a) communication and (b) education.

(Dr. M. T. Barlow [BP], Convincing the management of the values of IP, Presentation in Meeting of Patent Practitioners Association in Finland, 23 November 1998.)

The support of the senior management is absolutely necessary for the success of innovations as well as successful implementation and maintenance of the IPR strategy and policy. In order to receive the support of senior management, one must understand how senior management works. They are interested in information on financial success, business risks, strategy, and the company's reputation. In order to be heard, IPR management must speak the same language.

Also, other corporate cultures besides business are represented in technology companies: technological and corporate law cultures. IPR management must understand all of them. The management of immaterial assets has a similar role, and it is a suitable reference and, where applicable, a model for IPR management in many aspects. In practice, this means teaching IPR management lawyers in understanding technology, engineers in understanding legal matters, and both these groups in understanding finances and business.

In practical training activity, this means increasingly moving from one's own profession toward the region shared by these three organizational cultures in the following fields:

- From professional doing to a successful advisory role
- From the legal perspective to the business perspective
- From focusing on details toward an understanding of more extensive entities

- From invention orientation toward commercialization of the application
- Shifting attention from an individual inventor to the technology
- From internal processes toward proactive activity targeting the operating environment

In order to reach these objectives of IPR training, training provided by the industry experts is needed in the four following areas:

1. Ensuring good basic knowledge through formal training, taking into account the background factors of each individual
2. Ensuring knowledge of the technology, with good collaboration with R&D and the specializations of people in different technologies as key aspects

Understanding business: theoretical understanding of and practical competence in business strategies, understanding the sources of competitiveness of one's own business operations, and learning about the valuation of intangible rights

Personal development by deepening competence, finding ways to make operating methods more efficient and networking

5.7.2 *Patenting in View of Porter's Five Forces*

In Mintzberg's strategy classification, Porter belongs to the Positioning School. Especially, his Five Forces model is usable in many respects, both in patent strategy work and also practical work with patents when composing an individual patent application.

Figure 5.3 shows this five competitive forces model. According to the model, the competitive situation prevailing in an industry is under the influence of five forces: competitive rivalry within the industry, suppliers, buyers, substitutes and new entrants.

The significance of the analysis and understanding of these five competitive forces becomes evident when thinking about the typical product development process and its protection through patenting.

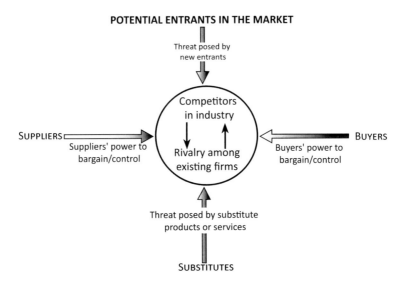

Figure 5.3. Five competitive forces model.

1. The patents of competitors in the industry are something that is usually already taken into consideration at the beginning of a chemistry-related product development process. Naturally, this already requires careful monitoring of competitors before the R&D project, as well as during it.

2. Once the own product or process has been established, any protections of raw material suppliers, which increase their negotiating power, often emerge. Typically, it has been possible to make agreements with the producers of raw materials on the use of existing technology, but one should also notice that the producers of raw materials and subcontractors develop their operations and aim to decrease the competition through patents.

3. Buyers of the developed products also want exclusive rights to their applications and aim to patent their applications, thereby limiting the market for the new product. Customers are the best experts in their respective fields, and also want to obtain a competitive edge compared to others.

4. A new product or process can make the industry an appealing one and introduce completely new players. Follow-up of new entrants is more difficult than follow-up of established players.

5. The properties or upsides of one's own product can be achieved with a substitute. This competitive situation is difficult to predict, which is why the aim in patenting is to protect the favorable properties of the product or the benefits offered by the product to the customers.

This idea can be applied in composing an individual patent application, by considering the scope of protection with regard to the entire competitive field, not only the visible product features or process conditions.

Similarly, a more extensive survey of the competitive situation can be applied to the product or family of products protected with patents.

5.7.3 Evaluation of Patent Portfolio

IPR assets are becoming an increasingly significant strategic tool. At the same time, the management of these assets is becoming a key factor in companies. This, on the other hand, increases the load of IPR units and the costs caused by IPR.

Thus, the evaluation of IPR assets becomes a significant tool for managing these problems. One of the best presentations on the evaluation of patents is by Lex van Wijk [28]. In it, he divides the target into four areas: employee competence, internal structure, external structure, and alignment of strategies. Employee competence is the sum of the competence, know how, and skills of the company's employee. Internal Structure includes all the company's databases, written procedures, intellectual assets such as codified technologies and intellectual property rights, and its infrastructure. External structure includes a company's relationship with customers, its reputation, the company's understanding of the needs and wishes of customers, and the company's position vis-à-vis competitors. Alignment of strategies includes a company's in-house business interest support to commercialize the technology of the invention, role of cost savings in providing competitive advantage, and revenue impact (e.g., by product/service differentiation) of the invention.

The evaluation of patents should be performed by a team with representatives from the R&D, production, sales, and IPR units, at the least. Patents should be evaluated regularly, once every one to two years. The patent application process has several decision phases; the decision to patent an invention, to expand the patent to abroad and the extent of the maintenance of granted patents (annual fees), and extensive evaluation according to the table is not necessary in only one of these phases.

In general, the aim in IPR activity should be to pursue procedures that are as simple as possible, in order to save resources. It is useful to apply the 20/80 rule so that full evaluation is only carried out on 20% of the patents that are considered the most valuable, while the remaining 80% undergo a briefer evaluation.

5.8 Summary

Two matters complicate the concept of strategy in connection with IPR activity: the diverse nature of the concept of strategy, and that a patent, by its nature, is a right to exclude. When discussing strategy, it is necessary to ensure that the parties' views of the contents of this activity are uniform. In its simplest form, IPR strategy reviews the benefits and disadvantages of patenting, keeping in mind the fact that the aim of patenting is to obtain a competitive advantage. In addition, it is necessary to bear in mind the alternatives to patenting, the most important of them being publishing new innovations or keeping them business secrets.

Creating an IPR strategy is not just a one-off project; rather, the IPR strategy should also be seen as a continuous development activity. It is possible to compare one's own position with regard to others and the advanced state of one's own strategy with regard to others to the levels of the value hierarchy of IPR activity. The initial level concerns protecting oneself from the threat of competitors' patents. At the second level, the focus of activity is on building a patent portfolio and managing costs. The third level concerns proactive IPR activity, while at the fourth level operations are integrated as part of other business activities and take the company's activities into account as a whole. Finally, at the visionary fifth level, which few reach, IPR activity takes into account trends from outside the company and consumer behavior.

The patent landscape is a way of illustrating the relationships between the holders of various patents in a field of technology and the individual technical solutions (patents) held by them. As technology develops, the patent landscape is filled with solutions protected by different parties, and one needs to find a path for one's own business. The task of the IPR strategy is to react to obstacles caused by other parties' patent portfolios with one's own protections.

Empirical studies have shown that the utilization of patents is more significant in the chemical industry than in several other fields of industry. On the other hand, in many industries the focus of protection of R&D results and innovations is largely on matters other than the exclusive right provided by a patent and the related right to exclude. It has also emerged in empirical studies that companies' patenting is increasingly strategic by nature. This means that patents not only aim at obtaining exclusive rights; rather, the aim of patenting is to obtain negotiating power, cross-licensing, obtaining licence revenue, and gaining reputation.

Nanotechnology patents are associated with certain special features that make it a unique area compared to previously emerged technologies. Firstly, patenting has been involved since the very beginning of the development of nanotechnology, and as a result, there are a number of central, broad patents in the field. This has not been the case in the development of the previous new technologies, IT, and biotechnology.

Another special feature is the link of nanotechnology to other fields of technology. Key application areas of nanotechnology include electronics, energy technology, biotechnology, and other material technologies. There is an abundance of applications also in other fields of industry, such as food and packaging and medicine, in particular diagnostics. Also contributing to how nanotechnology tends to transcend sector boundaries is the diverse nature of parties involved in the industry: they include large corporations, their smaller subsidiaries, SMEs, universities, and research institutes.

A third special feature of nanotechnology patents is patenting by universities and research institutes, as their position in negotiations over the use of patents and licenses differs from companies with production. The transfer of an innovation from development stage to production typically requires a decade or more of resources. Hence companies that manufacture products have a great need to launch their products in the market, whereas the interests of nonmanufacturing patentees may differ. Often, research institutes aim to obtain new research contracts with their patents in addition to licence revenue.

As the field of nanotechnology IPR players is so unique, dispersed and undeveloped, it is not possible to present a single way of creating a patent strategy and the patent policy required in its implementation. Therefore, this chapter presents some key tools for composing a patent strategy.

As with the chemical industry, competitor analysis and a survey of technological solutions discovered by others is the starting point of in-house R&D in nanotechnology. The special nanotechnology classifications made by patent offices facilitate making patent surveys. However, R&D activity is long-term work, and it is not sufficient to find out the existing competitive situation; one also needs to foresee future changes in the field.

Porter's model of five competitive forces is a usable framework not only in seeing the existing competitive situation but also its possible changes. In addition to competitors, Porter's model surveys the activities of raw material suppliers or subcontractors, buyers of the products and their applications, new players aiming to enter the field and products that substitute the properties of one's own product.

IPR assets are becoming an increasingly significant part of a company's operations, while on the other hand, as the significance of protection increases, the costs and the load on IPR units also increase. Hence, the evaluation of IPR assets, aiming to detect the most valuable patents in the business, becomes a significant tool for managing these problems. The competence of in-house personnel, the organization's operating methods, the structure of external competition, and the suitability of the patent for the company's strategy are associated with the usability and value of patents.

The evaluation may concern an individual patent or invention, and the same pattern of thinking can also be applied to evaluating the patent portfolio of one's own or a competitor. The practical evaluation should be carried out in a team representing competence in the technology in question, the actual invention, the planned business activity and patenting.

IPR strategy, which shows the long-term direction for the activity to reach the objectives, is discussed above. The IPR policy, on the other hand, aims to simplify the decision making and procedures related to patents and other industrial property rights.

Attention must be paid to a number of factors in composing the IPR policy. The list is a long one, and the factors that need to be taken into account can be divided into three areas: how the activity will be organized, what the decisions need to include, and other matters. Creating the IPR policy with the help of such a list of things to be taken care of is a practical way to establish an IPR policy.

At the same time, composing an IPR policy involves the orientation of the parties involved in the decision-making process and the contents of the decisions, and the establishment of the policy requires extensive expertise from senior management, R&D activity, key business functions and, naturally, an understanding of practical IPR-related matters.

The aim of the company's IPR activity is also to obtain a competitive advantage, and the IPR strategy and policy are tools for it. In order to function appropriately, IPR activity needs to develop and also rise up through the levels of the value hierarchy of IPR activity. The support of the senior management is absolutely necessary for the success of innovations as well as successful implementation and maintenance of the IPR strategy and policy.

In order to receive the support of senior management, one must understand how senior management works. They are interested in information on financial success, business risks, strategy, and the company's reputation. In order to be heard, IPR management must speak the same language. In addition, the IPR administration must understand legal matters, finances, and business.

References

1. Ackoff, R. L. (1981) *Creating the Corporate Future*, John Wiley & Sons, New York, p. ix.
2. Mintzberg, H. (1987) The strategy concept I: five Ps for strategy, *California Management Review*, Fall, pp. 11–24.
3. Mintzberg, H., Ahlstrand, B., and Lampel, J. (1998) *Strategy Safari: A Guided Tour Through the Wilds of Strategic Management*, Free Press, New York.
4. Earl of Halsbury 20/2/85, House of Lords ref: Pitkethly, R., *Intellectual Property Management*, www-page entered last 7.7.2010. Available at http://users.ox.ac.uk/~mast0140/SEC/SECIP.ppt (Patents Pros and Cons 3.2)- (IPR strategies in literature 3.3)
5. Kortum S., and Lerner, J. (1999) What is behind the recent surge in patenting, *Research Policy*, **28**(1), pp. 1–22.
6. Hanel, P. (2006) Intellectual property rights business management practices: a survey of the literature, *Technovation*, **26**, pp. 895–931.
7. Knight, H. (2001) *Patent Strategy: For Researchers and Research Managers*, 2nd ed., Wiley, New York.
8. Davis, J. L., and Harrison, S. S. (2001) *Edison in the Boardroom: How Leading Companies Realize Value from Their Intellectual Assets*, Wiley, New York.
9. Granstrand, O. (1999) *The Economics and Management of Intellectual Property: Towards Intellectual Capitalism* Elgar, Cheltenham, UK, and Northampton, MA.
10. Porter, M. E. (1980) *Competitive Strategy*, Free Press, New York. (Appropropability of patents (3.4.)).
11. Levin, R. C., Cohen W. M., and Mowery, D. C. (1985) R & D appropriability, opportunity, and market structure: new evidence on some Schumpeterian hypotheses, *The American Economic Review*, **75**(2), Papers and Proceedings of the Ninety-Seventh Annual Meeting of the American Economic Association, pp. 20–24.
12. Cohen, W. M., Nelson, R. R., and Walsh, J. P. (2000) Protecting their intellectual assets: appropriability conditions and why US manufacturing, firms patent (or not). *NBER Working Paper*, WP 7552.
13. Pitkethly, R. H. (2001). Intellectual property strategy in Japanese and UK companies: patent licensing decisions and learning opportunities, *Research Policy*, **30**(3), 425–442.

14. Aoki, R., and Prusa, T. J. (1996) Product development and the timing of information disclosure under U.S. and Japanese patent systems, *Journal of the Japanese and International Economies*, **10**(3), 233–249.

15. Blind, K., Edler, J., Frietsch, R., and Schmoch, U. (2002) Motives to patent: empirical evidence from Germany, *Research Policy*, **35**(5), 655–672.

16. Arundel, A., van de Paal, G., and Soete, L. (1995). Innovation strategies of Europe's Largest Industrial Fims. Results of the PACE survey for information sources, public research, protection of innovations and government programmes, Maastricht University, UNU-MERIT, Working Paper Series 2008-058.

17. Duguet, E., and Kabla, I. (1998) Appropriation strategy and the motivations to use the patent system: an econometric analysis at the firm level in French manufacturing, *Annales d'Economie et de Statistique* 49/50, pp. 289–327.

18. Cohen, W. M., Goto, A., Nagata, A., Nelson, R. N., and Walsh, J. P. (2002) R&D spillovers, patents and the incentives to innovate in Japan and the United States, *Research Policy*, **31**(8–9), pp. 1349–1367.

19. Schalk, H.J., Täger, U.C., and Brander, S. (eds) (1999) *Wissensverbreitung und Diffusionsdynamik im Spannungsfeld zwischen innovierenden und imitierenden Unternehmen*, Ifo-Institut für Wirtschaftsforschung, München.

20. OECD (2003) Committee for Scientific and Technological Policy, *Preliminary Results of OECD/BIAC Survey on the Use and Perception of Patents in the Business Community*. Working Party on Innovation and Policy, 19–20 June.

21. Lemley, M., A. (2005) Patenting nanotechnology, *Stanford Law Review*, **58**, p. 601

22. Li, X., Lin, Y., Chen, H., and Roco, M. C. (2007) Worldwide Nanotechnology development: a comparative study of USPTO, EPO, and JPO patents (1976–2004), *Journal of Nanoparticle Research*, **9**(6), 977–1002.

23. Preschitschek, N., and Bresser, D. (2010) Nanotechnology patenting in China and Germany: a comparison of patent landscapes by bibliographic analyses, *Journal of Business Chemistry*, **7**(1), 3–13.

24. OECD (2007) Capturing nanotechnology's current state of development via analysis of patents, DSTI/DOC(2007)4, *STI Working Paper* 2007/4/.

25. Huang, C., Notten, A., and Rasters, N. (2010) Nanotechnology publications and patents: a review of social science studies and search strategies, *The Journal of Technology Transfer*, doi: 10.1007/s10961-009-9149-8.

26. Fernández-Ribas, A. (2010) Firms' Global Patent Strategies in an Emerging Technology, www-page entered last 7.7.2010. Available at http://hdl.handle.net/1853/32344

27. Munari, F., and Toschi, L. How Good Are VCs at Valuing Technology? An Analysis of Patenting and Venture Capital Investments in Nanotechnology, www-page entered last 7.7.2010. Available at http://ssrn.com/abstract=1158467

28. van Wijk, L. (2001) Evaluating inventions, *Patent World*, November, p. 29.

Chapter 6

How to Identify Patent Infringements in the Nanotechnology Sector

Marco Spadaro

*Italian and European Patent Attorney, Partner, Cantaluppi
& Partners — Padova, Rome, Milan*
m.spadaro@cantaluppi.com

6.1 Introduction

A patent is a right conferred by a public authority on behalf of the state for exclusive exploitation (monopoly) of a new, inventive, and fully disclosed invention in an industrial field.

The monopoly provided by the patent to its proprietor must be observed by third parties, in particular, competitors. In other words, the patented invention can be carried out only by the patent proprietor or by subjects who got the permission to do so by the same proprietor (license).

The invention, in its innovative characteristics, is defined by the claims, which establish the boundaries beyond which the public cannot enter without violating patent owner's property.

For example, if the invention is, say, the mechanism for accurate control of shutter speed of a digital camera, the patent owner will

Nanotechnology Commercialization for Managers and Scientists
Edited by Wim Helwegen and Luca Escoffier
Copyright © 2012 Pan Stanford Publishing Pte. Ltd.
ISBN 978-981-4316-22-4 (Hardcover), 978-981-4364-38-6 (eBook)
www.panstanford.com

have the monopoly of manufacturing and selling this mechanism, the shutter and a camera equipped with it, but will not have any exclusive right on other kind of shutters or any camera not equipped with the objects defined in the claims.

6.2 The Laws

The manners of exerting these exclusive rights are regulated by national laws.

With very few exceptions, each State has its own patent law. Patent owners can therefore resort to national law for seeking enforcement of their rights.

Global trade pushed nations to agree on common principles also in Intellectual Property Rights and today we have a more defined landscape with a minimum of common criteria that define these rights, duties, and the means for enforcing or defending them from possible abuses.

Just for sake of simplicity, we refer here to only two major supranational organizations: World Trade Organization (WTO) and World Intellectual Property Organization (WIPO).

The first one is the only global international organization dealing with the rules of trade between nations. Agreements are negotiated and signed by member nations and ratified in their parliaments. The goal is to help producers of goods and services, exporters, and importers conduct their business.

The second one is a specialized agency of the United Nations dedicated to developing a balanced and accessible international intellectual property (IP) system, rewarding creativity, fostering innovation, and contributing to economic development while safeguarding the public interest.

Within the scope of this chapter we also consider another supranational organization, the European Union and its relation with European Patent Office.

All these systems provide common legal structure for obtaining patent protection and regulating the relationship between the patent owner and society.

Now, let's see what the law provides and how it defines the right of exclusive use provided by patents and, by contrast, how an action from a third party, different from the patent proprietor can be determined to be an action of infringement.

In a global view, and in connection with the scope of this chapter, we will see the common legal basis that provides a link between the United States and Europe.

We have to keep in mind that the United States and Europe are two comparable markets, in term of potential users, but very different in terms of their legal systems. The United States is a federal system with a central government, with legislative power, issuing federal laws applicable in each of the states that make up the federation. Europe, in its geographical expression, is a group of independent, sovereign States, with different political and legislative systems. In Europe there are republics and kingdoms and there are different economic systems, with more or less regulated markets. Until the recent past, Communist governments ruled in a number of European States. Now Europe is still a complex network of political/economic agreements, the most imposing is European Union.

It is often difficult to keep traders within legal fences, and trade was and is at the basis of harsh contrasts between or among nations.

Nations are associated in many international organizations with the scope of establishing common rules of procedure and behavior. Today we talk about "globalization" as a matter of fact. However, the need for international protection of intellectual property urged when foreign exhibitors refused to attend the International Exhibition in Vienna in 1873, because they were afraid their ideas would be stolen and exploited commercially in other countries. In 1883, the Paris Convention for the Protection of Industrial Property was established and nowadays is still a fundamental pillar for the protection of innovation and Industrial Property, in general. Of course, a history of Industrial Property is not the scope of this chapter, but we must know at least the general frame wherein innovators move in their trade.

World Trade Organization (WTO) is an International Organization to which both the United States and the European nations belong to. The treaty concerning Intellectual Property Rights and

Patents is contained in the Trade Related Aspects of Intellectual Property Rights (TRIPS). This treaty provides a commonly agreed set of rules for trade and recognition f intellectual property rights.

6.3 The United States of America and Europe

In the United States, the right to a patent is stated in the Constitution. Article 1, Section 8 states: "to promote the science and useful arts by securing for a limited time to the inventors the exclusive right to their respective rights and discoveries." The exclusive right is defined as "to exclude others from making, using, offering for sale, or selling the invention throughout the United States or importing the invention into the United States." The right of exclusive conferred by a patent is described by the above list of actions, which can be done only by the patent proprietor or a licensee. The US patent is valid throughout the United States, therefore the right of monopoly is valid in all the States, independently from the fact if the patented product is marketed in only one or few States.

In Europe there are two main situations: trade in Europe as a whole and trade in the European Union. Trading in Europe is, more or less the same as trading among other countries in the world, dealing with borders and different laws. However, within Europe, European Union represents a special area of common market, wherein special trading rules apply. Trading within European Union is as trading in one single nation, with no borders and a set of common rules. Let's now concentrate on European Union and patent monopoly.

One statutory principle of trading in European Union is keeping competition alive, for the benefit of the public. Fair competition is a principle of high importance within the European Union and abuse of dominant position is actively prosecuted by European Authorities. In this context, the patent monopoly can be seen as a stain in the picture of free, and fair competition. However, patents are a fully recognized tool for fostering innovation, hence competition and economic growth and general progress.

Some special considerations must be taken in designing a patent strategy within European Union.

As said before, European Union can be seen as a single state (Super State) wherein goods can freely circulate. Although the EU has a single body of law for trademarks, such a single body of law does not exist for patents. Currently, EU legislators are working on a Community Patent and on a Community Court for deciding nullity and infringement in patent cases. For the time being, patent owners must resort to national courts for seeking protection against infringers. We could say we are halfway to full harmonization, as legal provisions exist for cross-border injunctions against infringers, determination of jurisdiction, and recognition of decisions.

As a matter of fact, the only true supranational body entrusted with patents, the European Patent Office, can only decide on whether or not a patent is granted; it is not competent on deciding on infringements.

The EU is continuously working on its legislative production in order to ensure a better and more and more homogeneous certainty of rights in Industrial Property. Laws are now available on jurisdiction and enforcement.

6.4 Exclusive Rights and Infringement

Determining the act of infringement is still a matter of national law. We could say that in the territory of the European Union, each State has very similar provisions of law on the definition of the act of infringement and, for sake of simplicity, we can take the definition given above for United States Patent Office as a reference also for Europe. Both the United States and the European Union are WTO members; hence a common set of principles and rules is present.

In a more general view, patent owners may refer to the TRIPS Agreement in order to assess the minimum standard of right that a state adhering to WTO law must ensure.

In this frame, a patent owner can resort to the following exclusive rights

(a) where the subject matter of a patent is a product, to prevent third parties not having the owner's consent from the acts of making, using, offering for sale, selling, or importing for these purposes that product

(b) where the subject matter of a patent is a process, to prevent third parties not having the owner's consent from the act of using the process, and from the acts of: using, offering for sale, selling, or importing for these purposes at least the product obtained directly by that process.

6.5 The Product and the Claims

We have touched one of the main problems in protecting innovative products with patents: harmonization.

As said above, the scope and extent of protection of a patent and their direct influence on the efficacy defending the innovative product against competitors are determined by the claims.

Therefore, we must be sure that our product is well covered by the claims and that anyone, from competitors, to researchers and judges, can understand the correspondence between the patented product and the claims (we say "read the product in the claims"). Patent attorneys regularly observe that the inventor says: "Hey, is this my invention?" after having read the draft of his patent application. Claim language is sometimes somewhat abstract and exoteric. This is the result of the effort to claim more than what the invention is. The main goal of a good claim is to ensure to cover the product. A further step is to claim more embodiments of that product. The next step is to claim equivalent embodiments of the product. The highest step is to claim the inventive concept and future developments. The latter step is usually the patentee's wish but it can lead to difficult or even catastrophic situations, where the patent can become unenforceable because the claims cannot be understood.

Another big problem is the outcome of substantial examination in different Patent Offices.

This problem is even more likely than a cumbersome claim written by a poetic patent attorney.

Patent Examiners are persons, who as any one of us, look at the world with their own eyes. In more than a few cases, claims are not a precise description, but they start with a "this is what you want" from the Applicant and end with a "this is what you get" from the Examiner. Examiners read the claims, interpret them through the description of the invention, filter across prior art, and give their own interpretation. Patents are not science, but neither scientific law nor legal science. Projects are on the way for providing a harmonized system of examination and mutual recognition of allowable claims.

For the time being, waiting for better moments, we will discuss claim interpretation and infringement problems as far as special features of nanotechnology are concerned.

6.6 Recognizing Infringement

We have bored the reader with general principles and some legal issues, but this was necessary to set the discussion in the proper frame.

The first thing we have to do in order to recognize if our patented product is infringed is to take an accurate survey of the market.

Where is our nanotech product? Cosmetics? Car manufacturing? Telecommunications? Pharmaceuticals? More than one market segment?

When we feel the sting of competition, it could be too late to recover decline of profits. This aspect can be particularly critical in consumer's products rather than in specialty products. A last generation LED TV set, or the most-wanted beauty cream or mobile phone have a high selling potential. However, innovative products have a high initial price and many people wish that product, but do not want to pay it for its value. Competitors are very active in offering "the same" product at lower price. If the competitors can offer only lower quality products, consumers will likely continue to buy the original product from the innovator, but if the quality of the "infringing" products is acceptable, or are proposed as a cheaper equivalent, then it is time to defend innovation. If I buy a fake Rolex, I am conscious that my watch is "like" a Rolex, but it is not *a* Rolex. But, if I can find on the market a LED TV set at lower price, maybe it is not as excellent as the original, but I know that it is *that* technology. Patent owners must take into account that accepting the presence of infringers on the market can be very deleterious for their sales and reputation. Allowing infringement can boost other competitors to gain market shares too. Then, it can be too late. Consumers can appreciate also competitors' products, as well as (or especially) lower prices. Starting litigation with a high number of defendants and in many countries can be unaffordable.

An important expedient is to make people know that your product is patented. In United States, marking is compulsory. Not doing so, can hinder or make impossible to recover damages against the infringer.

So, closely monitor the market, competitors' activities, and sale trends.

If you realize that another product like the patented yours is on the market, take action.

The first step could be a warning letter, wherein you deem the competitor's product is infringing your patent and invite him to clarify his position or to stop continuing to infringe your patent.

The second step could be suing the alleged infringer, when you are sure that the competitor's product is infringing your patent. Strategic thought can suggest different action, depending on the circumstances.

Now we arrive at the core of this chapter.

When can I suppose or be sure of infringement? In a basic, simplified view we have two types of infringement: (a) literal infringement (or direct infringement) and (b) infringement by equivalents (or indirect infringement). The first type is the exact copy of the technical features claimed in the patent and implemented in the patented product. The second type is a similar product, wherein the technical features claimed in the patent and implemented in the patented product are not exactly the same, but are substituted by other technical features performing same as the patented ones. If the first type of infringement can be detected with a good degree of certainty (stupid infringer), the second one is very tricky to spot (smart infringer) and trying to pursue this infringer can be frustrating.

We have already said before that the claims are the crucial part of a patent, because they represent the legal boundary of the protected territory (the fence of the property). Reading claims on the patented product and on competitors' products is critically important, since this reading will be confronted with the interpretation made by the competing party. The players in the patent game are various: patent proprietors, public, competitors, patent attorneys, attorneys at law, technical experts, and judges to whom the final decision pertains. All those people are involved in a patent litigation and all of them will give an interpretation of the claims. So, first be sure your product is covered by the patent claims, second, be sure competitors' product is read in the patent claims, either literally or by equivalent, third, be sure your patent is valid and withstands any possible counterattack on its validity by the competitors, fourth, be sure your attorneys and technical experts understand the technology.

Here, we discuss some of the problems arising from infringement in the special field of nanotechnology.

6.7 Infringing Nanotechnology

We will discuss how to reckon that a marketed product is infringing your patented product, how to state it to the alleged infringer, how to explain it to the experts, and how to explain it to the judges. Finally, what the law tells us is how to reckon that a marketed product is infringing my patented product.

The first thing is to understand if my patent is a product patent or a process patent. As a matter of methodology, we will refer to TRIPS provisions for the general concept covering both United States and EU, then the specific laws. There are two basic kinds of claims: product claim and process claim. The first one directly covers a product, independently from its use or destination. For example, patent claiming a nanoparticle protects that nanoparticle per se, no matter for which use it is intended. Any manufactured product containing that nanoparticle falls within the patent protection. The second one covers the process for manufacturing a product. This patent has a lesser degree of protection, since it covers the process as far as recited in the claims, and the product directly obtained by that process. A fancy example, a patent claims a process for manufacturing a gold nanoparticle covered with a particular substance A. The innovative aspect is a process step achieving an accurate control of $\pm$ 5 nm of the thickness of that particular substance in the range 30–120 nm. The invention is layering substance A on the nanoparticle at a pH of 6.7 $\pm$ 0.2 for 3 to 12 min. The patent does not cover any gold nanoparticle covered from 30 to 120 nm of substance A, but will cover only a population of gold nanoparticles covered from 30 $\pm$ 5 to 120 $\pm$ 5 nm of substance A, when prepared with a process wherein the layering step is run at a pH of 6.7 $\pm$ 0.2 for 3 to 12 min. However, the presumption of infringement and the burden of proof can be differently conceived by different laws. The general principle provides that the patent owner has to prove infringement. In case of a patented process, the burden of proof is shifted to the alleged infringer (defendant) if the product is new or if, in spite of any reasonable effort, the patent owner could not determine the process actually used by the alleged infringer.

One of the main problems in defending own patent rights against infringers is to recognize your own invention in the competitors' products. Recognition must be accurate because an inaccurate one can bring to the uncomfortable situation to sue an alleged infringer and to reveal later in the court proceedings that your consideration was wrong. This will in many cases lead to an obligation to refund damages of the third party that was believed to be an infringer. In nanotechnology, this is a very sensitive problem.

In the above example, a judge, an attorney-at-law, a patent attorney, could read a properly drafted, therefore enforceable claim. However, a person skilled in the art of nanoparticles, such as a competitor or an expert called to give a technical opinion in litigation, is aware that nanoparticles are of polydisperse nature and cannot correctly be considered a single entity, but the determination of their dimension must be made on a statistical basis. The determination of the dimensional value must take variance into account. Experimental error is not the only factor to be counted for, but dimensional analysis is also method-dependent. Last but not the least, also dimensions must be defined. Does the claim refer to geometric, hydrodynamic, gyration dimension? Each of these dimensions has its own method of determination. The claim could or could not be limited by the above factors. Such will depend on the Examiner's criteria about clarity of the claim, but if the description of the patent is deficient in specifying all the elements for a correct determination of the technical elements of the claim, in this case the kind of the dimension to be found, the method of determination, the statistical treatment, and perhaps other elements, enforceability of the claim are tricky problems. Unfortunately, patents covering nanotechnology suffer of additional problems with respect to patents covering other technologies, albeit they also share common difficulties.

The common difficulty is to detect and recognize in the market, today in the global market, a product infringing our patent.

A patent covering a bicycle handlebar, wherein the claimed features are determined by the shape of the handlebar is easy to defend. It is sufficient to look at the marketed bicycles. Those with a handlebar identical to the patented one are infringing the patent. Less easy is determining infringement for similar products. Here we have the principle of infringement by equivalency. Careful comparison must be made between the patented product and the allegedly infringing one. In this case, not only the form of the products, but also the function or technical effect must be accurately scrutinized.

The more sophisticated is the patented product and the more complex is the technology, the more difficult is ascertaining infringement. Consumer products can be very complex. A DVD recorder is a now more similar to a computer than to a "recorder," wherein software and hardware play important roles in the performance of the apparatus.

Let's go back to the "queen" claim: product claim. To see whether a product claim is infringed can be an easy task if the claim *clearly* sets out the innovative characteristics *and* these characteristics can be easily recognized in a marketed product.

Let's take as example one of the most difficult products to be analyzed in their composition: cosmetics. A cosmetic composition is typically a homogeneous mixture of a number of ingredients and can be in solid, semisolid, or liquid form. The "active" ingredient(s) (the one(s) that is (are) responsible for the cosmetic effect) are dispersed in a matrix of components with different functions. Resolving the mixture in its different components, detecting, and quantifying each of them can be a hard task from a technical point of view. Cosmetic companies use nanotechnology in their products, especially nanoparticles.

Referring back to the problem of ascertaining infringement, not only process patent can be difficult to enforce, but also product patents. For example, EP1468672 claims an "oil-in-water emulsion in which the oil globules of the emulsion have an average size of at most 500 nm, containing at least particles of ionic polymer and at least one UV radiation-screening system, **characterized in that** the screening system comprises at least one UV-A-screening agent of the 4,4-diarylbutadiene type."

Its US counterpart, US 7,381,403, has a more limited and detailed claim for a "fluid, homogenized, UV-photoprotective oil-in-water emulsion ,microscopically and microscopically stable for a period of time of at least 30 days in the absence of phase separation, [in which] the oil globules of [the emulsion] which[have] an average size of at most 500 nm, containing [at least] particles of at least one ionic polymer and at least one UV radiation-screening system, [characterized in that the] said screening system comprises at least one 4,4-diarylbutadiene UV-A-screening agent."

For the purpose of comparison with respect to the EP patent, US claim is written here with underlined added words and bracketed deleted words. Regardless of the language style, typical of the two US and EP systems, we note that US patent has the additional feature for the emulsion to be *microscopically [sic] and microscopically stable for a period of time of at least 30 days in the absence of phase separation.*

A reader skilled in patent practice understands that this patent is in the field of nanotechnology and that the feature of stability is provided by the nanoparticles. However, if we read the claim of the European patent, we see that this claim is typically a two-part claim, where the inventor declares which part of the invention is prior art and which one is the innovative contribution coming from the invention. This boundary is clearly set by the words *characterized in that.* Then the right of exclusive is on the feature *the screening system comprises at least one UV-A-screening agent of the 4,4-diarylbutadiene type.* US claim provides the same extent of protection, given by the words recited after the term *said.* As a matter of fact, the expression *screening system comprises at least one 4,4-diarylbutadiene UV-A-screening agent* is equipotential.

So, what is the weight of the additional feature *microscopically [sic] and microscopically stable for a period of time of at least 30 days in the absence of phase separation* in the US patent? In the writer's opinion, this feature has no considerable weight for the reason that the fact that the claimed emulsion is *microscopically stable for a period of time of at least 30 days in the absence of phase separation* is an intrinsic feature provided by the *oil-in-water emulsion in which the oil globules of the emulsion have an average size of at most 500 nm.* This feature is part of the prior art, as declared in the patent specification: "particularly stable fine and fluid emulsions containing and stabilized by particles of ionic polymer have been developed in EP-0-864,320, the oil globules of these emulsions having an average diameter of less than 500 nanometres. These emulsions have particularly satisfactory sensory qualities (feel). These emulsions may be used for photoprotecting the skin and the hair against the effects of UV rays since they can contain UV-A screening agents and/or UV-B-screening agents."

Hence, the exclusive right is in the addition of a *4,4-diarylbutadiene UV-A-screening agent* to an *oil-in-water emulsion in which the oil globules of the emulsion have an average size of at most 500 nm, containing at least particles of ionic polymer.* Recognition of a product infringing the patent is possible as far as it is possible to ascertain the following elements:

- The product is an oil-in-water emulsion,
- Oil globules have an average size of at most 500 nm, and
- One of the sunscreen agents is a 4,4-diarylbutadiene derivative.

Although this patent is classified in the field of nanotechnology, determination of its infringement does not show specific characteristics of nanotechnology but, more generally, presents the problems of detecting by way of proper instrumental analytical chemistry the characteristics of complex systems, such as emulsions and of organic chemistry. It may happen that the invention, although conceived in the nanotechnology field, appears in the claims of the granted patent with its essential features, upon which infringement is to be determined, as not characteristic of nanotechnology. Indeed, cosmetics is a peculiar field of technology, where all products are complex mixtures and isolating and determining a specific component can be very difficult or even impossible.

Another interesting example is EP1741717, which claims: *a metal nanostructure comprising: (1) a nanometric metal core comprising gold, silver, or an assembly or alloy of gold and silver, and (2) one or more molecules attached to one or more surfaces of said nanometric metal core, characterized in that each of said one or more molecules has the structural formula W-X-Y-Z, wherein W is an atom selected from sulfur and selenium bound to said nanometric metal core, X is a hydrophobic spacer, Y is a hydrophilic spacer, and Z is hydrogen or is selected from the group consisting of amines, thiols, carboxylic acid halides, carboxylic acid anhydrides, isocyanates, isothiocyanates, sulfonyl halides, aldehydes, imines, epoxides, ketones, phosphoric esters, alcohols* [other chemical structures are provided].

This main claim immediately poses the following question: what is a nanometric metal core? Apart from the common understanding in the nanotech community, competitors and patent owners must know where the invention lies. Without entering the discussion on how the patent application was examined, the main claim of the patent leaves the question open. In this case, can the patent owner and competitors rely on an official definition of nanometric, which is commonly shared in all the Contracting States where the patent was validated? In defining the invention, we must be sure that the terms we use are clearly shared and understood in a well-accepted language in all countries.

However, claim 2 states that in the nanometric metal core *each* dimension is from 1 to 100 nm. This claim could be pleonastic or instead be absolutely necessary to define the actual boundaries of the nanometric structure. Since a claim is infringed when all its elements are present in the infringing product, it is better to have a clear definition of the elements. This is to avoid long, cumbersome, and expensive discussions on infringement before a Court. Certainly, the problem is not solved by an exact definition. What if the claim recited 1 to 100 nm and the competitor's product contains a structure (nanometric) with dimension of 101 nm? This problem is not specifically of nanotechnology, but of all patents wherein one feature of the invention is represented by a numerical value or interval. The only option left in this case is to convince the Court to adopt the doctrine of equivalents.

Claim 3 recites "said metal nanostructure has a pyramidal or branches shape;" therefore it adds another, more limiting feature of the pyramidal or branched shape of the pyramidal or branched shape. According to the invention, this feature brings a technical advantage to develop more heat when carried out in thermotherapy of cancer. But this advantage does not spring from the invention, rather is a feature well known in the art. Other claims relate to particular embodiments which, apparently, do not add any special inventive flavor to the invention of claim 1. Therefore, if in the course of litigation for infringement, the main claim is declared invalid, the fallback claims could be useless for recovering a right of exclusive, since they add nothing inventive. Many patentees and their attorneys like patents with a high number of claims. This is not always an advantage. Claim what is claimable. So to say, your fallback position must be a valid trench.

A patent must specify the industrial applicability of the claimed invention. The inventors must show at least one industrial use for a claimed product. Such use can be the object of a specific kind of claim (use claim). If the product claim confers exclusive right of making and using this product, independently from its use or uses (absolute protection), what's the scope of a use claim, which confers exclusive right for the use of the product for the specific purpose recited in the claim? Use claim can be a very good fallback position, in case the product claim is invalidated. The use claim will maintain the exclusive right to exploit the invention in the field of interest of the patent owner.

In this patent, use claim relates to the specific application of the nanoparticle the inventors have explored: analyticals.

The application of the invention in cancer therapy is explicated in the claims directed to pharmaceutical compositions comprising the metal nanostructure.

Again, this patent is in the field of nanotechnology, but its enforcement does not strictly depend on a characteristics of this technology, since the nanoparticle used in this invention are prepared with well-known methods, the chemical compounds are well-known and the assembly of nanoparticles and chemicals is made in a conventional way. Rather, the invention stands in that the combination of nanoparticle with the specific compounds sorted out by the inventors provide metallic nanostructures with stable shape and have chemical functionalities on their surface.

Thus, this patent will give a right of exclusivity in what is defined in the claims, but detecting infringement will pose the following problems:

(a) Finding metal nanoparticles is not a sufficient evidence
(b) Finding metal nanoparticles of a size between 1 and 100 nm is not a sufficient evidence
(c) Finding metal nanoparticles of a size between 1 and 100 nm with pyramidal or branched shape is not a sufficient evidence

These features are prior art; finding metal nanoparticles as in any of (a) to (c) above having attached at least one of molecules defined in claim 1 is a sufficient evidence of infringement.

Therefore, the patented product is not that much protected by the nanoparticle content, but by the organic molecule bound on it. Detecting infringement is a matter of analytical organic chemistry. Let's see another example in a different technology.

EP1829204 claims that [a] diode quad is characterized by

— a substrate; said diode quad being characterized in that
— first and second electrodes formed on the surface of said substrate and separated by a first gap;
— fourth and third electrodes formed on the surface of the substrate and separated by a second gap, a portion of said fourth and said third electrodes being formed in said first gap;
— a first p-n junction carbon nanotube diode connecting said first electrode with said third electrode;
— a second p-n junction carbon nanotube diode connecting said first electrode with said fourth electrode;
— a third p-n junction carbon nanotube diode connecting said second electrode with said third electrode; and
— a fourth p-n junction carbon nanotube diode connecting said second electrode with said fourth electrode,
— and in that each carbon nanotube diode includes a doped nanotube array.

Other claims relate to the embodiments of said third and fourth electrodes are local oscillator inputs to the diode quad, to p-side connections of the electrodes, to the fact that four metallic nanotube antennas are connected to the four electrodes, that the diode comprises single-wall carbon nanotube.

US 7,359,694, corresponding to the above EP, claims the same device, in the same for, but the claim does not contain the limiting feature that each carbon nanotube diode includes a doped nanotube array.

Does a competitor have a different degree of freedom for the two patents?

Apparently, the European patent seems more limited, since, in the competitor's scope to go around the patent, it could be sufficient to make a diode quad with non-doped nanotube array. The feature of doped carbon nanotubes had to be introduced during examination of the patent application before European Patent Office in order to distinguish the invention from prior art. Indeed, the patent reveals that carbon nanotubes may be doped in order to achieve semiconductor characteristics, otherwise they will have metallic properties. Doping nanotubes is to form p-n junctions. The problem to be considered is whether using a non-doped (metallic) carbon nanotube will produce a diode quad with p-n junctions connecting the electrodes. We can make two hypotheses: (a) the embodiment with non-doped nanotubes works, and (b) the embodiment with non-doped carbon nanotubes does not work. In the first case, the European patent has a weak point with respect to its US counterpart by leaving the competitors the way to non-doped nanotubes. In the second case, the European patent is strong since it has blocked the only way to manufacture the diode quad. The US patent will be infringed only in the claim relating to doped nanotube array, but it will be a strong patent as well, since doping the nanotube array is the only way of carrying out the invention. In this latter case, detection of infringement is just a matter to ascertain doped nanotube array.

An evident and clear patent is EP1887052, which claims "[a] piece which is in part or in full of a pigmented ceramic, characterized in that the pigment is comprised of nanoparticles based on a metal of column IB of the periodic table of the elements or an alloy of the latter, coated with a layer of silica, the silica being a crystalline silica."

This claim is quite simple and will have the advantage to be readily enforceable. To detect infringement, we have to ascertain that the pigmented ceramic is comprised of nanoparticles based on a metal of group IB of the Mendeleev Table and that this nanoparticle is covered by crystalline silica. The patent explains that the piece is brightly colored.

US 6,86,394 represents a clear-cut case of nanostructure with clear definition. This patent claims a free-standing, helical semiconductor oxide nanostructure including a nanobelt having a substantially rectangular cross-section, wherein the nanobelt is about 5 to 200 nm in width and about 3 to 50 nm in height, and wherein the radius of the helical semiconductor oxide structure is about 200 to 5000 nm. This semiconductor structure could be embedded in a "conventional" technology, and the "nano" dimension should impart the special characteristics of the "nano" universe. This patent faces and solves the problem of manufacturing defects and/or dislocation in nanostructures. The helical nanostructure claimed in this patent is characterized by a specific dimensional range, and the method for its manufacture is disclosed in detail. Methods and equipment for determining dimensions of the nanostructure are also provided in the patent. These nanobelts are particularly free from defects. The nanostructure can be used in nanosensors, nanotransducers, and nanoactuators. The precise dimensional range and the methods of measurement make rather easy to ascertain if products made by third parties are within the terms of the claims.

As the last of the many possible examples, let's consider US 5,720,805. This patent deals with sunscreen compositions, product with a very large market. This patent claims: "A stabilized nanometer size cluster comprising Ti-Sn-O clusters complexed with a salt of an alpha hydroxy acid." The second independent claim covers "[a] composition for protection against ultraviolet radiation comprising a carrier and a stabilized nanometer size cluster comprising Ti-Sn-O charge transfer clusters complexed with a salt of an alpha-hydroxy acid, wherein the complexed clusters are present in an amount effective to absorb ultraviolet radiation." The third independent claim is to "[t]he method of making nanometer size Ti-Sn-O charge transfer complex clusters comprising hydrolyzing a titanium alkoxide with an SnX_2, in which X is a halide, at a temperature and for a time sufficient to form said clusters." Defending this patent should not pose particular problems as far as the nanometer size cluster complex with an alpha hydroxy acid is detectable in a sunscreen composition. This claim is rather broad since the definition "nanometer" covers virtually any dimension falling under this term, say below 1 μm. However, the patent specification provides a range of about 20 to 100 Å, to achieve optimum UV absorption. This range should not be considered as a limiting feature, since it is provided as a best embodiment. The patent explains that a higher size can also be obtained, but no contraindications are given. Therefore, a product by a competitor outside the preferred range should still infringe the claims.

6.8 Determining Infringement in Nanotechnology

The problem of ascertaining infringement of a patent in the nanotechnology field is the same as the general technology: determining identity or similarity between the competitor's product and the one covered by the patent. In certain technical fields this can be very difficult, if not impossible; for example, very complex compositions wherein the components characteristic of the invention cannot be detected with analytical techniques. Nanotechnology may have the additional problem of analytical limit of detection and accuracy, in order to ascertain infringement. Nanoparticles are a typical example, where determination can be statistical.

6.8.1 *How to State to the Alleged Infringer*

This point is exquisite matter for an expert attorney-at-law. When you detect an alleged infringer, discuss the strategy with your IP team. At this stage, the team should comprise at least a patent attorney and an attorney-at-law. This team sometimes can be embodied by one person. The combination of the two competences can give the best effect in stopping the infringer and avoiding or reducing loss of profits. Also, the action may depend on the laws of the State where the supposed infringement occurs. One of the most difficult situations is when infringement occurs by offering the product on the internet. Legislation is complex and how to make the first step is important. A warning or a cease and desist letter can be the first step, but also a direct action to a Court can be a good move. In any case, and in my opinion, the move depends also on the degree of certainty you have of the infringement.

6.8.2 *How to Explain It to the Experts and the Judges*

This point can be dramatically decisive for the defense of the patent. Nanotechnology is a complex matter and patent law also is a complex business. So, the square complexity needs good experts. First, be assured to entrust your IP to a specialized and experienced patent attorney and attorney-at-law. This is the team which has the competence to conduct the matter. We must consider that in litigation proceedings, the judge is the person who will issue the decision. In certain States, a jury can be called. Generally, the judge is a man of law, only in few cases the judge has technical knowledge. It is evident how important is that your IP team is knowledgeable in the subject matter.

6.8.3 *What the Law Tells Us*

The law contains few, but very simple words: a patent proprietor has the exclusive right to exploit the patented invention, period. These few, simple and clear-cut words imply lawsuits which last years, involve a considerable number of persons, are extremely time consuming and are tremendously expensive. In certain cases, the cost can bring litigants to bankruptcy (usually the infringer). Again, it is critical to entrust a competent team.

6.9 Am I (Patent Owner) an Infringer?

Before entering in a lively discussion on how to defend nanotech patents from infringers, it is important to discuss the *first* problem a patent owner has to face: is my invention free to operate? This point is sometimes underestimated by innovators, and they can later find themselves in big troubles either with a problem of patent dependency or, worse, in infringement.

Big swallows small. This is a law of nature often adopted in the relationship among humans. In our civilized world this is transposed not only in a metaphoric way. In patents, small is comprised in big. This means that an object is patented independently from its size and smaller versions of the patented object will fall within the boundaries of the claims. Patent examiners generally considers miniaturization not a distinctive mark and miniaturization is not sufficient to provide inventive or non-obvious character to an invention (Fiedler and Reynolds, Legal Problems of Nanotechnology: an Overview; 3S Cal. Interdisc. L.J. 593 (1993–1994)). If an invention falls within the claims of another invention patented by others, we can be in different situations: (a) we are infringers; (b) we are not infringers, but we depend on the third party patents; and (c) we are not infringers and do not depend on third party patents. In a dimensional perspective, nanotechnology is a matter of smaller things, which replicate already existing bigger things (a nanoparticle is a small "particle") and an invention with nanoparticles could be seen as comprised in a patent covering "particles." The problem is that a claim to a nanotechnology product, as a nanoparticle, is generally defined in structural terms, for example, dimensions and materials, not in terms of functional features. Thus, a patent claiming a silica particle covers also a silica nanoparticle. An exemplary case is the story between DuPont Air Products Nanomaterials L.L.C. and Cabot Microelectronics Inc. (Bleeker R., *A Small Charge of Infringement: Strategic Alternatives for Nanotech Patent Defendants*, Nanotechnology Law and Business, Winter 2007). DuPont believes their activity in the nanoscale field does not infringe Cabot patents, which deal with a technology in the microscale. The trial is still open and its outcome is not yet defined.

In a general way, many patents relating to "conventional" technologies, or downsized to microscale, could include also nanoscale products, hence be enforced against nanotech inventions. However, nanotech products in certain cases need special and specific technology to be manufactured. Also, nanotech products perform different functions just because of their nanosize. If the "traditional" patent does not explicitly disclose the specific preparation of the nanoscale product, this patent should not be allowed to claim a right of exclusive on the nanotech product (Troilo L.M., Nanotechnology Law and Business, Volume 2.1 (2005)).

However, absence of size limitation in the claims could lead the Court to conclude that the patent covers the nanotech product and declare the patent to be infringed. The doctrine of equivalents can lead to different judgements whether the allegedly infringing product performs the same functions or specific functions due to the nanoscale range (Voigt R. and Michelson E., *Nanotechnology-Related Inventions: Infringement Issues*, Nanotechnology Law and Business, Volume 2.1 (2005)).

We could conclude that nanotech innovators can defend themselves from "generic" or "traditional technologies," including microrange technologies, by invoking nullity of their patents in enabling nanotech products or processes. This may be a difficult task, since the interpretation of the claims of the earlier patents is made by Courts reading a technical matter through the eyeglasses of the legal language (claim construction).

6.10 Different Kinds of Infringement

We have discussed above how the determination of infringement of a nanotech patent can be difficult in view of the analytical problems in this field. We have also talked about the two basic types of infringement: literal and by equivalents. Literal infringement is made very easily, by simply copying the patented product. A "smart" infringer will try to conceal the illegal activity by modifying the product in order to "apparently" circumvent the patent, but performing the same functions by using the same features of the patented product. Ascertaining this infringer's activity may imply the use of the Doctrine of Equivalents, at least in those States where it is recognized. However, the situation of a device performing the same or similar function in a substantially different way than a patented device, even though it falls within the literal words of a claim is also possible. This is the "Reverse Doctrine of Equivalents." This theory is a double-edge blade for patent owners and innovators. On the one hand, earlier patents in nanotechnology can have claims so broad so to cover subsequent developments;

on the other hand, nanotech patents may be so specific to cover just the developed product. Doctrine of Equivalents and Reverse Doctrine of Equivalents will play a discriminating role in defining technological developments with respect to patents with broad claims (Vyas, P.; *Reverse Doctrine of Equivalence: Role in Evolving Technologies*, Social Science Research Network, 2009; Wasson A. *Protecting the Next Small Thing: Nanotechnology and the Reverse Doctrine of Equivalents*, Duke Law & Technology Review, 2004, No. 10). Reverse Doctrine of Equivalents will be available both to resolve disputes among nanotech innovators and between traditional and nanotech innovators.

Harris *et al.* (Strategies for Resolving Patent Disputes over Nanoparticle Drug Delivery Systems, Nanotechnology Law & Business, Vol. 1, Issue 4, 2004) predict serious problems in nanoparticle field, mainly due to early overlapping patents granted by United States Patent and Trademark Office (USPTO) and to the still indefinite claim language in nanotech patents.

6.11 Court Decision: United States of America *(Kumar)*

By way of example, the *Kumar* decision by the United States Court of Appeals for the Federal Circuit of August 15, 2005 discusses the problem of nanoparticle size and size distribution. Although this decision is mainly focused on procedural matter, it sets some principles that can be used by nanotech players in considering prosecution and enforcement of patents. The Court had to deal with an obviousness rejection based on an earlier patent that disclosed a size and distribution size range overlapping with the claims of the application under examination. The Court confirmed also in the nanotech field the principle that the prior art can be considered for obviousness if it allows the skilled person to make and use the claimed invention. In the *Kumar* case the problem of a correct and accurate teaching of nanoparticle preparation and determination of the size distribution was critical to distinguish Kumar's invention with respect to prior art. Conversely, the same problem is present to the patent owner who has to determine whether the patented invention is infringed or not. As we saw above, the method of determining the physical parameter which is the, or one of the, characterizing features of the patented product must be clearly indicated in the patent and must be a method which is acknowledged as standard in the technical field of nanoparticles.

Sometimes, the aspect of the disclosure of the invention is underestimated. One of the patentability requirements is that the invention must be disclosed in a clear and sufficient way for its accomplishment by a person in the relevant field. If we find in the situation in which we are sure that our patented product is infringed and we can show that the infringing product reads on our claims, but the patent does not provide all the information on how to prepare and characterize our product, we will run the risk to loose our case, because the alleged infringer will have a good argument against us. He will have a good point if he can demonstrate that the patent does not provide not only all the information on how to prepare the product, but also how to characterize it. In the field of nanoparticles this can be critical.

The problem of nanoparticle size characterization is discussed in Zattoni A. *et al.*, "Asymmetrical flow-field fractionation with multi-angle light scattering detection for the analysis of structured nanoparticles," *Journal of Chromatography A*, **1216** (2009), 9106–9112, and shows how different methods can provide different, noncomparable results and how manipulation of the sample within the analytical technique can alter the product.

This aspect can be one of the major and most tiresome points in litigation on infringement and can also have a dramatically different impact depending on the country where the litigation takes place. There are countries where the case will be conducted by technically qualified judges, and countries where this kind of judge is not foreseen. In the latter case, the judge can call a technical expert to give an opinion on some technicalities in patent litigation. It is not difficult to think how many different outcomes can be reached in this case and how expensive is the trial.

6.12 Court Decision: Europe (Germany)

I have outlined at least the minimum terms for a discussion on the specific problems for nanotechnology patents in enforcing them against infringers.

I will now examine examples of court decisions, which can be used to understand the problem of infringement in nanotechnology field.

As in many other technological fields, litigations most often terminate before a final decision is reached because the parties reach a settlement. In the nanotechnology field, not many court decisions exist, due to the youth of this technology and many other difficulties, such as the level of understanding of the subject matter by the players (judges, attorneys at law, and sometimes even patent attorneys).

I have already commented the *Kumar* decision and I would not enter in more detail.

Although no other decisions could be retrieved, a paper from Huebner (*The Validity of Nanotechnology Patents in Germany,* Nanotechnology Law and Business, Fall 2008) that focuses on a critical issue in nanotechnology can be correlated to the above mentioned *Kumar* case. Germany is (one of) the major markets in Europe in advanced technological fields and nanotechnology. I'm not aware of German Court decisions in infringement cases in nanotechnology, but the considerations presented in Huebner's paper make me draw some conclusions. We have already discussed how nanotechnology is firstly defined by the dimensional range of its field of existence. According to the EPO definition, "The term nanotechnology covers entities with a controlled geometrical size of at least one functional component below 100 nm in one or more dimensions susceptible of making physical, chemical, or biological effects available which are intrinsic to that size. It covers equipment and methods for a controlled analysis, manipulation, processing, fabrication, or measurement with a precision below 100 nm." Also, in view of the above considerations about the possible coverage by "conventional" technology patents on "nanotechnology" patents, we have come to the determination that nanotechnology can be seen as a "selection" of size for the materials in order to achieve previously unknown properties. That is to say, the inventors "select" specific small sizes to achieve these special properties. The European Patent Office recognizes and has well codified the so-called "selection invention" (see EPO, Case Law of The Board of Appeals and *C. Kallinger "Patenting Nanotechnology:*

A European Patent Office Perspective, 5 Nanotech. L. & Bus., 95 (2008)). According to European Patent Office, a selection invention must comply with three basic requirements: (a) the selected sub-range must be narrow in comparison with the known range; (b) this sub-range must be sufficiently far removed from the examples that illustrate the known range; and (c) it must provide a new technical teaching. The last point is very significant. The selected sub-range should not be just a further example of the effects of the well-known, wider range, but is another invention (purposive selection). Nanotechnology seems to fit into this scheme, since the properties of nanomaterials are so different from "conventional" materials, because of their dimensions. Unfortunately, the German Supreme Court (*BGH-Crackkatalysator I, 1990, 150 GRUR; BGH-Chrom-Nickel-Legierung, 1992, 842 GRUR; BGH-Inkrustierungsinhibitoren, 2000, 591 GRUR*) came out with the principle that "in accordance with the rules of arithmetic, the naming of a numerical interval represents a simplified notation of the numerous possible values which lie between the minimum and maximum values." By this principle, the disclosure of a range makes disclosed all possible sub-ranges. This line of reasoning brings to the unusual situation that a nanotech invention can be provided with inventive step, but is not novel.

Even though a patent on nanotech invention is granted by the European Patent Office, it could be non-enforceable in Germany, because it will be considered invalid.

Dr. Huebner suggests to have use claims, the broadest possible, in order to be able to enforce the patent.

However, the same German jurisprudence allows an escape. In an old decision (*Fluoran, 1988, 447 GRUR*) the Court established that only a teaching which can be actually practiced by the skilled artisan can be considered novelty destroying. So, even though a sub-range is considered to be disclosed by a broader range, novelty is recognized if the person skilled in the art is not able to obtain the nanomaterial or nanodevice.

This makes an interesting contact point with the Kumar situation.

6.13 Specific Problems in Nanotechology

In nanotechnology, the method of fabrication is much more important than in "conventional" technical fields.

In product patents, process claims are often considered not very important, since the process itself is not provided with inventive step (is obvious) and we call this kind of claim "analogy process claim."

In nanotechnology, a process claim can be of critical importance, since it could be the only way to protect the product itself. The European Patent Convention and the TRIPS Agreement too, provide that a patent for a process confers protection also on the product directly obtained by the patented process.

Hence, a claim to the process for manufacturing the nanoproduct or nanodevice is a powerful protection.

The combined experience of the Kumar case and of the numerical doctrine of the German Supreme Court lead us to conclude that the patent for a nanotech invention must be drafted with great care, especially as far as the method of fabrication is concerned. This allows to better distinguish the invention with respect to the prior art and to provide easier enforceability of the claims. Moreover, it renders the patent stronger against possible attacks for insufficiency of disclosure.

Enforcing a patent is not an easy play. Even though we have a "stupid" infringer, so to say a copycat, pitfalls in the patent may render it impossible to enforce. If we have a "smart" infringer, the doctrine of equivalents could help us, but, again it depends on how our patent was written.

An accurate description of the method for manufacturing the nanotech product is critical.

6.14 Conclusion

As a conclusion, defending nanotech patents presents the same general problems as for any other patent. Detecting infringement, ascertaining it, defending the patent against the classical coun-terattack on its validity. However, nanotech patents have some peculiar problems. The first one is to be able to precisely detect the patented product in competitors' activity. This detection can be very problematic if the claims are limited by a specific size or size range or other kind of technical problems affect analytical methods which can be applied for this detection. The second problem is to resist the alleged infringer's counterattack against the validity of the patent. Other chapters in this book deal with patentability, and we will not repeat this argument. However, in view of this chapter topic, we will mention that the technical effect and properties of the patented nanotechnology must be clearly explained and shown in the patent. This is necessary to rebut the possible objection that our nanotech invention is "just" miniaturization, which is per se non patentable. Another very important point is to provide ample and complete disclosure on how to make the patented invention. The objection from the alleged infringer that the patent description does not allow to obtain the claimed invention would render void any attempt to stop infringer's activity and the patent could be declared invalid.

From patent point of view, nanotechnology is young, but jurisprudence is in its infancy, so uncertainty is a factor to be taken into consideration.

Growing number of patent prosecution in the Patent Offices, increasing jurisprudence and, possibly, harmonization among the Patent Offices and, hopefully, the Courts would increase certainty degree for establishing the balance between rights of exclusive and freedom to operate.

The challenge is to build a common understanding among operators in the complex field of nanotechnology.

Chapter 7

Licensing Issues in Nanotechnology

Joanna T. Brougher

Harvard School of Public Health, 677 Huntington Avenue, Boston, MA 02115, USA
jbrough@hsph.harvard.edu

7.1 Introduction

Innovation is often driven by collaborations. Pursuing a collaboration is desirable because it can create opportunities for a nanotechnology company to develop new products or technology and expand into new markets. Such alliances can also benefit both parties by controlling financial expenditures and solidifying mutual commitment.

Licensing intellectual property is one way in which companies may collaborate. While licenses can be extremely beneficial to the parties involved, they nevertheless involve many complex issues, such as negotiations surrounding research and development, ownership, profits, remuneration, exclusivity, transferability, fields of use, territory, and duration. Such issues should be thoroughly discussed at the earliest stage of the collaboration. Otherwise, problems may arise during the partnership that can leave the parties in a vulnerable position.

Nanotechnology Commercialization for Managers and Scientists
Edited by Wim Helwegen and Luca Escoffier
Copyright © 2012 Pan Stanford Publishing Pte. Ltd.
ISBN 978-981-4316-22-4 (Hardcover), 978-981-4364-38-6 (eBook)
www.panstanford.com

In the nanotechnology arena, companies may face unique challenges when it comes to the licensing and transferring of intellectual property rights. Many challenges stem from the nature of the technology itself. Nanotechnology generally involves products that operate on an extremely small scale in which structures do not exceed 100 nm [1], and many non-experts struggle to grasp a scale of this magnitude. Moreover, nanotechnology spans across a myriad of industries and disciplines. For instance, materials and devices that incorporate nanotechnology may be used in biomedicine, electronics, information technology, environment and sanitation, energy production, lithography, data storage, optics, aerospace, and molecular robotic manufacturing processes [2]. Furthermore, nanotechnology carries with it potential risks that may not be known or even contemplated by the current scientific climate. These risks can develop quickly and without any warning. Due to the complexity of the field, nanotechnology is a unique area with unique hurdles.

Since nanotechnology-centered licenses deal with technology that is emerging and unpredictable, nanotechnology companies should do their best to anticipate future developments and risks that may arise before the expiration of licensing agreements. The following article seeks to highlight potential issues facing companies when licensing nanotechnology and presents strategies for negotiating license agreements. Understanding these issues can help companies on both sides of the license agreement to maximize their protection and rewards and minimize their risks and liability. The article will primarily focus on the US practice, although some international issues are dealt with at the end.

7.2 Reasons for Entering into License Agreements

There are numerous reasons why nanotechnology companies may enter into license agreements. Generally speaking, partnerships between two or more companies may help each company overcome its shortcomings. Both small and large companies may benefit from having another company assist in providing resources, whether such resources are financial or technological processes or personnel.

Such partnerships can also make it possible for both small and large companies to research and develop new technologies.

From the perspective of a large company, small or emerging companies can provide access to a specific product pipeline. Emerging companies are often founded to research and develop a single product. Large companies may choose to license a product from an emerging company rather than infringing intellectual property rights by developing its own product. Moreover, licensing a product from a small company can allow a large company to invest in a specific product without the unwanted assets and risks associated with purchasing the small company. Such risks include, for instance, unknown liabilities, claims which have not yet been asserted, potential infringement of third party intellectual property, and employee integration and loyalty, to name a few.

For smaller companies, the primary benefit of partnering with a larger company is the ability to access resources and distribution channels that might otherwise be prohibitively expensive. Established companies may be able to provide the necessary capital to bring the smaller company's products through clinical development and onto the market. Additionally, a large company may provide access to additional resources such as laboratory space, access to leading technology, a well-established supply chain and sales force, and customer channels. With access to such additional resources, small companies may be able to develop products in a highly efficient and cost-effective manner. Establishing a relationship with a large and reputable company may also help emerging companies gain credibility in their field.

7.3 Overview of Intellectual Property Licensing

The idea of licensing intellectual property stems from the principles of property ownership. Once a patent has been granted on an invention, that invention can be treated as a personal property belonging to an owner [3]. Just like real estate, intellectual property can be transferred, willed, or mortgaged. The original owner of the intellectual property is the inventor whose name appears on the patent. When multiple inventors appear on the patent, those inventors are joint owners of the patent who can each transfer or sell the patent rights without the consent of the other joint owners [4]. For instance, if one joint owner decides to grant an exclusive license to a manufacturer to produce the patented product, that exclusive license would bind each of the joint owners regardless of whether they gave their consent or even know about the exclusive license.

Owners of patents have several options when it comes to deciding what to do with their inventions. In some cases, they may choose to retain the patent and pursue the development and commercialization of the product by themselves. Due to the costs and skills required for successfully manufacturing and marketing a product, however, many patent owners may instead choose to transfer those responsibilities to third parties with greater expertise and resources. This transfer can be affected via a license or an assignment of rights.

7.3.1 *A License: Transferring Less than the Entire Ownership Interest*

One way to transfer patent rights to a third party is through the granting of a license. A license involves transferring less than the entire ownership interest; the patent holder retains ownership of the patent while permitting the licensee to make, use, or sell the invention. In essence, a license is a contractual agreement in which the patent owner will not sue the licensee for patent infringement if the licensee makes, uses, or sells the claimed invention, as long as the licensee fulfills its obligations delineated by the agreement.

There are generally two types of licenses. The first is an exclusive license. An exclusive license prevents third parties from competing with the licensee over the rights set forth in the license agreement. A license can also be nonexclusive, in which more than one licensee may be given some rights to the same intellectual property. Regardless of whether the license is exclusive or nonexclusive, a license may also place restrictions on the licensee's use of the patented product. For instance, the license may be limited in scope, such as time, geographical area, or field of use.

7.3.2 An Assignment: Transferring the Entire Ownership Interest

A license should not be confused with an assignment, which is another means of transferring intellectual property rights. An assignment of patent rights occurs when the patent owner conveys an entire ownership interest or a percentage of the ownership interest in the patent to a third party. Essentially, an assignment is a transfer of the ownership of one's property to another party, including the rights, title, and interest in that property. To be legally binding, the assignment must be in writing and must be recorded with the United States Patent and Trademark Office (USPTO) within three months of its execution [5].

Under some circumstances, inventors are under a legal obligation to assign their ownership rights in an invention to a third party. In the context of university research, for example, employees may be required to sign an express agreement asserting that they will assign all patent rights, title, and interest to any inventive activity to their employer. Such agreements ensure that individual employees do not profit from the employer's time and resources. In such a case, a court will generally uphold the assignment agreement as long the agreement is reasonable in duration and restricted to inventions relevant to the employer's business. Moreover, if an employee is hired to perform specific tasks and develops an invention in the course of his duties, that employee is generally required by law to assign his rights to his employer, even in the absence of an agreement. In certain other contexts, dividing the assignment of patent rights between employers and employees can be extremely complicated.

7.4 Best Practices When Entering into A License Agreement

While licensing can be a useful strategy to expand into new nanotechnology markets, there are a number of issues that nanotechnology companies should consider before entering into a license agreement. Such issues include maintaining confidentiality, conducting due diligence of the licensor's and licensee's assets, and determining the scope of the intellectual property to be licensed.

7.4.1 Non-Disclosure Agreements to Protect Confidential Information

Entering into a license agreement often requires companies to disclose confidential information. Disclosure allows both parties to the agreement to know and understand exactly what technology they will be licensing. Disclosure, however, also provides a competitor with confidential information about a company that the competitor would not otherwise have access to. Without properly protecting that confidential information, companies may be hesitant to enter into licensing discussions with competitors for fear that the competitor will later use that information to gain a competitive edge. A non-disclosure or confidentiality agreement is a proper tool for adequately protecting confidential information while still allowing companies the freedom to disclose the confidential information.

Adequately protecting confidential information requires a company to first define what information constitutes confidential information. Not all information is confidential. The definition should not be drafted too narrowly since it would limit the protection only to what information is specifically designated by the agreement. In addition to providing a definition, the non-disclosure agreement should further restrict or prohibit the sharing of that confidential information with third parties. By providing such a restriction, the non-disclosure agreement helps ensure that only those parties central to the license agreement (e.g., those with a need to know) are the ones with access to the confidential information. Additionally, the non-disclosure agreement should provide for proper handling of the confidential information to avoid accidental disclosure or misuse of the information. Finally, the non-disclosure agreement should include a provision about the duration of the agreement as well as a provision regarding the return of all confidential information and materials upon termination of the agreement. Thus, before disclosing any confidential information, parties to the license agreement must make sure that the information is properly protected.

7.4.2 *Due Diligence to Uncover Potential Issues*

Since companies entering into a license agreement will share assets with one another, it is also important for each party to the agreement to perform due diligence on themselves and on the other party. Performing due diligence will allow both parties to uncover potentially unforeseen problems early in the process and will allow the parties to take adequate precautions to prevent problems from later arising.

Performing due diligence may reveal certain issues with the technology or assets being licensed. For instance, a search of the assets may reveal that a patent in the portfolio being licensed is expired or abandoned. If a patent is expired or abandoned, the licensee may be deprived of potentially desired and paid-for assets. A search of the patent assignments may further reveal that the licensor does not actually own the entire rights of the patent. Without proper assignment, the licensor may not be able to license the assets to the licensor, thereby once again depriving the licensor with potentially desired and paid-for assets.

Additionally, an examination of the claims may reveal that the scope of the promised assets is less than that which was previously negotiated and agreed upon. A patent having claims with a narrow scope may often have a lower value than a patent having claims with a broader scope since it is easier for a competitor to design around the claimed invention. As such, due diligence can be used as a means to reevaluate and renegotiate provisions and terms of the licensing agreement, or may be a way to dismiss an opportunity if enough red flags are raised early on. Conducting a thorough due diligence is therefore critical to protecting oneself against potential problems with the licensing of the assets.

7.4.3 *Properly Define the Scope of the Agreement*

One of the initial factors that parties to a license agreement may want to consider is the scope of the technology being licensed. In a field like nanotechnology that spans across different industries, scope is particularly important. Some licensors of nanotechnology may only want to extend a license to one particular downstream application of nanotechnology, such as medical device coatings. On the other hand, some licensees may want to receive a much greater scope, such as the entire field of biomedicine. In many instances, the scope may be limited only to those inventions that were conceived or reduced to practice in furtherance of the license agreement. Regardless of how broad or narrow the scope is, it is critical for both the licensor and the licensee to agree upon the territory covered by the agreement. Such determinations should be made before a dispute arises rather than having to fight over the scope of the licensed technology in a costly and time-consuming litigation.

7.5 Potential Issues in Nanotechnology Licensing

When dealing with licenses involving nanotechnology, one should not neglect the usual issues encountered with conventional licenses, such as remuneration, exclusivity, transferability, fields of use, territory, and duration. However, when granting or accepting a license involving nanotechnology, one must contemplate certain unique technological factors which have the potential to affect all involved parties to a significant degree. Such factors include the challenges posed by reverse engineering, government or university ownership, unforeseen negative impacts of technology, the amount of competitors within the same technology space, and the difficulty of policing infringement.

7.5.1 *Protecting IP Rights Under Trade Secret Law*

Nanotechnology is a unique field in that it can be very difficult for competitors to discover the underlying technical principles of an invention simply by analyzing its structure or function. This is because nanotechnology operates on a much smaller scale than other kinds of technology. As such, nanotechnology inventions are often protected from reverse engineering.

Given the inherent difficulty of reverse engineering in certain areas of nanotechnology, patent law may not always be the best available method for protecting nanotechnology inventions. In exchange for patent protection, technology must be disclosed to the world, and upon expiration of patent rights, patented inventions fall into the public domain. Thus, if disclosure is not desirable for any number of reasons, an alternative to pursuing patent protection is to treat the nanotechnology invention as a trade secret without publishing details of the invention in any form.

Trade secret protection involves protecting ideas simply by keeping them secret. It therefore avoids the effort and expense associated with filing a patent application. Protection remains as long as the underlying technology is kept secret. One example of a long-standing trade secret is the formula for Coca Cola$^{®}$. Trade secret protection, however, requires continuous diligence, since once the technology is revealed, it is no longer protected. Thus, when the likelihood of reverse engineering a nanotechnology invention is small, licensors and licensees should consider protecting their inventions as trade secrets.

When licensing nanotechnology, it is possible to convey nanotechnology rights without revealing the trade secrets themselves via a type of license agreement known as a "material transfer agreement." Under these agreements, the licensor shares the materials with the licensee but does not disclose their composition. A material transfer agreement usually involves a contract between two parties that governs the transfer of tangible materials from one party to the other [6]. The agreement defines the respective rights of the parties, including IP rights, duties, and obligations of the receiving party with respect to the materials and their derivatives. Under a material transfer agreement, therefore, the recipient has the right to practice the technology but is prohibited from analyzing, reverse engineering, altering or otherwise modifying the materials, and further cannot sell, transfer, disclose or otherwise provide access to the materials to another party without the prior written consent of the transferor. As a result, the underlying structure and function of the licensor's invention remains secret.

While confidentiality and non-compete provisions may give nanotechnology owners similar protection, this route always carries the risks of inadvertent disclosure or misappropriation of secrets. Protecting trade secret rights under material transfer agreements thus offers a more protective alternative that allows the recipient to use the technology without requiring the transferor to disclose the novelty or uniqueness of the material.

Another advantage to pursuing trade secret protection as an alternative to patent protection is that it is usually easier to obtain an injunction for misappropriation of trade secrets than for patent infringement. Under patent law, a permanent injunction will only be issued after a stringent four-factor test is applied [7]. That four-factor test requires a plaintiff to demonstrate that (1) it has suffered irreparable injury; (2) remedies available at law are inadequate to compensate for that injury; (3) considering the balance of hardships between the plaintiff and defendant, a remedy in equity is warranted; and (4) the public interest would not be disserved by a permanent injunction. Even after the plaintiff has satisfied all four factors, the court may still refuse to issue a permanent injunction, thereby allowing wrongdoers to continue infringing intellectual property rights. Under trade secret law, on the other hand, courts consider whether (1) the plaintiff has prevailed (or will likely prevail) on the merits of his misappropriation claims; (2) the plaintiff will suffer irreparable injury if the permanent injunction is not issued; and (3) the balance of the equities weighs in favor of the entry of the permanent injunction [8]. Applying trade secret law's three-factor standard, courts may be more willing to issue a permanent injunction. Due to its increased potential to prevent competitors from misappropriating technology, trade secret protection may be a viable alternative to patent protection for owners of nanotechnology looking to establish intellectual property rights.

7.5.2 *Ownership and Control of the Licensed Intellectual Property*

The Bayh-Dole Act, also called the University and Small Business Patent Procedures Act, is United States (US) legislation that addresses intellectual property arising from federal government-funded research. Enacted on December 12, 1980, the Bayh-Dole Act gives US universities, small businesses, and non-profit organizations the ability "to retain title to any invention developed under such support," or to control of intellectual property resulting from government funding [9]. Even if the development of an invention has been funded by the government, the Act allows a university, small business, or non-profit organization to pursue ownership of the invention rather than granting automatic ownership rights to the government. The Act therefore is responsible for reversing the presumption of ownership.

7.5.2.1 University ownership

Although the Bayh-Dole Act allows universities to retain ownership of patents derived from federally funded research, clear title to an invention may not always be easily determined. Problems that may hinder clear title generally arise when employees of the university collaborate with non-employees such as students, consultants, visiting professors, or government employees to research and develop an invention. Without clear title, a company's ability to practice the licensed invention may be affected. As a result, nanotechnology companies seeking to license technology from a university should conduct additional due diligence to determine whether a clear and traceable title to the invention exists.

Determining title to an invention generally requires a look into three sources: (1) inventorship; (2) assignment and employment agreements; and (3) the university's patent policies [10]. Inventorship is often tricky to establish. For journal or scholastic publications, individuals may be included as authors for various reasons including, contribution to the idea behind the publication, contribution to writing the publication, contribution to data collecting or analyzing, or even professional courtesy. In contrast, to be an inventor on a patent requires an individual to have contributed to the inventive concept of at least one claim in the patent [11]. As a result, those individuals listed as authors on a publication may be incorrectly listed as inventors on a patent. Similarly, individuals who should be listed as inventors may be incorrectly omitted from a patent.

Once inventorship is established, the affiliation of those inventors may determine whether those inventors can assign the invention to the university. Assignment and employment agreements, for instance, typically obligate employees of the university to assign their rights to an invention conceived within the scope of their employment to the university. These agreements are generally executed by the employee as a condition of employment. A university's patent policies, on the other hand, may obligate non-employees to similarly assign their rights to an invention to the university. Since a non-employee may never expressly consent to these policies, consideration should be given to whether these policies are enforceable against the non-employee. Moreover, the non-employee may have additional affiliations outside the university that may affect his ability to assign to invention to the university. Prior to entering into a license with a university, potential licensees should have an understanding of the various factors that may impact and hinder clear title to an invention.

7.5.2.2 Government march-in rights

In addition to revoking the government's automatic ownership of intellectual property developed with federal funds, the Bayh-Dole Act further permits, in certain circumstances, the federal government to step in and acquire the technology. This is known as the government's "march-in" right [12]. Under the march-in provision, the Bayh-Dole Act essentially allows the government, as the funding agency, to have an ongoing right to "require the contractor [university], an assignee, or exclusive licensee of a subject invention to grant a non-exclusive, partially exclusive, or exclusive license in any field of use to a responsible applicant or applicants, upon terms that are reasonable under the circumstances, and if the contractor, assignee, or exclusive licensee refuses such a request, to grant such a license itself." In order words, the Bayh-Dole Act allows the government to ignore the exclusivity of a patent and practice the invention itself, or have a third-party practice the invention on the government's behalf. Patent owners are therefore susceptible to losing rights to their technology to a potentially unknown party.

For the government to enforce its march-in right, however, certain criteria must be satisfied. Two of the most important criteria include (1) a failure on the part of the licensor to take "effective steps to achieve practical application of the subject invention," and (2) to satisfy "health and safety needs" of consumers. While the government has never exercised its march-in rights, one example in which the government contemplated using its march-in rights occurred during the anthrax attacks of 2001. At the time of the attacks, Bayer had patent rights to the drug Ciproflaxin, which was used to treat anthrax. As the anthrax scare continued and fear of a full-blown epidemic increased, the federal government contemplated marching in and acquiring Bayer's patent rights to the drug Ciproflaxin in order to benefit the health and safety of the public.

Owners of nanotechnology developed with federal funds thus face the possibility that their technology may be subject to government march-in rights. Since a great deal of nanotechnology research and development involves the military, national defense, and public health, that research is often funded by various US government agencies. Such government agencies include, for example, the National Science Foundation, the National Institute of Health, and the Department of Defense. Any technology developed using government funds granted by these agencies is generally subject to march-in rights.

The idea that the government can march-in and acquire patent rights presents a significant risk for both licensors and licensees. If the risk of the government marching in and acquiring the technology is high, a licensee may be cautious or unwilling to enter into a license agreement because it risks losing licensed technology. Under such circumstances, if the extent of use of the licensed technology is uncertain, a licensee may only be willing to pay a lower fee for the license. Additionally, the licensee may request a termination clause that provides contingent relief from certain obligations, such as paying royalties, in the event that the government takes ownership of the technology. Such terms in the agreement may serve to protect the licensee's interest in using the technology.

Relatedly, licensors may find it difficult to license their technology because licensees are wary of government march-in rights. Under these circumstances, licensors risk losing potential revenue. To offset this risk, a licensor may ask a licensee for upfront payments. These upfront payments would assure the licensor of receiving at least some revenue from the licensed technology. To further protect itself from the possibility that the government will march-in and acquire the licensed technology, a licensor may request that the license agreement include a covenant not to sue. The covenant would serve to protect the licensor from a possible lawsuit by the licensee in the event that the licensee loses the right to use the technology. Such a provision would protect the licensor's interests in generating revenues from the licensed technology.

7.5.3 *Unknown and Unforeseen Side Effects*

Nanotechnology involves new and exciting discoveries in a wide variety of fields, including biomedicine, electronics, information technology, environment and sanitation, energy production, lithography, data storage, optics, aerospace, and molecular robotic manufacturing processes, as noted above. Nanotechnology, however, also carries with it the possibility of unknown and potentially harmful side-effects. These side-effects may include toxicity, environmental harm, and health risks. Nanoparticles, for example, may present certain health risks. Nanoparticles are sometimes present in foods, creams, and other consumer products and maybe inadvertently inhaled, ingested, or topically absorbed, a health risk known as "nano breach." Following a nano breach, nanoparticles may accumulate within the lymph and nervous systems, around hair follicles, and in tiny skin folds. The health risks associated with nanoparticles has given rise to robust research and development in the growing field of nanomedicine [13].

There are very few sources that provide guidance or address safety issues relating to nanotechnology. Neither the US Food and Drug Administration nor the US Environmental Protection Agency has fully formulated nanotechnology standards. ASTM International, an organization that develops and publishes consensus technical standards, is one of a number of US and international organizations that have published guidelines relating to the characterization and handling of nanomaterials [14]. Moreover, some cities including Berkeley, California and Cambridge, Massachusetts have passed laws regulating the use of nanomaterials. Such laws require individuals to provide information on any known health or safety risks posed by the nanomaterial and report on how the nanomaterial will be handled, stored, and disposed off [15].

With nanotechnology's potential to cause unknown and harmful side-effects, owners of nanotechnology intellectual property may face a risk of liability for side-effects stemming from their technologies. This risk of liability presents difficulties for both licensors and licensees who must consider the probability of problematic side-effects before signing a license agreement.

From a licensee's perspective, a party may be hesitant to enter into a license agreement for fear of potential liability for unknown side-effects of the nanotechnology. Licensees may choose to include limited indemnity provisions in the agreement to protect themselves in the event of a third-party lawsuit. However, to further protect themselves from potential liability, licensees may demand a broader scope of indemnity to cover unforeseen side-effects of the nanotechnology. In fact, licensees may demand that the indemnity be so broad as to cover, for instance, all unforeseen negative effects of the nanotechnology on public health and the environment. Under such broad indemnity, the licensor may be required to defend the licensee and hold it harmless in the event that a court finds liability.

While licensees would certainly benefit from an expanded indemnity provision, licensors may be hesitant to offer such a provision. In many cases, licensors may want to limit indemnity to strictly intellectual property-related areas and may not be willing to indemnify licensees for any health problems or physical injuries caused by the nanotechnology. To minimize the risk of liability, a licensor may choose to incorporate the standards set forth by ASTM or a similar organization into the licensing agreement. The standards would serve as evidence of best industry protocols and would govern a licensor's conduct. As long as the licensor abides by the standards, the licensor may be able to insulate itself against liability for unforeseen side-effects of nanotechnology.

7.5.4 *Crowded Technology Space and Cross-Licensing Strategies*

Licensors and licensees of nanotechnology may face additional hurdles from industry competitors. In the nanotechnology space alone, to date there are over 45,000 published patent applications and over 20,000 issued patents [16]. Many of these applications and patents have a similar or overlapping scope. This is understandable in the field of nanotechnology, which spans a number of disciplines. As a result of overlapping intellectual property rights, owners of nanotechnology may often find it difficult to operate in the marketplace without infringing upon another's technology.

When parties have overlapping patents such that practicing one patent would infringe the other party's patent, both parties can opt to enter into a cross-licensing arrangement. A cross-licensing arrangement is a mutual sharing of patents between parties without an exchange of license fees in which each party promises not to sue the other. When entering into a cross-licensing arrangement, parties begin by pooling the relevant patents together and dividing the patent rights so that each party takes exclusive or non-exclusive rights to a particular field of use covered by the combined patents. For instance, one party may choose to work in the chemistry space while the other party may choose to focus only on electronics.

Cross-licensing arrangements can offer several advantages for owners of intellectual property in the field of nanotechnology [17]. First, such agreements can act to lower licensing fees. If two companies in the crowded nanotechnology arena cannot practice their own invention without infringing on the other company's patent, both companies benefit by entering into an agreement that allows them each to develop and commercialize their patented technology. Second, a cross-licensing arrangement can act to encourage quicker and cheaper settlements of infringement lawsuits. When parties enter into a cross-license, they do so with a promise not to sue one another. Such an arrangement allows each party to practice the other party's technology without fear of being sued for patent infringement. As a result of the arrangement, parties that own patents with overlapping scope will not sue one another to defend their patent rights. Even if the parties did not enter into a cross-licensing arrangement before practicing their respective technologies, cross-licensing could still remain a viable alternative to litigation.

Cross-licensing can further promote innovation by preventing competitors from blocking one another's products. This may be a particularly viable strategy for latecomers looking to enter into a field with well-established players. In the field of nanotube technology, for instance, there are several established players with patent rights to basic nanotube technology. If a latecomer enters the market, the company may be blocked from practicing basic nanotube technology without infringing the rights of previously established competitors. The latecomer, however, may be able to force a cross-license with a competitor if the latecomer has patented technology that can block the existing player from expansion and further development. In the case of nanotubes, the latecomer may be able to block the existing player by obtaining patent rights to the downstream applications of nanotubes, such as electronics. In this scenario, the latecomer and the existing player would enter into a cross-license arrangement that allows each company to access the other's patent rights. This arrangement would allow the latecomer to practice the basic technology while the existing player would gain access to more specialized technology. The terms of a cross-license in this situation may be favorable to the latecomer as a result of the latecomer's superior bargaining power.

7.5.5 *Policing and Enforcing Patent Rights*

The difficulty in policing and enforcing patent rights within the nanotechnology arena presents yet another concern for licensors and licensees. There are several reasons why nanotechnology presents such a unique problem. First, infringing activity of nanotechnology occurs on a very small scale, which can be difficult to observe, analyze, and police. Additionally, discoveries and improvements within nanotechnology occur almost daily and across a variety of disciplines. Continually monitoring potentially infringing activities occurring on such a small scale and in such frequency can be overwhelming.

Adding a mutual cooperation provision to a license agreement involving nanotechnology may help address some of the difficulties facing companies in monitoring and enforcing patent rights. A mutual cooperation provision is essentially a clause that requires that parties to an agreement, usually the licensor and licensee, to cooperate to protect each other's patent rights. Under a mutual cooperation provision, the licensor may be required to disclose features of the licensed technology to the licensee at the onset of the relationship. As the relationship continues, both the licensor and the licensee may be required to continually share critical developments with one another, including improvements, advancements, and/or modifications of the licensed technology. Such communication will allow both parties to follow the development of the particular technology.

Understanding the changes and modifications made to a particular technology will also help a nanotechnology company monitor developments made by competitors. Receiving continual updates regarding the licensed technology can allow both parties to follow activities occurring within or outside the scope of the licensed technology. If a third party, for instance, develops a product having certain features or functions that are within the scope of the licensed technology, the licensor and licensee may have a potential infringement claim. If, on the other hand, a third party develops a product that is outside the scope of the licensed technology, the licensor and licensee may be able to ascertain that the particular development likely does not infringe the patent. In the event of infringement, these provisions may require both parties to participate in gathering evidence and pursuing legal action. Such cooperation can lead to the identification of infringing activity early on, before any significant financial or other harm comes to either licensor or licensee. Moreover, when both parties cooperate in a patent infringement action, the likelihood of success increases. Since both parties to the agreement are actively working toward understanding the current status of the licensed technology, mutual cooperation provisions may act to alleviate the burden of only one party monitoring the market.

Even though mutual cooperation provisions may act to alleviate the burdens associated with monitoring patent rights, there are still hurdles to enforcing those rights. One hurdle is Rule 11 of the US Federal Rules of Civil Procedure [18]. Rule 11 requires a meaningful form of infringement due diligence on the part of both counsel and client prior to initiating a patent infringement action. In other words, before an infringement claim can be brought before the courts, the attorney and the client should reasonably believe that the accused product infringes at least one claim of the patent. In the case of nanotechnology, as noted above, it is often difficult to determine if a third party is infringing because of the size and complexity of the inventions. As a result, satisfying the Rule 11 requirement may be burdensome.

In addition to the Rule 11 hurdle, a recent Supreme Court decision has made it more difficult to receive a court injunction to stop an alleged infringer from continuing with its infringing activity. In *eBay v. MercExchange*, decided on May 16, 2008, the Supreme Court held that a permanent injunction will no longer be issued automatically upon a finding of patent infringement. Instead, the Court requires that a four-factor test, as previously mentioned, be applied. That four-factor test requires a plaintiff to demonstrate that (1) it has suffered an irreparable injury; (2) remedies available at law are inadequate to compensate for that injury; (3) considering the balance of hardships between the plaintiff and defendant, a remedy in equity is warranted; and (4) the public interest would not be disserved by a permanent injunction. Even if a plaintiff can satisfy each of the four factors, there is no still no guarantee that the Court will grant an injunction.

7.6 International Issues Surrounding Nanotechnology Licensing

While the above information has primarily focused on licensing nanotechnology in the United States, international licensing can also be a viable means of developing products and expanding a company's market share. In fact, for licensors, international licensing arrangements may be the only viable way to reach foreign markets due to the costs and risks related to establishing foreign operations. For licensees, international licensing can be an effective and economical way for foreign companies to acquire technology to help expand and strengthen their own market shares without exposing them to the costs associated with researching and developing new products. The benefits to both licensors and licensees may encourage continued and increased use of international licensing arrangements.

For companies interested in licensing nanotechnology, international licensing arrangements may be particularly beneficial because of foreign attempts to further nanotechnology innovation [19]. At least sixty countries have national nanotechnology projects or programs in place. Countries like Germany, Japan, and South Korea are making substantial investments across a broad range of nanoscale science, engineering, and technology. In addition, Russia and China have increased investments in nanotechnology. Other countries, such as Israel, Singapore, and Taiwan, have also increased investments in nanotechnology; however, they have focused their resources on a specific niche or technology department, rather than just general nanotechnology research.

International partnerships, nevertheless, may often be complicated by international laws, particularly those laws governing trade. Licensors who wish to license their nanotechnology internationally should therefore take precautions to protect their patents. These include, for instance, being aware of different patent laws, export control laws, and laws governing the transfer of the intellectual property.

7.6.1 *Different Patent Laws for Different Countries*

Since patent laws are territorial by nature, an initial inquiry for nanotechnology companies entering a foreign market through licensing is whether their intellectual property is protectable in that particular country [20]. For a product to be protectable in a foreign country, a company must file for patent protection in that country. Without patent protection, there may be nothing to prevent the licensee from making, using, or selling that technology without adequately compensating the licensor. Nanotechnology companies should be aware of the patent laws in each jurisdiction and what patent rights will be applied and how they will be applied.

There are several key differences between the patent laws in the United States and those of other countries. The first is the "first to invent" and "first to file" systems. The United States has a "first to invent" system. This means that if Company A developed a product first but Company B filed a patent application to that product first, Company A could still receive a patent to that technology, assuming Company A can prove that it created the product first. Most other countries, however, employ the "first to file" system. In the above scenario, Company B would get a patent to the technology and Company A would not be granted a patent because it did not file an application in time. As a result of this difference, nanotechnology companies looking to enter into international licensing arrangements should ensure that they filed their international application before a competitor.

Another key difference between the patent laws in the United States and other countries is the concept of novelty. In the United States, an applicant is entitled to a one year grace period between the time the applicant discloses the invention and the time an application is filed. The majority of other countries, however, employ the "Doctrine of Absolute Novelty." Under this Doctrine, no such grace period is provided. Generally speaking, disclosure of the invention prior to the filing of a patent application, with certain exceptions, would serve as a bar to patentability. As a result, companies should make sure to file their international applications prior to any disclosure of their technology.

Differences in allowable subject matter between the United States and other countries may provide additional hurdles to obtaining patent protection in the course of international licensing arrangements. For instance, while software and business method patents qualify as patentable subject matter with the United States, other countries do not recognize them as such. Since nanotechnology spans across a wide variety of disciplines and industries, there exists a possibility that certain nanotechnology-related inventions are not even patentable in a foreign market. Companies licensing their nanotechnology should make sure that they can obtain foreign patent protection in the licensee country to minimize the potential of losing control of their product.

Other jurisdictional differences that affect an international licensing arrangement may also exist. These include, for instance, less comprehensive intellectual property protection; licensor's requirements or limitations may not be given full effect; enforcement of judgments may be slow, uncertain, or expensive; and injunctive relief may be unavailable or uncertain. Understanding the differences that exist between the licensor's country and a potential licensee's country may help companies better navigate potential hurdles and negotiate license agreements.

7.6.2 *Export Control Laws*

Nanotechnology companies considering licensing their products internationally should further consider how export control laws will impact the international trade of their licensed technology. Export laws apply regardless of whether the technology is protected by patents or trade secrets. Export control laws are driven by foreign policy and are designed to protect national security. Companies dealing with the licensing of nanotechnology should be particularly careful when exporting their products since nanotechnology often involves matters of national security. As such, additional precautions may be necessary.

The Bureau of Industry and Security (BIS) is responsible for implementing and enforcing the Export Administration Regulations (EAR) [21]. The EAR regulates the export and re-export of most commercial items, but certain other US government agencies are also responsible for regulating exports. The US Department of State, for instance, is responsible for regulating the export of defense articles and services. "Dual-use" items, such as those that have both commercial and military applications, on the other hand, are governed by BIS. Certain nanotechnology-related inventions may be considered "dual-use" items since they may have both commercial and military applications.

Nanotechnology companies seeking to license technology having military applications or technology involving matters of national security may need to take certain precautions to ensure that valuable information is not unlawfully disclosed to others abroad. When exporting nanotechnology to a foreign country, a licensor should first explore whether the technology falls within a designated category and whether an export license from the US government is necessary. Whether or not a product requires a license is dependent upon several factors, including, the product's technical characteristics, the destination, the end user, and the end-use. Failure to obtain an export license can subject the licensor to penalties, and may risk invalidating patent protection covering such unauthorized disclosures.

7.6.3 *Choice of Law Provision to Govern the International Licensing Agreement*

One of the most critical issues in entering into an international licensing arrangement is the choice of law that will govern the agreement. The choice of law is important because it may impact the licensor's patent rights and ability to control its product in a foreign jurisdiction. The choice of law may further impact the terms of the licensing agreements. Parties to an international licensing agreement should negotiate upfront what law should govern.

Factors that may be used in making the determination should center on the type of technology being licensed and whether that technology would be better governed by US law, foreign law, antitrust law, or other public policy laws. Because of the potential military applications and ties to national security, US nanotechnology companies may prefer to have the United States serve as the choice of law. Any other governing body may compromise the US company's control over its technology and may potentially jeopardize national security and military matters. Due to the importance of maintaining US governing law in certain nanotechnology-related matters, a severability clause may be included in the license agreement to allow the choice of law agreement to stand on its own in the event another portion of the agreement is invalidated by a court.

In many jurisdictions, the choice of law selected by the parties may not necessarily be applied. In certain countries, the local law will automatically govern the agreement. Therefore, a US nanotechnology company entering into an international licensing arrangement should be aware of the potential risks associated with allowing a foreign law to control the commercialization or exploitation of the licensed product in the foreign country.

Parties to an international licensing agreement may opt to avoid either country's law and may choose to select an alternative body of law to govern the licensing agreement. The United Nations Convention on Contracts for the International Sale of Goods (CISG) may provide that alternative. The CISG is a treaty, adopted by over two thirds of the world's countries, offering a uniform international sales law. It applies to contracts between companies located in different countries and is designed to eliminate any ambiguity caused by different domestic laws concerning the international sales of goods. The potential for disputes may be reduced by applying the CISG. Despite the potential benefits of applying a uniform international sales law, the CISG may not necessarily be best route for all companies involved. Before electing the CISG as the choice of law, nanotechnology companies therefore should evaluate the advantages and disadvantages of having US law and domestic law govern. Regardless of what choice of law is selected, fairness should be considered in drafting the agreement, as courts often frown upon choice of law provisions that would cause an undue burden to one of the parties to the agreement.

7.7 Conclusion

As we have seen, license agreements often drive innovation and provide avenues for companies to expand into new markets. However, there are several unique challenges facing nanotechnology companies when it comes to the licensing and transferring of intellectual property rights. These challenges can leave companies susceptible to unknown future developments and potential risks that may arise prior to the expiration of existing license agreements. Nanotechnology companies on both sides of a license agreement should understand the potential issues surrounding their licensing strategies and should be prepared to handle the consequences. By being aware of the potential hurdles, nanotechnology companies can maximize their protection and rewards while minimizing their risks and liability.

References

1. National Nanotechnology Initiative. http://www.nano.gov/html/facts/whatIsNano.html.

2. J. D'Silva (2008) "Nanotechnology: Development, Risk and Regulation," in *Nanotechnology: Opportunities and Challenges*, ed. S. Zodgekar, ICFAI Press, New Delhi.

3. 35 USC §261.

4. 35 USC §262.

5. MPEP §302.

6. A Quick Guide to Material Transfer Agreements at UC Berkeley. http://www.spo.berkeley.edu/guide/mtaquick.html.

7. eBay, Inc. v. MercExchange, L.L.C., 126 S.Ct. 1837, 547 U.S. 388 (2006).

8. Nat'l Interstate Ins. Co. v. Perro, 934 F. Supp. 883, 889 (N.D. Ohio 1996) (denying, in part, granting, in part, permanent injunction).

9. 35 U.S.C. §202(c)(4); 37 C.F.R. §401.6.

10. R. A. Bailey and R. J. Hanson (2010) "The University Challenge," Patent World Issue # 221.

11. 35 U.S.C. §116.

12. 35 U.S.C. §203; 37 C.F.R. §401.14(a).

13. R. A. Freitas Jr. (2005) "What Is Nanomedicine?" *Nanomed. Nanotech. Biol. Med.* **1**(1): 2–9.

14. ASTM International Standards Worldwide. http://www.astm.org/

15. H. Bray (2007) "Cambridge Considers Nanotech Curbs City May Mimic Berkeley bylaws," *Boston Globe*.

16. USPTO.gov, search spec/nano. June 24, 2010.

17. P. Sutton, C. Pham, and J. Toke (2009) "Nanotechnology License Pitfalls," *J. Intellectual Property Law Practice* **4**(3):176–180.

18. U.S. Federal Rules of Civil Procedure Rule 11.

19. J. F. Sargent (2008) "Nanotechnology and U.S. Competitiveness Issues and Options," CRS Report for Congress.

20. K. Port, J. Dratler Jr., F. M. Hammersley, *et al.* (2005) *Licensing Intellectual Property in the Information Age*, 2nd ed., Carolina Academic Press, Durham, NC.

21. Bureau of Industry and Security. http://www.bis.doc.gov/licensing/exportingbasics.htm.

Chapter 8

IP Valuation: Principles and Applications in the Nanotechnology Industry

Efrat Kasznik

Foresight Valuation Group LLC, 260 Sheridan Ave., Palo Alto, CA 94306, USA
ekasznik@foresightvaluation.com

8.1 Overview of IP Valuation

8.1.1 *What Is Intellectual Property (IP) Valuation?*

8.1.1.1 Brief history of IP valuation in the United States

With its origins in the IP litigation of the 1980s and 1990s, the valuation of IP (primarily patents) in the United States was initially limited to damages calculations in legal cases involving claims such as patent infringement. With the introduction of tax planning involving IP, such as transfer pricing and patent donations, the valuation of intangibles became critical in non-litigation circumstances as well. Companies were required to include in their tax reporting the fair market value (FMV) of IP involved in transactions, such as the intercompany transfer of IP or the donation of a patent to a university. New accounting rules related to

Nanotechnology Commercialization for Managers and Scientists
Edited by Wim Helwegen and Luca Escoffier
Copyright © 2012 Pan Stanford Publishing Pte. Ltd.
ISBN 978-981-4316-22-4 (Hardcover), 978-981-4364-38-6 (eBook)
www.panstanford.com

business combinations in the United States, introduced in the early 2000s, expanded the need for IP valuations even more, as companies were now required to report the Fair Value (FV) of intangibles that were purchased with a target in a mergers & acquisitions (M&A) deal. These "compliance" situations — litigation, accounting, and tax reporting — carry with them a high degree of scrutiny by the court or regulating authorities, and require a third-party, IP valuation expert's opinion in the form of a report or testimony.

In parallel to the proliferation of tax and accounting rules, which mandated the valuation of IP in certain transactions, around the same time (late 1990s to early 2000s) the field of intellectual asset management (IAM) was starting to gain momentum with US corporations. Large companies with significant patent portfolios (like IBM, Dow, and others) were leading the way, and with the increase in sophistication of active IP portfolio management, came the need for valuation. The types of activities where a valuation became increasingly important include spin-offs, in kind contributions, licensing, patents sales, and other commercialization activities. Since many of these activities are not always subject to tax or accounting reporting, these transactions can be referred to as "non-compliance" activities. In these situations, due to the low to medium degree of scrutiny and the lack of reporting requirement, the valuation is often done in-house or between the parties, without the involvement of a third-party IP valuation expert.

8.1.1.1.1 IP valuation methodologies   There are three common methods for valuing intangible assets: the *market method*, the *income method*, and the *cost method*. These valuation methodologies were largely borrowed from the methodologies applied in the valuation of tangible assets (like real estate, machines, inventories, etc). As a result, they are more suitable for tangible assets and some of them are challenging to implement when intangibles are involved. Each of these valuation methods is briefly described below.

Market methods are useful where there is a market demand for an intangible asset and there are other similar intangibles that have traded hands under specified market conditions. In these circumstances, the market price for comparable assets may be used (with appropriate adjustments) as a measure of the market value of the intangibles at issue. The market method of valuation relies upon the availability of comparable external transaction data, which are sometimes difficult to find. In practice, although it is the preferred approach for many assets, it is often not possible to use market approaches if there is no observable active market for the intangible asset. The market approach is therefore not commonly applied

Income methods are the most commonly used methodologies in the valuation of intangibles. Income-based approaches involve calculating the value of an intangible asset based on the aggregate income stream that ownership of such intangible will provide. That income stream, net of any costs associated with its production, is discounted to its net present value (NPV) to determine the value of the intangible asset. The application of such methods requires the projection of economic income that is directly generated by the asset over its economic life. These projections are converted into NPV by using a present value discount rate, which represents the required rate of return over the intangible asset.

Cost methods involve an analysis of all cost components that went into creating the intangible asset, such as materials, labor, and overhead. Cost methods are hardly used when valuing intangible assets, primarily because these methods do not consider future economic benefits arising from the asset. The application of this approach is only appropriate for assets which are usually accounted for by the cost of reproduction, such as software and assembled workforce.

8.1.1.1.2 Types of intangible assets   The types of intangible assets that are covered in this chapter fall into several subcategories. One very useful framework for classifying intangibles can be found in the United States. Generally Accepted Accounting Principles (GAAP) definitions as found in Accounting Standards Codification (ASC) 805. The guidelines of ASC 805 are used for identifying intangibles involved in business combinations, such as M&A transactions [1].

Among the ASC 805 intangible asset categories, the most common groups of assets found in the nanotechnology industry are the following:

Technology-based intangible assets. Innovations and technological advances that are protected by contractual or legal rights. This group includes the following assets:

- Patented and unpatented technology
- Computer software and mask works
- Databases, including plants
- Trade secrets, such as secret formulae, processes, recipes, etc.

Marketing-related intangible assets. Assets that provide value to the marketing or promotion of products and services. This group includes the following assets:

- Trademarks, trade names, service marks, collective marks, certification marks
- Trade dress (unique color, shape, or package design)
- Internet domain names
- Non-competition agreements

Customer-related intangible assets. A customer relationship exists between an entity and its customer if the entity has information and regular contact with the customer, and if it stands to benefit from future contracts that are reasonably anticipated from that customer. This group includes the following assets:

- Customer lists
- Order or production backlog
- Customer contracts and relationships

8.1.1.2 IP valuation standards in the United States

8.1.1.2.1 IP valuation landscape in the United States Most IP valuations in the United States are done in *compliance* situations, either for financial reporting, tax compliance, or litigation damages. There is a fundamental difference between the IP valuation requirements in compliance versus non-compliance situations. In compliance situations, the valuation is mandatory, and is usually done after the deal has already been finalized. There is single point value that needs to be recorded (as opposed to a *range* of values that needs to be negotiated). On the other hand, most IP transactions today are done in non-compliance situations, where there is no mandatory reporting requirement. If an IP valuation is conducted under such circumstances, it is neither reported nor regulated under any standards.

Valuations done in non-compliance situations are usually done in-house for purposes of negotiations, and therefore there would not be a single point valuation, but rather a range of values that needs to be negotiated between a buyer and a seller. In recent years, a large volume of IP transactions involving the sales of patents, were carried out through IP brokers and various kinds of IP funds (such as patent aggregators, defense funds, etc.). There are rarely any formal IP valuations done in conjunction with these transactions, as most of them rely on legal claim chart analysis, and some heuristic rules of thumb as to the discounts that should be applied to future cash flows.

We turn next to discuss in more detail the IP valuation activities under the most common *compliance* situations: litigation damages, financial reporting (accounting), and tax reporting.

8.1.1.2.2 IP valuation for litigation damages   The litigation of intellectual property in the United States has seen a sharp increase since the 1990s, both in the number of cases filed annually and in the cost of litigation. According to the American Intellectual Property Lawyers Association (AIPLA), the number of patent infringement cases has doubled in 10 years — from about 1700 in 1995, to over 3300 in 2005 [2]. The number has since leveled somewhat, but is still significantly higher than a decade ago. The cost of patent litigation has sky rocketed as well; the average cost of litigating a patent case through trial, according to the AIPLA, is estimated at $1–$3 million, depending on the amount of damages and the size of the case.

The field of nanotechnology is at its early stages of patenting, and litigation activity has so far been limited. The first case filed in a potential wave of patent infringement litigations based on nanotechnology patents, *DuPont Air Products Nanomaterials v. Cabot Microelectronics* (filed in January 2007), represents what many feel will turn out to be a growing trend amongst companies. While it is difficult to predict how the litigation environment will evolve to absorb the changes brought by the nanotechnology field, all eyes will be on the top players and how they react to legal challenges related to their IP portfolio.

IP valuation analysis is implemented in IP litigation situations when it comes to damages calculations. The purpose of damages is to make the plaintiff whole, that is, compensate the injured party such that it would return to the position it would have been in *but-for* the infringement of its intellectual property. Generally speaking, there are four types of IP litigation in the United States: patents, copyrights, trademarks/trade dress, and trade secrets. Below is a brief overview of the damages calculation standards for patent litigation [3].

When plaintiff can successfully show infringement of a valid US patent, 35 USC § 284 states that: "Upon finding for the claimant the court shall award the claimant damages adequate to compensate for the infringement but in no event less than a reasonable royalty for the use made of the invention by the infringer, together with interest and costs as fixed by the court." The case law implementation of this statute allows for patent infringement damages based on lost profits, a reasonable royalty or a combination of both depending on the circumstances of the case

Much legal guidance, case laws and precedents exist in the United States as to how patent infringement damages are calculated. The first step in quantifying damages in a patent infringement matter involves the application of a four-part test set forth in *Panduit Corp v Stahlin Bros Fibre Works*, 575 F2d 1152 (6th Cir 1978). In the *Panduit* case, the court found that the conditions necessary for the calculation and recovery of lost profits are the following: a demand for the patented product, the absence of non-infringing alternatives, the existence of sufficient manufacturing and marketing capacity on the part of the plaintiff and the plaintiff's ability to quantify lost profits.

Should the *Panduit* conditions not apply, the damages calculations revert to reasonable royalty. The determination of reasonable royalty damages involves the construction of a hypothetical negotiation scenario between the parties at the date of the infringement. Such an analysis is generally conducted through a consideration of the 15 factors set forth in *Georgia-Pacific Corp v United States Plywood Corp*, 318 F Supp 1116, 1120 (SDNY 1970). These are economic factors that increase or decrease the reasonable royalty rate that would have been the result of the hypothetical negotiations

8.1.1.2.3 Trademark, copyright, and trade secrets damages   Damages in trademark, copyright, or trade secrets cases are estimated using a broader range of measures as compared with patent infringement damages. For example, the disgorgement of the defendant's profits may be an allowable measure of damages in these types of cases (when such damages are not allowed in patent cases) Copyright infringement offers an additional approach to damages not available in other IP matters: statutory damages that are set at a fixed amount of dollar per infringement.

8.1.1.2.4 IP valuation for financial reporting   According to US GAAP rules, intangible assets are only reported on the financial statements when they are paid for in a business combination transaction, such as an acquisition. As a result, "home-grown" IP assets, like patents and trademarks, are not measured or reported on the balance sheet of the company that created them. However, if that company buys another company that owns patents and trademarks, the acquired intangible assets of the target company will be valued at their *fair value* and reported on the acquiring company's books. That is an interesting anomaly that is frequently pointed out by members of the IP community; however, it is unlikely that any changes to that accounting treatment will take place in the near future, primarily because of the conservative nature of GAAP rules, and the somewhat speculative nature of IP asset valuations. As long as these assets are not priced by a market transaction, their value is not certain enough for financial reporting purposes.

That being the case, the valuation of IP for accounting purposes is primarily done in the context of business combinations (such as M&A deals) and is governed by GAAP pronouncements such as ASC 805 (first introduced in 2001 as Statement of Financial Accounting Standards 141). The valuation is done as part of the overall process of purchase price allocation (PPA), when the deal price paid for the acquired company is allocated among the various assets that are comprised in that company. All assets of the target company, tangible and intangible, as well as its liabilities, are identified and valued at their current fair value. When valuing assets at fair value, the appraisal needs to assume their "highest and best use," which refers to the use of an asset by market participants that would maximize the value of the asset, even if the intended use of the asset by the holding entity is different.

The standard of valuation applied in business combinations is "fair value," defined as the price that would be received when selling an asset (or paid when transferring a liability) in an orderly transaction between market participants. "Market participants" are defined as buyers and seller in the principal market for the asset that have all the following characteristics:

(a) Independent and unrelated parties
(b) Knowledgeable, having a reasonable understanding about the asset (or liability) and the transaction based on all available information, including information that might be obtained through normal and customary due-diligence efforts
(c) Able to transact for the asset or liability
(d) Willing to transact for the asset or liability, that is, are motivated but not forced or otherwise compelled to do so [4]

The fair value of all assets (net of liabilities) is then compared to the acquisitions price, and the residual amount is recorded as Goodwill, which by itself is considered an intangible asset. Goodwill represents the future economic benefits arising from other assets acquired in a business combination that are not individually identified and separately valued. The results of the PPA analysis, including Goodwill, are then reported in the new, combined financial statements of the two companies. In the following years, Goodwill is tested annually for impairment, a process that involves a valuation analysis. If the test shows indication of impairment then some of the intangible assets' fair value estimates may be revisited, and certain assets, or Goodwill, may be written off as necessary.

The US GAAP valuation guidelines state a preference for market-based valuation methods, when market data is available. In reality, the markets for intangibles are very thin, the assets are uniquely different from each other, and most transactions are not publicly reported. As a result, most IP valuations done for accounting purposes rely on income-based approaches. In a recent study done by the global accounting firm KPMG it was reported that income-based valuation methodologies are the most commonly used for fair value IP valuation [5]. The study looked at over 300 M&A transactions between 2003 and 2007, by industry groups. The results of the study show that the percentage of the acquisition purchase price allocated to intangible asset (based on the fair value valuation analysis) varies significantly across industries. The study further determines that, in the majority of the industries analyzed, the percentage allocation of the purchase price to Goodwill is typically over 50%.

8.1.1.2.5 IP valuation for tax reporting   The most common application of IP valuation in tax reporting and planning is related to *transfer pricing* between affiliated companies. The intangible nature of IP rights renders them easily movable and creates the potential for valuable tax planning. This often involves the migration of IP into favorable tax jurisdictions or into IP holding companies [6]. The role of the IP valuation expert is to calculate the *fair market value* of the IP portfolio when it is the subject of an intercompany transfer, such as the migration to another country. The IP valuation expert can also be asked to determine the *arm's length* royalty rates for licensing IP rights between affiliated companies; these are the royalty rates that approximate the rates that would have been established between two unrelated parties.

   Due to the complexity of global tax planning, the internal Revenue Service (IRS) in the United States established a comprehensive set of transfer pricing compliance and documentation requirements. IRS Section 482, and its accompanying regulations, lays out the general compliance framework for transfer pricing schemes involving US companies, including the list of acceptable valuation methodologies for transfer pricing situations. These methodologies generally fall into three categories [7]:

1. *Transaction-based methods* The arm's length intercompany royalties are determined based on the terms of *comparable* licensing deals between non-affiliated companies.
2. *Profit-based methods* The arm's length intercompany royalty rates are determined based on the relative contribution of

the intangible asset to the overall combined company's profit margin.

3. *Cost-sharing methods* Costsharing agreements are used to divide the cost of developing intangible assets between affiliated parties, based on their respective return from the future assets. The IRS set out a list of requirements for cost-sharing agreements' qualification, including specific documentation and reporting rules.

8.1.1.3 IP valuation circumstances in Europe

8.1.1.3.1 Litigation damages The laws of IP damages in Europe also rely on the premise of restoring the injured party to the position it would have been in, *but for* the infringement of its intellectual property, although the legal procedures could vary for the United States. For example: in the United Kingdom, damages hearings constitute a separate trial, which only kicks in once liability has been established; damages hearings are therefore less frequent than in the United States as the parties have a greater motivation to settle [8].

The concept of reasonable royalty damages, based on a hypothetical licensing negotiation between the IP owner and the infringer, is common in Europe as well. The courts apply a range of royalty determining considerations, similar to Georgia-Pacific, although European courts have not clearly established a set of relevant criteria yet.

8.1.1.3.2 Financial reporting As a result of the international harmonization of accounting standards, the treatment of IP valuation in business combinations is similar in Europe and in the United States. All member states of the EU are required to use the International Financial Reporting Standards (IFRS) system of accounting standards, as adopted by the EU for listed companies since 2005. The United States is expected to follow suit in the next few years, and transition into IFRS from the current US GAAP system. IFRS 3 is the standard that applies to valuation for business combinations (the counterpart of ASC 805 in the United States).

Interestingly enough, it should be noted that prior to the EU adoption of IFRS standards in 2005, the United Kingdom was one of the only countries where home-grown intangibles (i.e., intangibles that were internally developed by the company) could be measured and presented as assets on the balance sheet. That is no longer the case as IFRS rules mostly recognize intangibles as assets for accounting purposes if they were purchased in a business combination.

8.1.1.3.3 Tax reporting   Europe is a major hub of transfer pricing activity, as several European countries are favorite locations for setting up offshore IP holding companies. Currently, Switzerland and Ireland are the preferred locations for US-headquartered corporations, while EU-based corporations tend to prefer Ireland and Luxembourg. Switzerland, Ireland, and Luxembourg have good treaty networks and very advantageous tax rates. For instance, the tax rate in Switzerland is determined by negotiation with the local canton and is usually in the range of 4% to 8% [9].

The valuation of IP assets is critical at the time of transfer of the IP portfolio into the IP holding company, and needs to be done by a third party, independent valuation expert. Each IP asset should be separately assessed. The IP holding company needs to have to function as a real business; otherwise it might be treated as a "controlled foreign corporation," in which case its accounts would have to be consolidated with those of the parent company and it would lose the benefits of operating in a low-tax jurisdiction. In order to overcome this problem in the EU, the IP holding company has to have "functionality," which is demonstrated by a staff that actively manages the IP portfolio.

8.2 The Application of IP Valuation in the Nanotechnology Industry

8.2.1 *Nanotechnology IP Landscape and Technology Development*

8.2.1.1 Patent landscape overview in the United States

The field of nanotechnology is inherently interdisciplinary: a nanotechnology invention will likely have applications in multiple industries, from pharma and biotech to computer hardware.

Examples of some of the early nanotechnology products on the market highlight that diversity: French cosmetics company L'Oreal introduced a nanocapsule antiaging skin treatment in 1995; and US fabric manufacturer NanoTex introduced a line of nonmaterial-based fabrics that are stain resistant [10].

Patent activity in the nanotechnology sector is at its early stages of development. The patent landscape is rapidly evolving, as more patents are beginning to issue while the number of new patent applications has significantly increased over the last few years. Patent holders include companies of all sizes, from start-ups to multinational conglomerates, as well as research institutes such as universities and government laboratories. The nanotechnology patent landscape is expected to follow a path similar to that which occurred in biotechnology and semiconductors, but in an accelerated way.

When patents issue in an emerging sector with interdisciplinary characteristics, there is inevitably going to be overlap among patent claims issued by different inventors. This phenomenon is referred to as a "patent thicket" [11]. A 2005 study by research firm Lux Research shows the degree of "white space" (absence of patents) plotted against the degree of "entanglement" (claim overlaps) of nanotechnology patents in different sectors and components. It shows that, with the exception on nanowires where there is still a significant amount of white space and fewer instances of patent entanglement, all other major research areas that were examined (fullerenes, carbon nanotubes, dendrimers, quantum dots) show a great amount of overlap and lack of white space [12].

As in any emerging field, the United States Patent and Trademark Office (USPTO) is walking a thin line between protecting new innovations, and allowing freedom to operate to more players in the sector by not issuing patents that are too broad, as has been the case in the early days of the biotech industry. As a response to mounting complaints about USPTO examiners' lack of understanding of nanotechnology research and difficulty in tracking down prior art, the USPTO recently introduced a new cross-reference classification system to facilitate prior art searches for patent applications which cover multiple disciplines [13].

8.2.1.2 State of nanotechnology development around the world

Technology development is impacted by the state of research and development, and the national infrastructure that supports its commercialization. Lux Research issued a study in 2010 that ranks the status of nanotechnology ecosystems in 19 countries, based on an analytical framework that analyzes data from 2007 through 2009 [14]. The Lux Research framework maps out the nations using a two-dimensional matrix with the following axes:

Nanotech activity measures the absolute level of nanotechnology development, drawing on 8 metrics including nanotech initiatives, nanotech centers, government funding, risk capital, corporate spending, nanotech publications, nanotech patents, and active companies.

Technology development strength measures the robustness of technology commercialization infrastructure, drawing on 6 metrics including: high-tech manufacturing, R&D spending, intellectual capital, technology and science workforce, knowledge emigration, and technology infrastructure.

The mapping on the matrix ranges from "minor league" (low on both dimensions) to "dominant" (high on both dimensions), with "ivory tower" (high on activity, but low on development) and "niche" (low on activity and high on development) in between. It is interesting to compare the US ranking on the national mappings opposed to European countries, and in particular, Italy, the Netherlands, the United Kingdom, France, and Germany.

The United States earned an "Ivory Tower" ranking — with full marks in every activity metric. The United States launched the National Nanotechnology Initiative (NNI), which is well coordinated and funded, and helps support start-up and academic research. In addition, corporations and private funding invested billions of dollars and filed thousands of patents in 2009 alone. On the other hand, the United States ranked average on the Development axis. When compared to the European countries, Italy and the Netherlands are closely ranked as "minor league," with low ranks on both dimensions. France and the United Kingdom are in the middle of the map, scoring about average on both activity and development.

Germany seems to be faring out the best with a "dominant" ranking, scoring above average on all counts.

8.2.2 *Managing an IP Portfolio in the Nanotechnology Industry*

8.2.2.1 IP portfolio challenges in nanotechnology

The macro-level overview of IP landscape and national rankings leads to unique challenges at the micro level (company or research institution) when it comes to creating, managing, and commercializing a nanotechnology IP portfolio. While the challenges vary by company size and the specific industry in which companies operate, and while research organizations face problems that are different from those faced by corporations, there are still some common themes related to IP portfolio management than run throughout the industry. Three of these IP management topics stand out as broadly applicable to a wide range of nanotechnology IP holders around the world (in the United States and in Europe), as the industry follows the path from early stage to maturity:

- Patenting along the nanotechnology value chain
- Technology transfer from university to industry
- Mitigating litigation risk

Following is a brief discussion of each of these topics, including a discussion of the role that IP valuation can potentially play in supporting these challenges. We then conclude the chapter with some valuation case studies that demonstrate the versatility of the technologies that IP valuation experts analyze in the nanotechnology industry.

8.2.2.2 Patenting along the value chain

The concept of the *nanotechnology value chain* is closely related to the interdisciplinary nature of the industry. This concept is illustrated in a 2003 Intellectual Asset Management (IAM) Magazine article [15]. The authors claim that the key to a strong IP portfolio is found in patenting across the entire nanotechnology value chain, taking into consideration the following elements:

- The basic chemical composition
- The physical structure
- The tool for developing that structure
- The method or process for using that tool
- The article of manufacture (end product)

Vertical integration along the supply chain can be done by either licensing or acquisitions. Both types of transactions require the support of IP valuation analysis. With regards to licensing, there could be instances where the licensing is done between affiliated companies, and such activity would require the royalty rates to adhere to transfer pricing regulations for tax reporting purposes, a process that usually involves a third-party valuation expert. In acquisitions, the buying entity would need to engage in purchase price allocation and value the target's IP portfolio for financial reporting.

8.2.2.3 Technology transfer from university to industry

Nanotechnology relies heavily on federal funding, technology transfer, and university inventions. The transfer of new technology from university laboratories to the private sector has a long history and has taken many different forms. Prior to the 1980 enactment of the *Bayh-Dole Act*, companies did not have exclusive rights under government patents to manufacture and sell resulting products [16]. As a result, companies were reluctant to develop new products when competitors could also acquire licenses to the same technology. Government remained unsuccessful in attracting private industry to license government-owned patents. The Bayh-Dole Act permitted universities and small businesses to elect ownership of inventions made under federal funding and to become directly involved in the commercialization process. This new policy also allows for exclusive licensing when combined with diligent development and transfer of an invention to the marketplace for the public good.

With the passage of the Bayh-Dole Act, colleges and universities immediately began to develop and strengthen the internal expertise needed to effectively engage in the patenting and licensing of inventions. In many cases, institutions that had not been active in this area began to establish entirely new technology transfer offices, building teams with legal, business, and scientific backgrounds. As a result, many new technologies have been diligently and successfully introduced into public use. Another significant result of the Bayh-Dole Act is that it provides a strong incentive for university–industry research collaborations.

The technology transfer process from university to industry is primarily done in two ways: licensing and joint ventures. In both instances, there is an IP valuation analysis that needs to take place in order to facilitate the transaction. In licensing deals, the parties need to agree on the royalty rate, structure (lump sum, running royalty, or a combination of both), and other terms related to the license. In joint ventures, a university would typically contribute the IP, while the industry partner contributes the tangible components (usually cash or equipment). In order to figure out what share of the venture each side gets, the university's "in-kind" contribution of IP needs to be valued, a task that is usually assigned to a third-party valuation expert.

8.2.2.4 Mitigating litigation risk

The biotechnology and semiconductor industries experienced accelerated patenting activity, followed by litigation, over the last 20 years [17]. This pattern could emerge in the nanotechnology field, especially in light of the claim overlap due to the unusually high number of patent applications covering similar technologies. Bringing a product to market involves navigating patent thickets, and facing potential litigation down the road as the industry matures seems highly likely.

Legal experts following the nanotechnology industry vary in their assessment of the future of litigation in the industry. Some believe that cross-licensing is an effective tool for resolving patent thickets. The argument here is that by carefully carving the fields of use in a cross-licensing deal, each party can have an exclusive field of use that does not overlap with the other party [18]. The 2005 Lux research study goes on to predict that the nanotechnology industry will, for the most part, manage to avoid a "self-destructive IP war" through a flood of cross-licensing agreements and other types of IP licensing.

On the other hand, other legal experts are not as optimistic as to the mitigating role of cross-licensing when it comes to future patent litigation. Their argument is that cross-licensing has worked in the semiconductor industry due to the oligopolistic nature of that industry, where there are a relatively small number of large firms with similar products and similar IP portfolios [19]. The nanotechnology industry is inherently different: there are many patent holders in a variety of different industries. Two start-ups trying to commercialize the same product in the same market are more likely to try and litigate each other out of business, as opposed to cross-license, as has been the case in the biotech industry.

The role of IP valuation is fairly clear and straightforward here: if and when a wave of litigation hits the nanotechnology industry, it would give rise to a tide of economic expert who testify on damages.

8.2.3 *IP Valuation Case Studies*

The case studies below are based on real-life valuation analyses of nanotechnology patents that were carried out in support of tax reporting for two major US corporations. Each of these organizations donated intellectual property (including patents and know-how) to a university, and the valuation was conducted to estimate the *fair market value* of the technologies for tax deduction. These two cases demonstrate a *compliance* situation (tax reporting) and are part of the technology transfer process from university to industry (even though the technologies were donated from the industry to the university, the goal with technology donations has been for the universities to further develop the technology, and then license it back into the marketplace). These cases were also selected because they represent the interdisciplinary nature of the industry, and the wide range of applications that nanotechnologies have.

8.2.3.1 Advanced thermoelectric technology

Originating industry. Aerospace.
Technology description. Method for making advanced thermoelectric devices known as thermopiles. The patent involves a method for thermopile fabrication that produces advanced thermoelectric

devices supporting applications over a broad range of areas. The method uses a unique semiconductor microfabrication technique in conjunction with a printed circuit board, integrated circuit, and pick and place/flip-chip technology.

Application areas. Power generation for micro-electromechanical systems (MEMS) was found to be a major emerging application area for this technology. MEMS are the integration of several micro components on one chip, which include integrated circuits, sensors, and actuators. At the time of the valuation (2003) only a handful of organizations, primarily universities and research institutions, engaged in exploring solutions for powering MEMs. It was further determined that the patented technology would be appropriate for meeting the power supply requirements of low-power MEMS devices, in areas such as information processing, biomedical, telecommunications, consumer electronics, and military/aerospace.

Valuation standard. Fair market value for tax reporting.

Valuation method. Income method based on discounted cash flow analysis (relief from royalty).

8.2.3.2 Nanocomposite plastic technology

Originating industry. Chemical.

Technology description. Process for making plastic nanocomposite materials with improved barrier qualities. The technology has originally been developed to enhance gas barrier properties of beverage packaging plastic. It is based on dispersing nanometer-size clay platelets in a resin matrix, to create Polyester and Polyamide nanocomposites. Besides improved gas barrier, these nanocomposite materials were also found to impart higher strength, stiffness, dimensional stability, and heat resistance, which could be utilized in non-packaging application areas as well.

Application areas. The main application areas that could benefit from the patents include beverage packaging, baby food packaging, medical packaging and containers, medical devices, and automotive body parts. The valuation further focused on the beverage packaging application, and on the beverage groups that were most likely to benefit from the nanocomposite's improved properties: beer and fruit juice. The analysis focused on the supply chain from the raw material to the bottle. Since the patent portfolio included foreign filings as well, the analysis covered the United States and international markets in Mexico, United Kingdom, Germany, and France.

Valuation standard. Fair market value for tax reporting.

Valuation method. Income method based on discounted cash flow analysis (relief from royalty).

References

1. Based on Ernst & Young publication, *Financial Reporting Developments, Business Combinations – Accounting Standards Codification 805*, Chapter B4, Revised October 2010.

2. *Cost of Patent Litigation*, AIPLA Mid-Winter Conference, Jan. 25, 2008. Available at http://www.aipla.org/Content/ContentGroups/Speaker_Papers/Mid-Winter1/20083/Showalter-slides.pdf

3. Chelton, D., Tanger, M. P., John, D., and Tony, S. *Accounting for Damages in Intellectual Property Litigation,* PwC, Atlanta, Washington DC, London and Sydney.

4. All definitions taken from the ASC Master Glossary.

5. *Intangible Assets and Goodwill in the Context of Business Combinations, An Industry Study*, KPMG – Corporate Finance Advisory, May 2009.

6. Chelton, D., Tanger, M. P., John, D., and Tony, S., *Accounting for Damages in Intellectual Property Litigation,* PwC, Atlanta, Washington DC, London and Sydney.

7. Based on the discussion in "Fundamentals of Intellectual Property Valuation," Chapter 18, by Weston Anson

8. Chelton, D., Tanger, M. P., John, D., and Tony, S. *Accounting for Damages in intellectual Property Litigation*, PwC, Atlanta, Washington DC, London and Sydney.

9. Larry, C., and Sean, F. (2010) *Offshoring your IP holdings*, New Legal Review, May 14, 2010.

10. Joff, W. (2003) *Patent Challenges for Nanotech Investors*, Intellectual Asset Management Magazine, September/October 2003.

11. Stephen, B. M., and Leon, R. *The Nanotech IP Landscape: Increasing patent Thickets Will Drive Cross-licensing*, Foley & Lardner LLP.

12. Lux Research and Foley and Lardner, *The Nanotech Intellectual Property Landscape*, 2005.

13. *Nanotechnology: Thinking Small in a Big Way*, JonesDay publication, 2008.

14. *Ranking the Nations on Nanotech Hidden haven and Fallen Threats*, Lux Research, 2010.

15. Joff, W. *Patent Challenges for Nanotech Investors*, Intellectual Asset Management Magazine, September/October 2003 (sidebar titled: "Developing the nanotech patent portfolio," by Rachel E. Schwartz and John E. Cronin of the ipCapital Group, Inc.)

16. University of California Office of Technology Transfer, *The Bayh-Dole Act – A Guide to the law and Implementing Regulations* (http://www.ucop.edu/ott/faculty/bayh.html)

17. *Nanotechnology: Thinking Small in a Big Way*, JonesDay publication, 2008

18. Stephen, B. M., and Leon, R. *The Nanotech IP Landscape: Increasing Patent Thickets Will Drive Cross-licensing*, Foley & Lardner LLP.

19. Ruben, S., Kirk, H., and Chris, D. (2005) The nanotech intellectual property ("IP") landscape, *Nanotechnology Law and Business Journal*, 2(2), Article 3.

Chapter 9

Investing in Nanotechnology

Po Chi Wu

School of Business and Management, & School of Engineering,
Hong Kong University of Science and Technology, Hong Kong, HKSAR
wupc@ust.hk or wu.pochi@gmail.com

9.1 The Nanotech Challenge

Throughout history, every wave of new knowledge, new technology, has resulted in a split in society, between those people who embrace change and learn how to benefit from that new knowledge and those who do not. Shifts in power also follow. Even established major players that fail to respond appropriately to innovative and disruptive technologies can find their business models severely challenged. Some of the innovators go on to form businesses and create wealth for themselves and for society. These are the entrepreneurs, sometimes considered foolish and unrealistic, but bold enough to follow their instincts and dreams. Whether they are technical experts or not is often less important than their ability to sniff out opportunities and their courage to act on them before others do. Somehow, they are able to see how to develop and combine the elements that are necessary for building

Nanotechnology Commercialization for Managers and Scientists
Edited by Wim Helwegen and Luca Escoffier
Copyright © 2012 Pan Stanford Publishing Pte. Ltd.
ISBN 978-981-4316-22-4 (Hardcover), 978-981-4364-38-6 (eBook)
www.panstanford.com

a business. In terms of general direction, they know what must be done and acknowledge that they will undoubtedly encounter many difficulties, especially unforeseen ones. More importantly, they accept the reality that the only way to resolve those challenges is to keep trying, often failing repeatedly, as they build their own experience base.

For a new technology to really take root and become a force for change in the industry, a small group of inventors and innovators is not enough. An entire ecosystem, a supporting infrastructure, must also be created. Extensive value chains must form, especially to develop the human resources that will be needed for these new businesses. Look at what the state of North Carolina proposed to do in 2006 (taken from *A Roadmap for Nanotechnology in North Carolina's 21st Century Economy — Findings and Strategic Imperatives of the Governor's Task Force on Nanotechnology and North Carolina's Economy*). These are the elements that the report suggests are necessary to build the ecosystem in North Carolina, if the state is to focus on nanotechnology as a strategic priority in economic development:

1. Establish a North Carolina Nanotechnology Alliance.
2. Through the development of multiple centers of nanotechnology excellence at North Carolina's universities, develop a diverse critical mass of nanotechnology research, development, education, and outreach expertise in the state.
3. Establish a not-for-profit nanotechnology "imagineering" group staffed to identify emerging nanotechnology opportunities and execution agents.
4. Create an information clearinghouse about nanotechnology in North Carolina.
5. Convene an annual North Carolina Symposium on Nanotechnology.
6. Ensure that nanotechnology is explicitly considered in education and workforce development activities.
7. Strengthen teacher knowledge of advances in nanoscale science.
8. Integrate information about nanotechnology into the North Carolina Biomanufacturing and Pharmaceutical Training Consortium.

9. Explicitly integrate the environmental, ethical, health, legal, safety, and other societal implications of nanotechnology.
10. Emphasize education of policy makers, the public, the business community, and the scientific community on issues related to nanotechnology.

Note that virtually all of these elements refer to knowledge — education, research, and building communities to promote sharing. All this will take significant amounts of time, perhaps a decade or two. The scope is broad, and includes people from academia, industry, and government. This is critically important in nanotechnology, where innovations, almost by definition, result from a convergence of multiple scientific and technical disciplines. The ability to determine precise molecular and atomic alignment may be a scientific breakthrough. It becomes commercially valuable only when the result is a novel solution to an important practical problem. Only then might an investor show some interest to invest in the project.

Innovations in other fields of science and engineering are typically simpler and their supporting ecosystems are already established. Apple's new iPAD has many innovations, but these have been developed and manufactured by experienced component suppliers all over Asia. Steve Jobs created the vision for the product and the business model, then Apple's engineering team went to work actualizing the dream and figuring out what parts were needed, from whom, at what quality, in what quantities, and when. Chip designers know how to work against those kinds of requirements, so do the display makers, and the other hardware and software providers as well.

A major challenge with nanotech innovations is that they usually comprise new materials that have very promising technical characteristics, often never seen before. How do you integrate those new materials with other conventional parts? How do you attach a carbon nanotube to a silicon chip? How do you measure and control electrical impulses across such interfaces? Not only does the technology have to be developed, it has to become manufacturable in large volumes, at a reasonable cost. All that takes time and a lot of trial and error on the part of a lot of people in the value chain. The company with the one innovation just isn't going to be able to move the industry.

9.2 How Investors Think About Nanotechnology

- It's a complex technology, hard to understand. Who cares?
- What practical applications are really needed in the market-place?
- Who are the customers?
- Who are the competitors?
- Where is the supporting ecosystem?
- Where are the people with the talent, experience, and drive to build these nanotech businesses?

Investors who are looking at nanotechnology typically consider this last concern to be extremely critical — that currently available human resources are still relatively scarce. Senior management is being recruited from leaders in other industries, often from information technology and other electronics businesses. While these people often come with outstanding track records, the key question is how transferable their skills are and how quickly they can learn how to drive these new business models. The perceived risk is high.

Figure 9.1 highlights some of the major issues that investors consider when assessing uncertainty and risk.

In fact, in the context of nanotechnology, the answers to all the questions listed above usually leave most investors feeling uneasy about their perceptions of the risks involved. Despite the name "risk capital," which is the predecessor to "venture capital," venture capital investors are remarkably risk-averse. This is as it should be. They are handsomely paid to achieve returns on investment that are significantly higher than a public stock market index fund. Compared to a commercial banker, who deals with fixed interest financial instruments, a venture capitalist takes huge risks, much greater than a commercial bank would accept. However, in the context of the venture capital world, if one investor can find ways to mitigate and manage risk in a particular investment opportunity, while other investors cannot, he may decide to invest and the others won't.

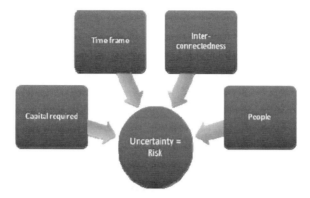

Figure 9.1. Major issues to consider when assessing uncertainty and risk.

The job of a professional investor is to make money from his investments. It is not to support research just because it is innovative and might someday lead to some potentially valuable outcome. In this sense, investors are basically "reactive," as opposed to being "proactive." Practically every entrepreneur seeking funding for a new seed-stage start-up company will confirm this, as very few venture capitalists now invest at the earliest stages of a company's life and development of new technology. When faced with an investment opportunity, investors have two fundamental questions:

1. Who are the customers?
2. How does the money flow along the value chain?

Investors also care a great deal about the ecosystem. Without that infrastructure, money can't flow. Successful investors understand what drives change and how to influence that change. Figure 9.2 highlights the major factors that contribute to change.

Innovations in nanotechnology are often significantly more complex than in other industry areas because of the scope of the technologies involved as well as their potential impact on societies around the world.

Despite the public excitement for imaginative science fiction loosely based on concepts from nanotechnology, commercialization of the technologies is still at an early stage. Only in the last 15 years have governments made significant commitments to research

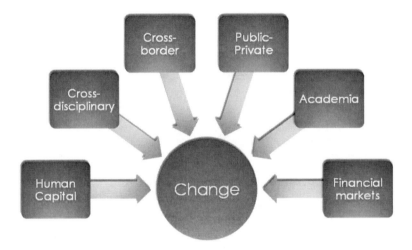

Figure 9.2.   Major factors that contribute to change.

and development. Major universities have now developed multi-disciplinary programs that combine resources from engineering, chemistry, physics, mathematics, and biology. Academic research in the laboratory has yielded some amazing new materials that show unexpected properties such as superconductivity. While the science may be worthy of a Nobel Prize, the bigger challenge, in some sense, is how to manufacture these materials in large quantities and consistent quality. The very precise, sophisticated tools that work in carefully controlled, world-class academic laboratories are not the same as those that are needed in the rough-and-tumble environment of a manufacturing plant. So, the first issue is one of "unprecedented scale," at both ends of the spectrum.

The human resource issue is only part of the "Intellectual Capital" that is the critical asset that businesses require. The other part is "Intellectual Property" — the know-how, patents, copyrights, trade secrets, and even brands that support commercial success. While US researchers still file and are granted the most patents every year, the progress of the Japanese and the Chinese is accelerating, supported by governments eager to build proprietary advantages for their economies. For the Chinese, especially, this is a field that offers great opportunities that are consistent with strengths of the local researchers — patient, aggressively hard-working, precise, ambitious — and the initial phases of the work are not particularly capital intensive. The big money is needed when large-scale manufacturing kicks in. By the way, China, as "factory of the world" has proven to be clever and capable in solving these kinds of challenges, perhaps better than any other country. From this point of view, China has tremendous advantages, especially over Europe. It has a huge and rapidly growing domestic market, and its products and services may serve markets in other developing countries as well. Its mass and rapid growth are driving an accelerating rate of change in all the countries in the world.

The most important fundamental aspects of globalization are the free flow of knowledge and capital across national boundaries as well as technical specialties. These irreversible trends are changing the world of business, and are particularly relevant in nanotechnology because success requires cooperation and collaboration. Inventions in nanotechnology require platforms of other technologies and extensive infrastructure before commercial outcomes can be achieved. This simple fact is not well understood among technologists but is all-important to venture investors, who care greatly about entrepreneurs "controlling their destiny." If an entrepreneur's intellectual property depends on or may be adversely affected by assets held or being developed by other competitors, investors are very reluctant to invest. The risk is perceived to be high and the outcome of any dispute is likely to be costly and definitely uncertain.

Reflecting the early stage of commercialization, the supporting ecosystems are also in their infancy. Production implies a full value chain of raw material sources, analytical tools, precise equipment to test and manufacture, other component manufacturers, product de-signers, not to mention marketing, distribution, and sales channels. Two major product areas that achieved success for companies and investors are photovoltaic solar cells and carbon nanotubes used as additives in tires and other rubber products. These two examples illustrate a couple of other points. Solar cells are manufactured using existing semiconductor knowledge and existing production facilities, so solving issues of scale-up was relatively straightforward. Investors poured money into those start-ups that had the most efficient and cost-effective technologies because they had been developed over several decades at the very best academic research labs and supported by hundreds of millions of dollars of government funding.

The rapid growth of the solar cell industry was a direct result of government subsidies in many countries, starting in Europe (Germany and Spain are particular examples), then in the United States, that offered rebates and other financial incentives to consumers and companies that installed solar panels for generation of electrical power. Regulations for the electric power industries also had to be modified to permit owners to sell "excess inventory" back to the state or local power grid. While these policies provided a short-term boost to the industry, the growth of the industry has slowed. Of course, several multi-billion-dollar companies have gone public, creating significant wealth for investors and represent the foundation of a new industry.

The development of carbon nanotubes took a slightly different path. While the beginnings were also in academic labs, very quickly companies realized that isolation of pure products would be extremely difficult and costly. Consequently, they discovered applications using heterogeneous mixtures to take advantage of simpler, common properties of the mixtures. These properties were relatively crude, such as physical strength, heat conductivity, or electrical resistance. The more exotic and novel properties of particular nanotube configurations have yet to be commercialized.

The flow of money is critical to understand because this is how investors think. Figure 9.3 illustrates four major geopolitcal issues which are also major economic issues. Until money flows freely through a cycle of producers, customers, and financial markets, and back to developers/producers, there is no participation from smart investors. For successful businesses, the cycle is self-sustaining. They develop and sell products that customers want and pay for. When a business has enough customers to make a large profit, it will have funds to invest in further R&D to create new products. Success breeds success. Only companies that show exceptional promise before achieving profitability can hope to obtain outside financing from professional investors. These are the ones with exceptional Intellectual Capital.

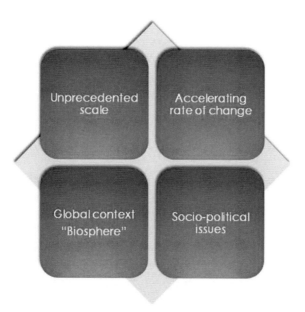

Figure 9.3. Geopolitical issues associated with money flows.

Other than consumers, all the other players in the cycle measure value in terms of money, specifically — leverage and multiples of investment. For consumers and end users the value of a new product or service is the benefit they receive. Everyone else is simply following the cycle and betting on its sustainability (or not). Investors are the last to participate, since they can benefit only when the perception of success is strong enough. Sometimes, just the perception is enough, but that is, of course, not really sustainable.

When investors think about the risks, they consider impediments to this free flow of capital. The first key question is: Are there or will there be customers for a company's products or services? Is there a market? Before the final product configuration is set, it can be difficult to determine who, exactly, the customer is, what benefits they will derive, and how much they will value those benefits – how much they will pay. Technologists who have not had a lot of experience with real product launches often underestimate the time and investment required to get to that point. A common weakness of unsophisticated business plans is an underlying assumption often referred to as "the better mousetrap theory," or "build it and they (customers) will come." Technologists usually love technology, and are passionate about their own inventions. When they talk to their friends, who are also engineers, they get a lot of positive feedback and encouragement. Even in Silicon Valley, this illusion is all too common. Unfortunately, while their friends and peers may be the early adopters of their new products and services, markets where potential customers are less savvy about technology may be far less receptive.

This issue of "comfortable acceptance" by potential customers is much more complex when dealing with nanotechnology, because of the underlying technologies. Recent research by marketers and psychologists who study the effectiveness of marketing campaigns shows that the vast majority of marketing campaigns are ineffective in determining what people will really buy. Conventional approaches such as focus groups are designed to elicit how people think and express how they think about new product ideas. The problem is that 95% of people's behavior is determined by habits, often unconscious. People's actual buying patterns are not consistent with what they say they like or will buy. People buy what they trust and feel comfortable with. This is the power of brands that have established reputations for desirable and trustworthy products and services. For consumer products, doing this takes time and a lot of investment. This undertaking is not something early-stage companies can afford or even are inclined to do because they lack the human and financial resources.

The history of venture investments in the area of new materials has not been characterized by great successes and this may be a pattern that is being followed in nanotechnology. Great scientific innovations, brilliant technological breakthroughs are not the issue. If the innovation is to be marketed as a new material, there can be and typically is resistance from a number of the players in the ecosystem. Every new technology breakthrough can shift the balance of power, as the early adopters can have tools that give them a competitive advantage. Figure 9.4 illustrates the potential for resistance from all participants in the ecosystem, especially for new technologies.

Figure 9.4. Points of resistance — participants at each point in the cycle can and often do resist change, especially change that results in shifts of power.

9.3 Current State of Investment in Nanotechnology

- Nanotech funding reached $18.2 billion in 2008, as government spending ballooned to $8.4 billion, corporate funding edged to $8.6 billion, and VCs provided $1.2 billion.
- International nanotech activity is still paced by the United States, Japan, Germany, and South Korea, but Russia and China made significant gains.
- Nanotech activity in the energy and environment sector is hot on many metrics; it accounts for 29% of all nanotech government spending in 2008, 13% of corporate spending, and 41% of venture capital. However, it barely makes a dent in total emerging nanotech revenue, amounting to just 0.6% or $876 million of the total in 2007 and 2% or $57 billion of the total in 2015.
- Nanointermediates captured 30% of the $2.5 billion that national governments invested in nanotech for renewable energy last year. Likewise, they represented nearly half of the $1.2 billion pie in nanotech VC funding in 2008. The

biggest beneficiaries are developing nanointermediates for energy storage and solar cells, but plenty of growth remains in emerging applications.

Investment in nanotechnology-focused venture capital in Europe has historically been a small fraction of that of the United States. An annual average is of the order of €20–40 million, and this level appears to have been maintained in 2008 with four deals alone accounting for investments of €24 million. The first quarter 2009 has already seen a substantial investment, with Oxford Nanopore (UK) receiving approximately €16.3 million from a strategic industrial partner and an undisclosed private investor.

Increased investment will occur when the perceived likelihood of investment success increases. That means that more companies will need to have public offerings of their stock, followed by several years of enthusiastic institutional investment.

With that in mind, two successful exits in 2008/09; Nanoco and OptoGaN, provide some encouragement. (*Venture Capital Funding of Nanotechnology*, a study by ObservatoryNano). Over the middle part of the decade, global venture capital investment in nanotechnology was averaging around €900 million per year. However, this investment was heavily focused on the United States. Whilst Europe performed better than the rest of the world, it received just €34 million in 2006, representing 3.6% of the global total. This proportion of the total was reasonably consistent over the period 2002–2006. Nine European deals were seen in 2007, down from 16 in 2006. The total value of European VC investment also decreased from 2006, to around USD 40 million. The largest deals in Europe in 2007 included Elmarco (CZ, USD 14 million), Heliatek (DE, USD 8.7 million), and Beneq (USD 4.1 million).

Lux Research developed the Lux Innovation Grid, a framework for analyzing company performance and evaluating the start-up space overall — both in nanomaterials and other emerging technology areas. Lux also noted that nanotechnology investments in healthcare had proven to be the most successful, having account for a majority of IPO value — USD 1.68 billion, with the total of nanotechnology IPO values at USD 2.57 billion. Three US companies; Nanosolar, A123 Systems, and Neophotonics had collectively received around three quarters of a billion dollars in investment. Nanosolar has a case for being one of the best funded companies in any sector, let alone nanotechnology. Total investment received by the company is USD 500 million, including around USD 300 million in August 2008. Partners for this funding round included strategic investments from energy companies AES Corporation and EDF, in addition to the Carlyle Group and a number of other investors.

The year 2009 brought news of the first nanotechnology IPO in Europe for some time. UK-based Nanoco Technologies was able to list on the Alternative Investment Market via a reserve takeover of Evolutec Group. Evolutec Group was originally a bioscience firm which had planned to develop novel therapeutic products. However, following results from clinical trial, the company decided to suspend its activities. The largest shareholder in Evolutec then established the following strategy; it would maintain its listing, with the intention of purchasing a company and reversing it into the Evolutec.

Nanoco is a bulk manufacturer of a range of functionalized quantum dots, which have applications in lighting, displays, solar power, and biological imaging. The company has a current market capitalization of around GBP 173 million on revenues of $1,600,000 and a loss of $241,000 for the six months ended January 31, 2010.

Compare Nanoco with InVisage, a Menlo Park-based company that has developed a proprietary product, QuantumFilm^{TM}, based on quantum dot technology, for application in digital cameras. InVisage is backed by three prominent venture capital firms who recently invested $30 million in the company. Nanoco was formed several years earlier, and did garner some investment, but has focused more on the research and biomedical diagnostics markets. InVisage is focused on a "killer app" for its technology that the company claims could revolutionize the quality and cost of digital photography.

Finland has also seen a nanotechnology exit in 2008. In December, high-brightness LED developer OptoGaN was acquired by Onexim, a private investment fund. The acquisition created a joint venture, OptoGaN Group, which is planned to mass produce solid-state lighting based on GaN chips. The new joint venture also involved the Ural Optical and Mechanical Plant (UOMP) and RUSNANO, the Russian Corporation of Nanotechnologies. GaN lighting components will be produced is St. Petersburg, and then incorporated into lighting products at a plant adjacent to the UOMP facility in Yekaterinburg. OptoGaN had previously received investment from VNT Management and Finnish Industrial Investment.

Clearly, a lot of investment in nanotechnology has been made around the world and a lot of progress in commercialization is evident. However, the nature of the technology, which results mostly in new materials, lends itself to development by large corporations, rather than start-ups and venture capital. Large corporations can bring all the resources of their supply chains to form the needed infrastructure. Most importantly, they can position new commercial products as "sustaining innovations" that can help them maintain their market share and existing customer base. Most nanotechnology innovations are not likely to be "disruptive innovations" as defined by Clayton Christensen.

However, a small number of adventurous venture capital firms have been investing in the area for the last 10–15 years. The following sections show how venture capitalists have been thinking about this "brave new world."

9.4 The Venture Biosphere

Any approach to solving environmental problems requires global awareness and international cooperation. Acknowledgment of this reality has come only recently, perhaps in the last 10 years. How well is it being accepted? How are practical solutions being implemented?

The same kind of thinking actually applies in many industries, from high-technology products, including nanotech, to low-tech service offerings. In this global economy, how do companies sustain competitive advantages? For start-ups, the situation is even more challenging. With limited financial and human resources, how can early-stage companies hope to survive and develop a presence in international markets?

Much hope has been placed on what we can learn from the example of start-ups in Silicon Valley, specifically those that are fortunate enough to have secured venture capital financing. What are the factors in Silicon Valley that help and support the rapid growth of start-ups? Governments all over the world have spent hundreds of millions of dollars, building fancy industrial parks often labeled by local media as "The Silicon Valley of XYZ." In Taiwan, there is the Hsinchu Science-Based Industrial Park, formed in the late 1970s. This is one of the earliest and most successful of such industry infrastructure projects. In China, "The Silicon Valley of China" often refers to the Zhongguancun area of Beijing, near Peking and Tsinghua Universities, the top two academic institutions. Singapore has truly world-class Science Parks, so does Malaysia. Of course, there are similar facilities in Korea and Japan, as well as in Europe.

What do most people think of as the most critical aspect of the Silicon Valley model? Usually, the focus is on the availability of venture capital. A handful of venture capitalists have celebrity status, and vice versa. Al Gore, former Vice President of the United States, is now a partner at Kleiner Perkins Caufield & Byers, one of the top investment firms in the world.

Without money, a company often cannot attract top talent, the brilliant minds that will create the new technology products and services. There can be no doubt that financial resources are required. Money is like fuel to drive growth. However, what is less obvious is the need for a whole infrastructure of financial services and products that may only be indirectly influential for any particular company. For nanotech companies, this environment is even more complex. I call this the *Venture Biosphere.*

9.4.1 *What Is the Venture Biosphere?*

A biosphere reflects the interactions, the biorhythms of its members, the evolution of individual human behavior, and the evolution of social behavior. This involves a dynamic balance of creative destruction — of births and deaths. For individual members, survival is everything — and very uncertain. The natural biosphere/ecosystem reflects *random,* that is, *stochastic,* interactions among its members. The *Venture Biosphere* reflects interactions driven by human needs and desires in the context of economic development. Intentionality on the part of individual humans and human society is the key difference between the Venture Biosphere and the natural one. The Venture Biosphere is driven by *psychology* as much as economics.

The "Venture Biosphere" concept is offered to supplement the term "ecosystem" which seems to suggest a homeostasis, a stable condition, whereas the "Biosphere" concept implicitly includes voracious appetites and an intense struggle for survival.

The usual term, "ecosystem," usually consists of those companies that have a *direct* influence on a particular company. Fifty years ago, management theory and practice emphasized the importance of "supply chain" thinking. A manufacturer/marketer of a television set needed to be very aware of the status of its suppliers of all the components that go into the production of a television set (upstream), as well as the downstream partners that enable distribution, marketing, and sales of the products. The assumptions underlying their financial planning would not have included issues such as the cost of petroleum, which was relatively stable then, or the cost of recycling toxic heavy metals, for example. Life was a lot simpler, because we didn't know there were other issues that needed to be addressed.

Relationships were more stable in those days too, as supply contracts tended to be longer. In the early days of air travel by jet, before faxes, communication took much longer. Change occurred at a much slower pace. There weren't the same drivers of change, the same urgency of the 24/7 workday many companies face today. China was closed to the outside world. India was very backward, still reminiscent of the British colony that it had been for a hundred years.

Between 1960 and 1980, dramatic improvements in telecommunications, travel, and increasingly rapid growth of new economies, that is, in Japan and Korea, led to a major shift in thinking. If supply chains are seen as one-dimensional, then adding more complexity in terms of more players in each category can be seen as more two-dimensional. A company had to think of more alternative suppliers, for example. Competitors, present and future, became increasingly important concerns, especially from foreign sources. The 1960s and 1970s were truly a golden time for the United States and Western Europe. The first big shock came with the petroleum crisis in 1977.

The Silicon Valley model for entrepreneurship and venture capital started in the 1970s. Private investment has probably been around since the invention of money. What was new was the invention of new financing methodologies for businesses, the end points of which are the purchase of illiquid stock (that cannot be traded in a public market) in start-up companies and the hope of hugely successful initial public offerings several years later, at which time the investors could receive impressive returns on investment. In the intervening years, these venture investors applied their own experience and personal relationships to guide and facilitate the rapid growth of their portfolio companies. By definition, they themselves embodied the "ecosystem" concept.

Fast-rewind to 10 years ago, to around year 2000. The venture capital industry itself had matured and evolved as an institution. While it had become a major force in developing major technology businesses, like Intel, Cisco, Google, etc., some industry experts would even say that some of the original advantages have been lost because the industry as a whole had turned inward more than outward. The "Internet Bubble" of 2000–2002 showed that sound business principles have evolved for good reasons. Almost everyone would agree that too many companies were funded for poor or inappropriate reasons. The adage "You can fool some of the people some of the time, you can't fool all of the people all of the time" is a fundamental truth about human nature.

Because the investment bankers saw the public's fascination with Internet companies, they literally created "new products," new investment opportunities to soak up capital from naive investors. There were wishful dreams, like Webvan, which offered home delivery of groceries, ordered by customers from their personal computers at home. Hundreds of millions of dollars were raised for this venture, including money from very well-established venture firms. Many of these companies were not generating revenue, and very few even had business strategies that showed the possibility of generating profits. Again, wishful thinking prevailed, and most plans depended on "advertising," the assumption being that advertisers would "pay for click-throughs." Of course, this idealistic metric also resulted in very creative ways to generate totally misleading data on the number of real customers viewing those websites. Hackers of a different stripe appeared.

What made the bubble possible? Fascination, the temporary suspension of disbelief, can be very powerful. What kind of story can induce this phenomenon? Human nature has a number of perversities that often do not serve us well. One is that we are really attracted to things that appeal to our emotional natures, the right hemisphere of our brain, and, at the same time, challenge our intellectual capabilities, the left hemisphere of our brain, but only a little bit. The best video game developers understand this very well. How else to explain how Zynga, the San Francisco developer of games like Farmville, reports annual revenue in 2009 of almost $500 million from customers buying "virtual goods," for themselves and for their friends in their social network! Real money in exchange of virtual goods. Truly amazing!

How does this insight apply to nanotech investing? One of the bigger challenges is for lay people to understand what nanotechnology is really about, as technology, as products, as an industry. The dream, the future vision of an amazing product or service, can be appealing, for example, a micro-machine so tiny that it can be swallowed by a human so it can repair some damage at an internal organ — like in the movie *Fantastic Voyage*, which was released in 1966. At the same time, something like that is such a stretch of the imagination that the rational mind knows this will not become reality any time soon.

Most of nanotechnology is about materials that are new specifically because atoms and molecules have been organized in novel ways through clever human manipulation. Properties of these materials are surprising precisely because their behavior is NOT intuitively understandable, and not predictable by the rules we are familiar with. So, many people don't know how to think about them. The balance between credibility and fascination is delicate and not always under the entrepreneur's control. The imagination takes off first, but the rational mind literally provides the "reality check." When people don't understand *how* the technology will be used, how it is manufactured (in volume), how it will interact with conventional components, how waste material will be disposed, then the questions start. Too many questions in the investor's mind

— and the entrepreneur will lose his audience. Say hello to spending a tremendous amount of time and energy educating and say goodbye to the hope of quick funding.

9.4.2 What Makes the Silicon Valley Ecosystem Work So Well?

Acknowledging and celebrating the value of "learning from failure" is one of the secrets of Silicon Valley. What is this process and how did it evolve?

Over the last 40 years, an amazingly effective and efficient set of relationships has developed as an infrastructure to support the process of getting technology companies to grow very big as quickly as possible. The technology driver was literally the integrated circuit, or silicon chip — not a native son of California, but quickly adopted as a favorite son. This set off the vision of a new wave of industrialization, to be realized through massive increases in computing power. The results we all know. The chip, as an enabling technology, attracted people with diverse talents and interests because they all shared the vision of automating knowledge, information, and democratizing this capability. Everyone's imagination was fired up. They were passionately driven to discover new uses for the chips. These were visionaries, prepared to accept the lonely path of being innovators and entrepreneurs.

The silicon chip is itself a metaphor. To be useful in a product, an integrated circuit has to interact with many other kinds of electronic components as well as mechanical parts. These pioneers understood the value and necessity of working together. Since no one could possibly know what the future might be, they were comrades exploring what could be created. *This attitude of collaborative exploration is key.* The scientific method shows us that, during the course of many experiments, we can learn as much, if not more, from unexpected results, some of which are labeled, at the time, as "failures" because those outcomes were not the ones that had been expected or hoped for.

The next key to successful innovation is being open to new possibilities. Just because we don't achieve the outcomes we were seeking doesn't necessarily mean that what is found doesn't have value. Also, value is very much in the eyes of the beholder. One man's failure is another man's treasure. The diversity of interests and personalities that are drawn to California means that a lot of different perspectives are constantly interacting and diligently studying what is evolving and why. This is how the community as a whole learns and benefits from the efforts of every individual.

Then, there was the money. The first investors were wealthy individuals who had made their fortunes in traditional industries. Fascinated with the new technologies, and persuaded by the visions, these people understood how to put together the resources to make successful businesses. Legends were born out of blood, sweat, and tears – of frustration as much as joy. This was also the beginning of "professional" venture capital. Success of this new industry was founded as much on the model of "value-added investors," investors who could bring a wealth of resources and experience to support the most promising of entrepreneurs who dared to dream big, as well as on the sophisticated capital markets, that is, Wall Street.

This network has expanded to become an ecosystem that has some unique characteristics, reflecting the powerful, visionary personalities that populate the community. Very importantly, the social structure of Silicon Valley is a meritocracy and relatively open. The barrier to entry is very high, as people are judged constantly, and only the best are invited to participate in the most elite social and professional circles. The group has high regard for demonstrated intelligence and ability. Coming from Stanford or Harvard helps, but the diversity of cultural backgrounds has been a strong positive factor as well. These people know how to talk to each other; they share common experiences, hopes, and dreams. Importantly, they are open to trusting each other, based on objective data, which, in America, is more readily available and reliable than in any other country. They share the feeling of being driven — by ego, by desire for money, or simply passion for what they are doing — for fun. They are willing to sacrifice almost anything, including personal relationships, like marriage, to achieve their goals.

Trust is the second critical aspect of the success of Silicon Valley and one of the hardest to replicate in another location, even in the United States. Where does it come from? The culture puts a premium on individual accountability, responsibility, and transparency because the goal is optimizing achievement, doing great things — understanding and accepting that many will fail so that a few can succeed. All this happens naturally, without planning or conscious intent. It's a matter of mutual respect, a fundamental aspect of the energy that makes relationships work collaboratively — and unbelievably competitive attitudes.

In the last 10 years, following the Internet bubble, a lot of the mystique has gone. Nasty public lawsuits have literally stripped away some of the secret dealings. With tremendous success has come greed, pressure from investors, occasionally improper and illegal behavior on the part of unscrupulous individuals, intense focus from media, and institutionalization. In investment firms, the relatively small number of very experienced senior technologists and executives who are founders has been supplemented with scores of very bright, ambitious young people who come with impressive backgrounds and real expertise, but with little exposure to the practical realities of how industries really work.

Ironically, this approach is inline with the mantra of venture capital, which is "Focus, focus, focus." However, the sharpening of focus has also resulted in a shift of vision, to narrower perspectives of opportunity. The work has become more about how investment money is most effectively deployed and less about how new industry breakthroughs can be engineered, with a lot of help from friends …and a little money.

This trend reflects what has been happening in other industries as well. The power of the financial markets and the demands of stock traders who can move billions of dollars essentially instantaneously have distracted everyone from the fundamental realities and challenges of running the underlying businesses. CEOs often worry more about how their company's stock price is moving than with developing new products and selling them successfully. Of course, the structure of their personal compensation, which is tied to the performance of the stock, is a major factor in their strategic thinking.

All of these issues have been challenging in the last decade. After 9/11, the United States has assumed a defensive posture. People are fearful about an uncertain future, and many businesses have lost some of their competitive edge. At the same time, the

Figure 9.5. Global intimacy → Uncertainty.

challenges from rapidly growing economies like China and India make everyone nervous. There are many valid reasons for concern: foreign exchange, trade imbalances, wars, petroleum and other natural resource pricing, domestic budgets, and local and global political changes. The world is a lot more complicated than it appeared to be 50 years ago.

"Uncertainty — in the economy, society, politics — has become so great as to render futile, if not counterproductive, the kind of planning most companies still practice: forecasting based on probabilities." Peter Drucker, 1992 (*Wall Street Journal*)

How can we prepare ourselves and our companies to deal with these challenges and the uncertainty they bring? This question is depicted graphically in Figure 9.5.

When Chinese companies project growth at 100% or more per year, and their budgets can be off by 50% in 6 months, what does the term "strategic planning" even mean? How can we plan for uncertainty? The accelerating rate of global change, driven by large players, that is, China and India, is compelling the need for innovative ways to plan for the increasing levels of uncertainty. In this context of "global intimacy," how is "value" defined?

For nanotech companies, this particular aspect of planning is especially challenging. Because of the especially long time frames required to go from discovery, to product development, to manufacturing technologies, to marketing, and sales of products, lack of predictability and stability means that growth paths can be treacherous. Nanotech projects require a lot of capital, an extensive supporting infrastructure, experienced managers, and they will all operate more smoothly when circumstances are stable.

Very few leaders will have the vision, courage, and the discipline to embrace this phase of strategic planning, which is even more challenging and will require both more deep and thorough thoughtfulness as well as broader scope of consideration. Even fewer will have the ability to communicate their insights effectively within their companies and throughout the global community.

How will this differentiate the best in leadership? For all corporate management, the key question for survival is, What constitutes "sustainable competitive advantage? This question, by the way, is uppermost in the minds of investors as they assess investment opportunities. The investment process is all about balancing "present value" and "potential future value."

The more we know, the more we know we don't know. Traditional adjectives that define "value," such as "rare," "non-imitable," and "non-substitutable" may not have the same impact as before, as in a more stable environment. How does a company identify and benefit from resources in a "globally intimate environment," many of which it may not even be aware?

What is needed? A fundamental re-positioning of resources, processes, and values. Professor Clayton Christensen, of the Harvard Business School, and author of "The Innovator's Dilemma" clearly describes how major corporations have failed to respond appropriately when faced with disruptive innovation. His decades of research have shown that the decisions made by the top executives at these corporations were sound, in the context of their existing business models. Those decisions were determined by the resources, processes and values of the corporation. A multi-billion corporation simply cannot apply the same operating principles to an innovation as it does to its mainstream business.

One approach that is used in dealing with environmental challenges may be applied to bigger problems, and other industries, especially nanotech businesses — *Integrated Systems Design.* In green-building construction, the focus is on dynamic interrelationships and clear communication, allowing for flexibility and adaptation, instead of rigid rules-based operations. A key concept is to get all the stakeholders to discuss together their objectives and the trade-offs necessary to achieve practical solutions – long before a final design concept or plan is presented as the goal. There is also an implicit understanding that flexibility and adaptability must rule as conditions change, be they government regulations, economic imperatives (prices of raw materials), even climate change. We live in an open system, and the designers have to be willing to communicate as they collaborate, and understand that the final product has to reflect objectives that are greater than the concerns of any one of the participants, including the financiers.

All of this reasoning speaks to applying the Venture Biosphere approach in the design of businesses, industries, and society. We have to deal with uncertainty by embracing change as a given, by planning in a very different way than we have done in the past. Where we used to set ambitious, lofty goals that we assumed, literally, we could achieve, we now have to be more realistic about what we can control and what is not directly manageable. As outcomes become less certain, our near-term goals can focus on what we can learn from each step, more from an attitude of exploration rather than from goal orientation. Each move will test our assumptions constantly, with the understanding that many of our initial assumptions, even those critical to our activities, might be perceived as incorrect or irrelevant as the project proceeds. We have to change our priorities, to broaden our perspectives, to achieve greater awareness and sensitivity to what is going on around us.

This is an information management challenge. How do we deal with what we don't know (but could help/hurt us)? This is a "design challenge," not just technical, not "business management = operational implementation." Business and political leaders have to learn how to behave in fundamentally new ways. In this context of global hyper-competitiveness, different skills will be required:

- Emphasis on interpersonal communication skills
- More qualitative and less quantitative data
- Greater diversity of vision
- More insight from looking at "boundaries" — peripheral vision
- More emphasis on fundamental human/society interaction issues

Perhaps this will mean that overall growth will be slower, but the benefits are likely to be greater mindfulness and more attention paid to all the stakeholders, to employees, to customers, to the public at large, and to the planet.

9.5 Fundamentals of The Venture Capital Process: Advice to Entrepreneurs

9.5.1 *What Does a VC Really Look for When Reviewing a Business Plan?*

Due to the sheer volume of opportunities presented to a VC, and his time constraints, the VC process is one of elimination. At least 50% of the deals are eliminated within 5 minutes of review. Another 25% can be eliminated within 1 hour of review. Of the remaining 25%, some amount of "due diligence" is performed, always with the goal of minimizing the investment of time and effort by looking for serious flaws in the plan. In practice, less than 5% of the opportunities will be evaluated with much depth. Very few VCs study a plan with the attitude of looking for a reason to invest. The plan must be "compelling," strong enough to temporarily "suspend disbelief," to set aside doubts and fears.

9.5.1.1 VC's first key issues (instant "decline to invest") – "weakest links in the story"

- If the business plan is a "snapshot of success," then what's missing from this picture?
- Where are logical inconsistencies and unresolved conflicts?
- Where does the inexperience of the team show — "weakest link is the people?"
- Where and how can doubt and fear arise?

Fear, uncertainty, and doubt — FUD, the three deadliest opponents you'll ever meet, within yourself and from external sources. Your job is to eliminate or at least reduce their influence as much as possible. At the end of the day, there is a leap of faith – *your* belief in your passion and your conviction will have to bring others to share your faith.

9.5.2 *Reasons for Failure*

There is only *one* fundamental reason for failure to achieve objectives — not enough effort was made. Period. That means that *everyone involved* shares in the "failure." The good news is that every experience is an extremely valuable opportunity to learn. Everyone can improve themselves and their ability to interact in positive ways with others.

There are usually many factors that contribute to this condition of "not enough effort being made":

- The chosen goals were not appropriate (not realistic or achievable)
- The activities were not organized in alignment with the goals (progress not measurable)
- Expectations were not communicated clearly (clarity of purpose)
- The necessary resources were not available (tools, money, people)
- External challenges were not addressed (preparation, flexibility)

Everyone, the boss, the employees, the company, contributes to this "failure," which is really a learning experience – but aren't all of life's experiences learning opportunities?

The philosophical position regarding "failure" is to consider every effort as a positive learning experience, with the desired outcome and true objective being to "learn by experimentation." This is what life is really all about.

Improvement does not come from "fault-finding," or intense analysis of the past. This places attention/energy in negative directions, in separating everyone from the value objectives. This increases the power of the limiting beliefs that created the situation in the first place. The process already carries an implicit assumption that there are "problems to be solved," which is itself a limiting belief.

9.5.2.1 Success mode

Improvement does come, however, from intense effort to focus on what is really important and on what is actually happening on a day-to-day basis ("focal points"). The real challenge is dealing with constant change and the dynamics of human interaction, that is, how you feel about the changes. This is a process geared toward creating solutions, as opposed to "problem-solving." When people focus on positive directions that align their energies with each other and with the corporate objectives, the result is infinitely superior. Creative energy is released and encouraged; interpersonal interaction is cooperative and collaborative, rather than competitive.

9.5.3 *Practical Considerations: What to Do about your Plan (Teamwork Is Required, with Leadership)*

Take the point of view of a complete stranger to your plan. As you read your plan, reflect on the following questions:

- How can you tell if the energy of the team has coalesced into a common goal (compelling vision statement, with passion, energy)?
- How can you tell if the vision is sufficiently developed that your plan communicates it clearly, easily, and simply? (Strong, clear executive summary)
- How many facts (and which ones) are enough to build a strong, logical argument that success is all but "inevitable"? (Consistent, accurate, and relevant facts [market data, cost estimates, etc.])?
- How can you tell if the business plan is sufficiently developed to be "reasonable" (inferences, deductions, financial projections)?
- What are the scenarios for partial success, changes of direction, or failure?
- What are risk reduction strategies (VCs are notoriously risk averse)?
- What will you do when your plan is successful?
- How do you build on your success (future products, partnerships, exit strategies)?
- What will the customers want next (full circle from where you began this process)?

9.6 How to Raise Money from Venture Capital Firms

There are numerous books written on this subject, as well as websites that are full of advice on this subject. If you have the time, it is certainly worthwhile to peruse some of these resources. Not all are equally useful, of course. One of the classic publications is

"Pratt's Guide to Venture Capital," which includes some chapters on preparing for the VC interview as well as an extensive list of venture firms, with some details on their areas of investment.

Briefly, the steps are as follows:

1. Find someone who knows you to introduce your company to VC firms (in order from least to most effective sources of introductions)

   - Paid consultants, brokers, "middlemen"
   - Friends and acquaintances
   - Company employees
   - Company advisors (legal, accounting, banking)
   - Directors, investors
   - Senior management

2. Prepare the appropriate materials

   - Business plan
   - PowerPoint presentation
   - Executive summary
   - E-mail/cover letter intro
   - References/testimonials

3. Network for initial contacts (informational interviews, not fund-raising)

   Do some research to learn what specific areas each VC is interested in. To whatever extent possible, choose to contact VCs who appear to be "value-added investors." The biggest initial hurdle is just getting an appointment, an opportunity to present your case.

4. When making in-person presentations, think *dialogue*, more than monologue

   - PowerPoint presentation/hard copies for attendees.
   - Detailed responses to anticipated questions (presentation slides, written materials).
   - Do some research to learn what specific companies each attendee is involved with.
   - Do ask questions of the VCs to elicit information they have that may be useful — market data, names of competing firms, possible strategic partners, etc.
   - Ask for next steps to take in the evaluation process.

5. Suggest follow-up meetings at your company, to meet other staff, see demos of future products, something that was not discussed in detail at the first presentation (often at the VC's office).

6. Follow-up with additional materials as requested or with additional information that you think enlarges on certain topics that were covered in the presentation. Keep calling periodically to check in on the VC's progress. At some point, the VC will either decline or stop accepting calls from you, which is one of the ways he will indicate lack of interest.

7. When using "middlemen," consultants, or brokers, understand that their highest value is usually realized when an appointment for a presentation is set. They can help with the presentation, but a VC typically prefers to deal directly with the principals, that is, company management. After all, those are the people he wants to know. They are the ones who will execute the plan. They had better be able to articulate that plan convincingly. VCs absolutely prefer a slightly rough, "homemade" business plan prepared by the entrepreneur himself, than a slick, "professional plan" prepared by a paid consultant. The VC is looking for "integrity" in the opportunity, that the entrepreneur who proposes a plan truly has the ability to execute it. An experienced VC can sense a "prepared plan" very quickly and many will be inclined to dismiss such proposals.

   PowerPoint or other presentations that are "professionally prepared" do tend to be clearer and more persuasive than "homemade" ones. Here, the "integrity issue" is less important, since the VC understands that this is a marketing piece, designed to attract attention and deliver a quick, persuasive message. Of course, management still has to deliver this message as convincingly as possible. If the VC senses that the entrepreneur is "acting," working hard to deliver a performance according to a script written by others, that VC is not likely to invest.

8. The order in which information is presented can be important, as well as the quantity at each step. First step – get the VC's attention with a crisp, simple statement of why your company's product and technology have unique value, that is, why you deserve consideration as an investment opportunity. Second

step — find out how the VC processes opportunities — that
is, is he more concerned with technological superiority, size
of market, competitors, management, etc.? Let his interests,
priorities, and concerns guide you — understand his viewpoint
and the way he wants to work. Understand his past failures as
well as successes.

Put another way, give the VC enough to chew on, but not so
much that he holds up his hands in an attempt to avoid being
inundated with too much information. If a VC is interested, he
will naturally ask for more information or more meetings.

9. If middlemen or brokers are involved, company management
   should insist on maintaining direct contacts and relationships
   with the VCs. Don't expect the middlemen or brokers to follow
   up, provide additional information, make further appointments,
   etc. You can, certainly, ask them to do so, but company
   management should dictate and guide the process. The direct
   relationship is extremely important. The VC is also evaluating
   company management by the way they manage the process of
   fund-raising, the level of professionalism, the commitment, the
   responsiveness, etc.

   Treat the VC contact as you would your most valuable customer.
   Your pitch is the most important sale of your company's life
   – literally. Get to know him, cultivate a friendly relationship.
   Understand his viewpoint, his prejudices, his preferences. Take
   a long-term view to the process. VCs are important people to
   know. Even if they decline this investment opportunity, there
   will always be opportunities in the future. Next time, you won't
   be a complete stranger!

10. One of the most valuable aspects of the process is understanding
    the real reasons behind a VC's lack of interest, or his interest.
    You can learn a great deal about how you need to improve —
    your strategies, your product/technology, your communication
    skills, or some other aspect of your company. The challenge is
    that few VCs are willing to invest the time to give you that kind
    of feedback. If the introduction to the VC was made through
    another person, perhaps that person can help get you that
    feedback.

11. Track the process on a spreadsheet and manage accordingly. The more organized your campaign, the better your chances of getting valuable results, with a minimum of time wasted and uncertainty.

12. Be as thoroughly prepared as possible and polish your communications and interpersonal skills. The adage: "You never get a second chance to make a first impression" is particularly applicable to interactions with venture capitalists. These are people who make their living, whose reputations depend on the perception of their ability to make good (profitable) judgments. The first impression, the first contact, is extremely important. Some of the key issues related to credibility are illustrated in Figure 9.6.

Secret: Many people really would like to believe in you and your plan. People like to be associated with a "winner." You won't be able

Figure 9.6. Credibility is key — what must be communicated — clarity [of focus, execution] — consistency — comprehensiveness — reasonableness (matches expectations/experience of others).

to execute your plan without their help and you won't get their help if your plan doesn't make sense to them. Ask for their help and suggestions at the earliest possible opportunity!

Often, what is most critical for success is what business schools typically don't teach you — passion — integrity — conviction (persuasiveness = compelling) — because these are qualities of human character that are difficult to measure and nearly impossible to teach. They can, however, and must be, learned.

The learning comes from a focus on creating the solution, not merely "problem-solving." The desired solution is creation of value and wealth, to be shared "fairly," not only by the principals, but also by the public at large, including customers, partners, and investors. All your energy and attention must be directed toward this win-win-win goal, which is definitely achievable. True entrepreneurs are driven by this creative urge. They don't have the time and the luxury of entertaining and cultivating negative orientations and beliefs. Anyone who feels a strong "fear of failure" is burdening himself with a tremendous handicap, one that drains his energy and creativity. You will need as much positive energy as you can maintain, for yourself, your own health, as well as that of the company.

Chapter 10

Technology Transfer and Nanomedicine with Special Reference to Sweden

Claes Post

Business Development, LiU Innovation, and Department of Clinical and Experimental Medicine, Linköping University, Linköping, Sweden

10.1 Introduction

Excellent academic research has always been the base for innovation, patenting, entrepreneurship, and the establishment of new enterprises. This was the case for many companies around the world, like the Swedish pharmaceutical companies Astra and Pharmacia that were started during the early 1900s. Technology transfer during those years was performed mostly through direct research collaborations between academic scientists and companies, such as the Pharmacia collaboration between Professor Nanna Svartz and Salazopyrin as well as the collaboration with the Nobel Laureate Arne Tiselius, who was instrumental for various products based on dextran. The Stockholm University professor Nils Löfgren's novel local anesthetic substance lidocaine was key to establishing Astra as the leading company in the field of local anesthesia.

Nanotechnology Commercialization for Managers and Scientists
Edited by Wim Helwegen and Luca Escoffier
Copyright © 2012 Pan Stanford Publishing Pte. Ltd.
ISBN 978-981-4316-22-4 (Hardcover), 978-981-4364-38-6 (eBook)
www.panstanford.com

Today, technology transfer offices (TTOs) have been established at universities to handle intellectual property in the academic setting. This is performed by activities to scout, identify, and develop academic projects with commercial potential. The legal framework varies from country to country. In most countries, intellectual property rights (IPRs) are handled by TTOs at universities or similar institutions, but Sweden is an exception. Swedish scientists own their inventions and can freely choose how to handle IPRs related to their academic research.

To handle project opportunities from outside, virtually all companies have established licensing functions dedicated to in-licensing and acquisitions of projects or companies with complementary assets, as well as out-licensing. In particular in the life sciences industry, companies are actively pruning their products and project portfolios. Out-licensing of projects as well as spinning out start-up companies has therefore increased. Central to the licencing activities is that technology transfer is a people's business where formal and informal relations are built between the TTO function at the university and the licensing department at the company. Once the discussions advance beyond the initial non-confidential discussions, informalities are exchanged for legal frameworks with various contracts signed by the parties. In the cases where the initially shared non-confidential information has generated a mutual interest, the continued relation is framed by contracts that may lead up to a business agreement. Even though the process of technology transfer is formally carried out by legal entities, it is in the end run by individuals, leading to formal and informal contacts between scientific, legal, and general business development from both sides of the negotiation table.

In this chapter I will discuss the specific situation in Sweden in contrast to that in other countries. Sweden has a system of teachers' exemption, also called professors' privilege, where university-based scientists have the full ownership to their university-derived inventions. This runs counter to the situation of virtually all other countries, where universities own the IPR generated by their staff. The tech transfer process will be illustrated by a few case studies where nano-derived innovations in the nanomedicine field have been commercialized by either an external large company or a novel company created by scientists.

Nanotechnology is a broad term, which covers a wide range of methods, tools, and possible applications. There are a variety of definitions reported in literature, each generated for different purposes. For the purpose of this document, the definitions are based on those provided in the UK Royal Society and Royal Academy of Engineering report (*Nanoscience and Nanotechnologies: Opportunities and Uncertainties*, The Royal Society & The Royal Academy of Engineering, July 2004):

> Nanotechnology is defined as the production and application of structures, devices and systems by controlling the shape and size of materials at nanometre scale. The nanometre scale ranges from the atomic level at around 0.2 nm (2 Å) up to around 100 nm.

This chapter will for the major part be profiled towards nanomedicine, which the European Medicines Agency (EMA) defines as follows:

> Nanomedicine is defined as the application of nanotechnology in view of making a medical diagnosis or treating or preventing diseases. It exploits the improved and often novel physical, chemical and biological properties of materials at nanometre scale.
>
> (Reflection Paper On Nanotechnology-Based Medicinal Products for Human Use, EMA, June 2006)

Medicinal products are developed according to certain regulatory guidelines set at the International Conference of Harmonization, or ICH (ICH Steering Committee, 2005). These have been adopted in most countries (www.ich.org) and provide a framework that also guides medicinal products containing nanotechnology-based components. Patient safety is the prime focus of the guidelines, both among healthy volunteers and patients being enrolled in clinical trials and the patients to be treated once the product is approved.

10.2 Technology Transfer

10.2.1 *Definitions*

Although technology transfer can be difficult to define, some definitions are given below:

> Technology transfer is the process of transferring scientific findings from one organization to another for the purpose of further development and commercialization. The process typically includes
>
> - Identifying new technologies
> - Protecting technologies through patents and copyrights
> - Forming development and commercialization strategies such as marketing and licensing to existing private sector companies or creating new start-up companies based on the technology
>
> http://www.autm.net/Tech_Transfer.htm

> Exchange or sharing of knowledge, skills, processes, or technologies across different organizations.
>
> www.nsf.gov/statistics/seind06/c4/c4g.htm

> Technology transfer is understood to be the passing of theoretical and practical skills and know-how from the owner of a technology to outside users or beneficiaries of technology. Technology transfer is in comparison to technology cooperation an isolated and time-limited transaction. Technology transfer is not simply about the supply and shipment of hardware across international borders. It is about the complex process of sharing knowledge and adapting technology to meet local conditions. It strengthens human and technological capacity in developing countries.
>
> www.lineadecreditoambiental.org/html/glossary.html

> Assignment of technological intellectual property, developed and generated in one place, to another through legal means such as technology licensing or franchising, processes of converting scientific and technological advances into marketable goods or services.
>
> http://www.businessdictionary.com

10.2.2 TTO Networks and Guidelines

TTO structures have been implemented at most universities and other academic institutions. To develop this profession, which in many places is in its infancy, professional organizations have been established. The AUTM (Association of University Technology Managers; www.autm.net) in the United States is a forerunner in structuring processes and principles for technology transfer. National organizations have been launched in many other countries, for example:

- Association of European Science & Technology Transfer (www.astp.net)
- The Global Tech Transfer Summit Initiative (www. techtransfersummit.com)
- University Technology Transfer Association, Japan (http:// unitt.jp)
- Shanghai International Technology Transfer Network (www.sittnet.cn)
- Swedish Network for Innovation & Technology Transfer Support (SNITTS) (http://snitts.se)

The TTO conditions vary considerably from country to country. It is therefore unlikely that global guidelines for TTO procedures will or can be developed. Even so, in most countries TTOs have implemented similar processes and types of contractual agreements. Technology transfer from the individual inventor, usually a scientist at a university, to a commercial application involves complex structures, legal frameworks, and most of all, interactions with involved individuals. Trust and ethical standards are the keyword for inter-individual interactions, in particular when money is involved, as is the case when developing an invention into a product. To guide the US TTO organizations, AUTM has established guidelines to professionalize TTOs in that direction. The recommendations (http://www.autm.net/Nine_Points_to_Consider.htm) were established based on a meeting at Stanford University in 2006 between representatives of US universities. Even though this chapter is skewed towards the Swedish situation, nine points are generally applicable to technology transfer. The comments below each point are my summaries of implications.

1. **Universities should reserve the right to practice licensed inventions and to allow other non-profit and governmental organizations to do so**
   Comment: Researchers shall maintain the right to perform research, including publishing research data in dissertations and publications. Other researchers will have the right to verify such results without interfering with patent rights held by the licensee. At the onset of a license agreement it is important that the scope of reserved rights to the licensee is clearly defined.

2. **Exclusive licenses should be structured in a manner that encourages technology development and use**
   Comment: Exclusive licenses are often given when investments into their commercialization are significant and development takes a long time. At the onset of the license agreement, both parties have a common interest to develop the invention. This should best be described as critical milestones in the licensee's project planning, and how the fulfillment of the milestones is to be communicated to the licensor, in this case the researchers usually supported by the university TTO. Milestone achievements are usually connected to predefined payments to the licensor. In VC-funded companies, financial restrictions may lead to in-licensed projects' being put on low priority. This should also be taken into account in the contract.

3. **Strive to minimize the licensing of "future improvements"**
   Comment: The license should as much as possible be restricted to the existing patents. Conditions that also grant the rights of the licensee of future improvements of the original invention(s) can have significant negative consequences by locking in the university and its researcher for future research that may be interpreted as such improvements even though they should be defined as novel inventions. This is a difficult point in the negotiation since the company prefers as broad definitions as possible of the field covered by the agreement and the researchers and the university strives to maintain as much of the academic freedom as possible.

4. **Universities should anticipate and help to manage technology transfer related conflicts of interest**

   Comment: The TTO should be conscious about potential conflict of interest when technology is licensed out. In particular, problems can arise out of situations where senior and junior researchers together make a deal and where the parties may have different opinions about how to share the upside of such a deal. Problems may also arise in a situation where one of the researchers has a stake in a company that is considered a licensee, whereas others have other preferences. The TTO should professionally handle such situations by seeking the best and rationally sound solution. In Sweden, this is even more problematic since the university has no ownership stake in the IPR to be out-licensed.

5. **Ensure broad access to research tools**

   Comment: As far as possible, the license agreement should make clear that the license is for the commercialization of the invention and not for research purposes. If the commercialization is directed towards academic research as market, this may not apply.

6. **Enforcement action should be carefully considered**

   Comment: The universities are not geared to be involved in complex legal conflicts. Therefore the best alternative is that the licensee, in particular when the license is exclusive, takes the actions.

7. **Be mindful of export regulations**

8. **Be mindful of the implications of working with patent aggregators**

   Comment: When the patent aggregator is, for instance, a venture capital fund, patent aggregation may be an excellent way to create novel business opportunities, something that the university can support as institution. There are less attractive aggregators, "patent trolls," that lump as many patents together as possible to identify patent infringements as a means to generate litigation-based incomes.

9. **Consider including provisions that address unmet needs, such as those of neglected patient populations or geographic areas, giving particular attention to improved**

therapeutics, diagnostics, and agricultural technologies for the developing world

10.2.3 *Deal Sourcing*

Deal sourcing is key to the success of a TTO. Even though the number tells us nothing about opportunities for success, it is apparent that the higher the number of invention disclosures from high-quality academic institutes, the higher the likelihood of success. In the United States, the total number of invention disclosures has been estimated by AUTM (Fig. 10.1) with a constant annual increase of the numbers from 1991 to 2008. There is a concomitant increase in the United States in the number of start-up companies formed during the same period (Fig. 10.2).

Whether these increases have a causal connection is speculative, but quite likely. My estimate is that in Sweden, the total number of incoming invention disclosures from researchers to the respective university TTO is on the order of 1000 per year. Given the specific situation in Sweden with the teachers exempt, meaning that the university researcher has no obligation to inform the TTO about other paths of commercialization, the added number of inventions being processed by other organizations is very uncertain.

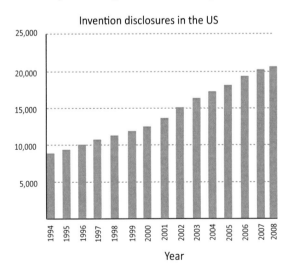

Figure 10.1.   Data from AUTM 2008 Licensing Activity Survey.

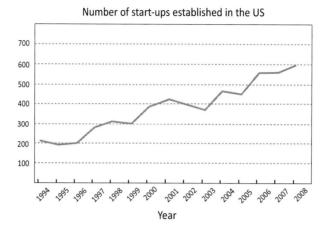

Figure 10.2. Data from AUTM 2008 Licensing Activity Survey.

10.2.4 *Swedish Context*

In Sweden, as in most other countries, the employer owns inventions derived from the work of an employee (Swedish Law, "Law on Employee's Rights to Inventions"). As mentioned above, teachers at schools and universities have a personal ownership of such inventions, the so-called teachers exempt.

TTO structures in Sweden have gradually been established at almost all universities. In 2009, the parliament adopted a governmental proposition to strengthen university-based research and innovation (Swedish Governmental Proposition, "Enhancing Research and Innovation"). Besides the main focus to strengthen certain strategic areas of research, universities were financed to strengthen the innovation support by setting up so-called innovation offices. Several universities have now, one year later, established this. At Linköping University, for instance, the Innovation Office provides support to the scientists at the university to identify, validate, and develop their projects towards commercialization. The Innovation Office will also support the scientist to create a company for the further development of the project. The competence mix at the Innovation office is intended to be supportive of the various aspects of this several-year process towards commercialization.

Technology transfer is a people's business, where projects with potential are identified by the personal interactions between the scientists and the tech transfer person. There is no shortcut for this, especially not in Sweden, where university scientists have the full ownership of their inventions.

One choice that is often made by the researcher is to publish the results, thereby forming prior art should a patent application be filed later. This is fine, as long as the decision to publish or perish for academic reasons is taken deliberately. Unfortunately it is very common that the publishing of potentially patentable data simply was a mistake that is discovered when analyzing possible prior art before filing a patent application. In countries where the university has full ownership rights to inventions from their staff, this is still a problem since sanctions usually are not used towards a researcher publishing despite potential IPR opportunities. In Sweden another reason to publish rather than taking the commercial path is that the individual not only owns the inventions personally but is also responsible for the costs required to develop the invention into a salable product. The product in this case may be anything from a filed patent application to an almost market-ready product. In other countries, the development cost should be carried by the university, but the budget for technology transfer is usually so small that most projects will not receive the optimal development budget from the TTO. Therefore, early licensing is usually considered the optimal way to exit projects.

Usually a deal is made whereby future profit is shared between the university, the department, and the inventors. If scientists choose to publish and thereby undermine IPR opportunities, the university rarely imposes sanctions. In the Swedish system, inventors can seek support from the TTO, still maintaining full ownership of the invention. New initiatives are emerging, however, where holding companies make deals with the inventors by agreeing that the holding company will support the commercialization, in return gaining an ownership stake in the company to be formed.

The TTO works in the intersection between research, education, and entrepreneurship. The latter is expected to generate university-derived spin-out companies or other businesses such as license deals. In the Swedish context with the aforementioned teachers' exemption, the university-based TTO will not participate as co-founder of companies. Instead, this role can be filled by the university holding companies that can play the role of co-founder of a company together with the founding scientists. Whether the teachers' exemption is good or bad for the country's development of innovation-based enterprises has been debated for many years. Neither the current government nor the opposition has indicated this as an issue, and we have adapted our operations accordingly. If the teachers' exemption is to be eliminated, such a move has to be well prepared by building structures, involving competent people, and accessing the required capital well in advance of such a change. During the last four years, more than 3000 papers have been published annually with at least one scientist from Linköping University. Assuming that 10% of those publications may have a potential for commercialization, this would imply that several manuscripts per day would need to be assessed for commercial potential. To make this possible and to avoid long handling times at the TTO, the resources would have to be considerably strengthened. Since this takes a considerable amount of time and money, it is not likely that the exempt legislation will be changed for several years to come.

In a Stanford Institute for Economic Policy research discussion paper, Paul A. David and J. Stanley Metcalfe raise concerns on how the public-partnership relationship between universities and the commercial sector may have negative effects on how the university sector could become impeded in fulfilling its other important functions (David and Metcalfe, 2008). In particular, they argue that the blind application of the Bayh-Dole Act in Europe will not and cannot be a successful way forward. Not even the Silicon Valley model could be successfully set up in the San Francisco area today. Silicon Valley was formed in a situation that was unique for San Francisco and that also was unique for its time.

The moves in Europe to overcome barriers between publicly funded academic research and commercially funded businesses can result in negative consequences by impeding some of the university's important functions and roles in society. The Swedish system, seemingly, is thus an exception in Europe. But Sweden is not an island; on the contrary, Swedish research institutions interact heavily with other countries in Europe within the EU research funding as partners in framework program funded research, where small and medium enterprises participate for the commercialization of the resulting research activities. Ownership issues do, therefore, have to be agreed upon between the parties in consortium grants, and Swedish scientists may not have the same ownership role as has previously been the case.

10.2.4.1 Tech transfer processes: a linköping University case study

Active scouting is crucial in the Swedish TTO system. There is no given formula on how to achieve high yield from the scouting. The base for the scouting is to set up a structured meeting schedule at various levels in the different faculties.

The university is operating at the intersection between education, research, and entrepreneurship. In Linköping, the TTO activities are organized within the Innovation Office (see Fig. 10.3), which was established recently because the state strengthened the financing of the Swedish universities to improve the TTO capabilities.

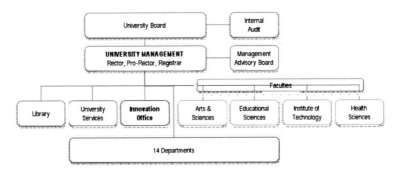

Figure 10.3. Organizational structure of Linköping University.

a. Infrastructure

At Linköping University we have an institute of technology, a faculty of art and sciences, a faculty of health sciences, and a faculty of educational sciences. At the various parts of the university, all types of communications means are used to penetrate as many sectors as possible. In addition, the university hospital is a research-based hospital, where inventions are also made. The legal system at the hospital is different from that at the universities, since those solely employed by the university hospital do not have the right to their inventions, whereas scientists with a double affiliation, both the university hospital and the university have the same position as those at the university, that is, full personal ownership of inventions. The forms of interaction vary from spontaneous meetings with a scientist at a coffee machine to lectures and seminars given at the group level to scientists, at a departmental meeting, or at larger gatherings at the faculty level. The homepage, university, as well as the university hospital are also used to spread the information about what the TTO may provide in terms of support.

During 2009, innovation and commercialization was discussed with some 600 researchers at the university and about 1500 students. The TTO structure and the Innovation Office also have an external interface with the local industry with the goal matching industries' needs with researchers' skills and problem-solving capacities. Approximately 1200 persons from industry, municipalities, and counties were exposed to the Innovation Office messages. From all these activities, 36 start-up companies were generated in 2009 and more than 40 collaborative projects with industry and society. The preliminary evaluation of 2010 indicates similar outcomes.

Even though the ambition is high, there are obstacles hindering the optimal deal-flow as well as the optimal outflow from the system:

- The TTO staff do not have sufficient time to meet as many researchers as they actually would like to
- Scientists have too little time to think about commercialization
- Scientists do not prioritize commercialization, because they have to bear the expenses themselves

- Discoveries, patentable or not, must be published or the researcher may perish because commercialization does not (yet) count as a career-supporting achievement

b. The Invention disclosure

The typical invention disclosure form (IDF), given below, is the starting point for discussions between the inventor and the TTO:

1. Give a short descriptive title to your discovery or invention.
2. Provide (in non-scientific terms, if possible) a short (4–5 paragraphs) description of the field/background of the invention.
3. Outline the essence of your invention and identify the public health need it fills or the current problems it solves (4–5 paragraphs).
4. In 4–5 paragraphs, speculate (be creative!) on all applications and possible commercial applications of your invention or discovery.
5. Who are the inventors (those who conceived the idea and put it into practice with their own creative input)? Describe in detail the specific contribution of all inventors.
6. Now identify all other colleagues who merit co-authorship credit for the associated publication, whether or not you believe them to be "co-inventors."
7. Is anyone outside the university aware of your invention or discovery? Identify such persons and describe dates and circumstances of how they found out.
8. Are you aware of any patent applications that are related to your invention or discovery?
9. List previous articles, presentations, or other public disclosures including all grant applications (Swedish Research Council, VINNOVA, Swedish Cancer Society, etc.) made by you or other researchers relating to your invention or discovery. Please attach copies (e-mail attachments, if possible) of papers, abstracts, etc.
10. Indicate any future dates on which you will publish articles or make any presentations related to your invention or discovery.

11. Is the subject matter of your invention related to any collaboration that involves your laboratory and other laboratories? If yes, identify the collaborator(s).
12. Is the subject matter based on research materials that you obtained from some other laboratory? If yes, attach any material transfer agreements (MTAs) under which you received the material(s).
13. What companies or academic research groups are conducting similar research? Identify companies that may have an interest in your idea.
14. What further research in your laboratory would be necessary for commercialization of your invention (estimate the time frame)?
15. What are your future research plans for the invention and/or for research in areas related to the invention?
16. Provide information about all inventors who contributed to the essence (conceived the idea) of this invention.
17. How did you come in contact with the TTO?

c. The way forward: patent and financial issues
Normally no patent application has been filed when the TTO first meets the inventor. In this case, we do a preliminary evaluation of patentability. This can be done internally in our organization, where we have patent expertise. Costs are usually carried by the TTO. In some cases we support the inventor to get access to grants from the Swedish state funding organization Innovationsbron, Swedish for "Innovation Bridge." These are early development grants up to a total of approximately €20.000. If the project is successful, we will further support the inventor to apply for a development grant from VINNOVA, with a total of up to approximately €200.000 towards the project, provided that it meets specified milestones. At this stage the project should be mature enough to be the basis for forming a company. The TTO supports the inventors in this, but takes no stake in the new company. If the university's holding company becomes involved, they may take up to a 10% ownership of the company on the condition that the holding company will commit time and management resources to it. Once a portfolio company has been taken into the holding company, it will have

access to various types of support. At most Swedish universities, holding companies have now been established. Karolinska Institutet in Stockholm has since several years established an innovation structure owned by Karolinska Institutet Holding (KIHAB), where KIHAB owns Karolinska Institutet Innovations AB (KIAB) and part of Karolinska Development AB (KD). KIAB scouts for projects with commercial potential, and KD has the exclusive first right to invest in spin-outs created by the researchers and KIAB. The system has broadened its scope beyond Karolinska Institutet and has invested in projects from other Swedish universities as well as from foreign institutions. A great advantage of the KIAB/KD/KIHAB system is that KD has managed to raise considerable funds to finance a broad portfolio of start-up companies, thereby spreading the risk of failure and increasing the chance of success among its portfolio of more than 40 start-ups. In a recent article by Jem Baker, the Karolinska innovation system is described in more detail (Baker, 2010).

The patent cost as outlined above is a deterrent to the founding scientists, since it normally is too large to fit in the private budget. Here's an example of the cost for a patent to be processed in 10 countries (the Swedish Patent and Trademark Office; www.prv.se):

1. Cost for a professional patent agent to compile patentability evaluation: €1000
2. Search from the Swedish Patent and Registration Office: €1200
3. Cost to compile the Swedish priority application, to be followed by PCT application, dialogue with patent offices for the prosecution of the patent, plus yearly fees: €18,000–25,000
4. Other costs
5. Approximate total cost to maintain the patent in 10 countries for 11 years: €80,000–120,000

Patent costs increase considerably over time (Fig. 10.4), the normal window of opportunity to successfully make a technology transfer is between the initial patent application filing and when the patent enters national phase 30 months after the priority date. The costs can be weighed differently, for example, by filing a less costly application initially and expand it later on to a regular application. For the TTO as well as the inventor, it is crucial to use a professional

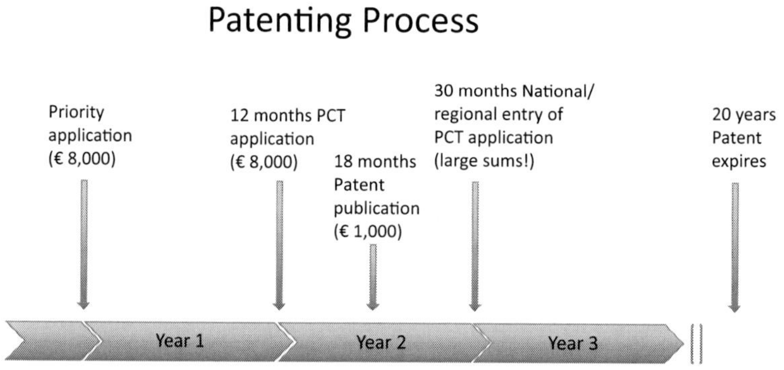

Patenting Process

Figure 10.4.   Costs and timeline of the patenting process.

patent agent in the process to avoid costly mistakes or even mistakes that can jeopardize the patentability. A common problem we as TTO encounter is the following:

- A scientist contacts the TTO to get support in evaluating a potential invention for patent opportunities. Written information on the invention is provided to the TTO by filling in an Invention Disclosure Form, IDF.
- Initial discussions on the potential invention conclude whether it's a yes (possible to patent/interesting potential) or no.
- During the deeper discussion to follow, the inventor mentions, "Oh yes, I gave a talk at this meeting." This happens all too often and usually puts an end to the discussion because the disclosure of inventions destroys the novelty that is required for obtaining a patent.
- Other problems relate to patentability, costs of patenting, prosecution of the patent, or simply lack of time and engagement from the inventor. We at the TTO may also be the bottleneck with limited resources to provide support to all researchers requesting our services.

d. Company spin-out

LiU Holding owns the local incubator LEAD. This will for the best cases become a new base for the further development of the inventor's company. In most cases the inventor leaves the active roles as manager at his or her company at this stage. LEAD is to my knowledge the only ISO-certified business incubator. Processes, education, and other quality-related aspects of the incubator are thus certified and the quality maintenance is assessed yearly. This gives the new and vulnerable company a better survival chance. Business incubators are now established at all Swedish universities, most of them publicly funded, even though privately held incubators are being opened.

10.3 Nanomedicine

The number of publications per year (PubMed), with "nanomedicine" as search term increases annually. Nanomedicine journals are now available, for instance, *Nanomedicine.*

Even though many articles in the literature not are indexed as "nanomedicine," it can be seen from Fig. 10.5 that the output increases considerably over the years. The figure for 2010 is up to mid 2010, and a significant increase over 2009 can be expected.

The proportion of publications that are nanotechnology related has increased in most countries, with the increase in the United States being relatively higher than in other countries (Fig. 10.6).

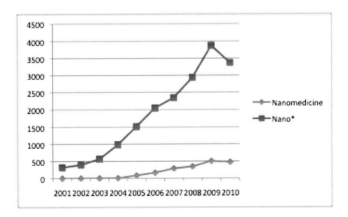

Figure 10.5. Number of publications (from PubMed) per year (*nano** or *nanomedicine* as search term).

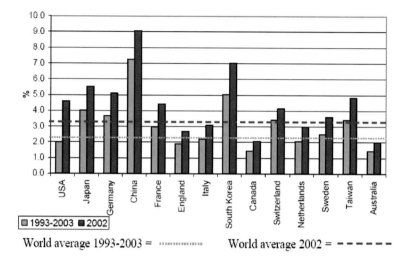

World average 1993-2003 = ░░░░░░░░░░░   World average 2002 = ‒ ‒ ‒ ‒ ‒ ‒

Figure 10.6. Percentage of the number of countries' publications which are nanotechnology related (Warris, 2004).

10.3.1 *Regulatory Aspects*

As mentioned above, the development of new medicines is regulated by the ICH Guidelines (ICH Steering Committee), and this applies also to nanomedicines or medicinal products containing nanotechnology-based components. The ICH guidelines are gradually being applied in more and more territories.

Nanotechnological products for medical applications are regulated either as medical devices or as medicinal or pharmaceutical products. Nanomedicine is such a new field that the authorities are contemplating adaptations and adjustments. Today, there are a number of medicinal products that are market authorized and have nanocomponents as defined constituents. Their addition may cause dramatic changes in the intrinsic properties of the material, such as the transformation of a material from particle physics to quantum physics. The mere reduction in size may also cause the particles to be handled differently by the human body compared with larger particles. Small enough particles cross cell membranes, and if the particles are loaded with drugs, novel pharmacological targets may be reached in ways that were not possible with today's pharmaceutical formulations.

Because nanomedicine is in the beginning of its development, it can be expected that we will see tremendous development with new opportunities when material science improves properties, scalability, and economy. In addition to improving properties of medicines, such as their solubility or stability, nanomedicine is expected to find new applications in the medical field:

- Deliver medicines into the body
- Target medicines in the body more accurately
- Diagnose and treat diseases
- Support the regeneration of cells and tissues

The European Medicines Agency (EMA) is actively following developments in this area and has declared that it may take action should there occur potential safety issues or variability in production quality or other aspects. In particular, safety concerns are high on EMA's agenda, since nanoparticles in medicines are a novel area with uncertainties both regarding the efficacy and potency *and* potential side effects. Recommendations from the Agency's Committee for Medicinal Products for Human Use (CHMP) have thus already led to the approval of a number of medicines based on nanotechnology, for example:

- Liposomes (microscopic fatty structures containing the active substance), such as Caelyx (doxorubicin), Mepact (mifamurtide), and Myocet (doxorubicin)
- Nanoscale particles of the active substance, such as Abraxane (paclitaxel), Emend (aprepitant), and Rapamune (sirolimus) (from EMA homepage)

10.4 Nanotech Case Studies

At the formation of the early spin-out companies, the university had rather rudimentary or no support systems, at least nothing that could be characterized as TTO structures. In the Swedish context this meant that the Tech transfer had to be handled for the most part by individual scientists. Given the situation that the scientists under the Swedish Teachers' Exemption furthermore have to finance the various parts of the commercialization, many potential inventions and companies have not come to fruition. Companies spun out from academia were also set up prematurely, only to later succumb for various reasons. Some should rather not have started, some have closed down due to a lack of funds, and others have not found a suitable entrepreneur to run the company.

10.4.1 *Biacore*

In 1983, an undergraduate student pioneered the key experiments that became the start of a successful commercialization. The student was Bo Liedberg, now full professor at Linköping University, and the company that later was formed by Pharmacia, based on these studies, was Biacore. Prior to the experiments by Liedberg and coworkers (1984), surface plasmon resonance had been used to study organized organic mono- and multilayers on metal surfaces. The experiments by Liedberg were key to the concept to use plasmon resonance for gas- or macromolecular sensing. The initial experiments used novel materials such as gold surfaces, where fibrogen was adsorbed on the surface and could be studied with novel methods compared with what was possible previously. Later in the development it became possible to study the binding characteristics of substances that interact with proteins immobilized on the gold surface (Liedberg *et al.*, 1984, 1995). The Swedish company Pharmacia became interested in developing a commercial application, making the study of macromolecular interactions feasible.

The instrumentation company BIAcore was launched in 1990 and BIAlite in 1994. Liedberg's key finding that interactions between macromolecules immobilized on a gold surface when interacting with, for example, a low-molecular-weight compound gave rise to data of interest in, for example, the drug-discovery process. The company project, originating from the experiments in 1983, is now a growing business arm of GE Healthcare, and their method is steadily finding novel applications. The tech transfer of the Liedberg project from the academic setting to Biacore generated a new technology application, and a new company grew out of it.

According to the scientists (personal reference) this successful spin-out was created based on an atmosphere of collaboration. It also strengthened the research base of the involved university scientists; they were rewarded with the scientific collaboration and successful publishing and financing of parts of their project costs. Had the university at that time owned the IPRs and had been resourced to do successful out-licensing, it may have developed in either of the following two ways:

- Either the TTO at the university would have been too resource scarce to pursue the collaboration and the projects may have failed without the personal engagement of the inventors, or
- The TTO may have been successful in defending the university's position and made a co-development with Pharmacia, sharing the future revenues among the inventors, their department, and the university, as is the usual model.

I am convinced that through their scientific contributions, researchers and inventors have been instrumental in establishing this novel technology. The technology has without doubt been, and still is, valuable to the study of protein interactions with, for example, drug substances.

10.4.2 *SPAGO Imaging*

A recent spin-out has been created based on the innovative research on nanomedicine applications from Professor Kajsa Uvdal's research at Linköping University. In brief, novel methods are devised that increase the sensitivity of imaging agents, opening avenues to improved sensitivity and level of detection of, for example, smaller tumor sizes than what was possible earlier (Ahrén *et al.*, 2010).

Kajsa Uvdal's research is focused on molecular surface physics, especially interactions between nanocrystals and the binding of organic and biomolecular materials. In collaboration with other research groups, Uvdal designed nanocrystals based on rare earth metals with the aim of developing novel types of imaging contrast media, findings that have been the basis for patenting.

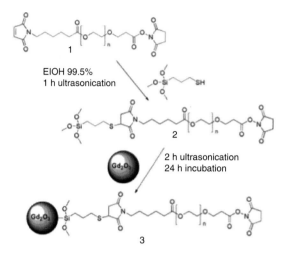

Figure 10.7. Synthesis of PEGylated Gd2O3 nanoparticles for MRI. Reprinted with permission from Ahren *et al.* (2010). Copyright (2010) American Chemical Society.

Uvdal's imaging project is now developed by SPAGO Imaging AB, a Swedish contrast agent company dedicated to the development of MRI contrast agents for the early and specific detection and diagnosis of cancer. The project is based on patents showing the potential of nanoparticulate rare-earth metal oxides as MR-contrast agents. In 2007, SPAGO Imaging budded off from Accelerator Nordic (http://www.acceleratorab.se) as a subsidiary. Since 2008 it is located in laboratories in Ideon Bioincubator at BMC, Lund, Sweden, with five full- and part-time employees. This case is an interesting example of how a project from excellent academic research recently was spun out to form the basis for a promising new enterprise that may improve human health care.

10.4.3 *Artificial Corneas*

In a recent article in *Science Translational Medicine*, Fagerholm and coworkers described a clinical study where patients with corneal blindness were given artificial corneal implants. They were produced from a cross-linked collagen type III matrix. The matrix used in the recently published paper by Fagerholm *et al.* (Fagerholm *et al.*, 2010) was shown to be a scaffold for remodeling into an integrated cornea with full functionality. It can also be anticipated that cellular components, growth factors, as well as drugs can be added to the matrix for various purposes. These additives may also be formulated as nanoparticles to extend half-lives as well as tissue-penetrating properties. This case is interesting from a TTO perspective in that Professor Griffith initially developed the technologies in Canada and now has moved to Sweden to implement them there. Furthermore, this is a project that can be cross-fertilized by the moves of researchers from one country to another.

10.5 Conclusion

Technology transfer is a complex set of activities where individuals, such as researchers, TTO professionals, lawyers, investors, licensing industry professionals, local support functions, entrepreneurs, and many other categories, have to come to various agreements to move a project forward from one context to another. What is common in all these situations is that trust is central for people-to-people interactions. Next to the trustful phase is that intentions must be translated into contractual agreements that truly reflect what has been agreed; otherwise the contracts will be obstacles in project development. Technology transfer is different from regular sale of a product in that university research often agrees to continuing its support for the project beyond selling it. The researcher's support is essential for the advancement of the project and often also for developing improved versions of the initial product envisioned at the time of agreement. The nanotechnology, or in particular the nanomedicine, field is no different in this respect. From my chapter it is also apparent that the conditions for technology transfer differ somewhat from country to country. The basics are the same, though.

Acknowledgements

I would like to thank Linköping University for giving me the opportunity to take of their time for the preparation of this chapter.

References

Ahrén, M., Selegård, L., Klasson, A., Söderlind, F., Abrikossova, N., Skoglund, C., *et al.* (2010) Synthesis and characterization of PEGylated Gd2O3 nanoparticles for MRI contrast enhancement, *Langmuir,* 26, 5753–5762,

Baker, J. (2010) Finding research diamonds and putting fire in the ice, *Medical Innovation & Business*, 1, 10–27.

David, P. A., and Metcalfe, J. S. Only connect: Academic-business research collaborations and the formation of ecologies of innovation. *SIEPR Discussion Paper No. 07-33, Stanford Institute for Economic Policy Research*, Stanford University.

Ett Lyft för Forskning och Innovation ("Enhancing Research and Innovations" [author's translation]), Swedish Governmental proposition, October 2008.

Fagerholm, P., Lagali, N. S., Merrett, K., Jackson, W. B., Munger, R., Liu, Y., *et al.* (2010) A biosynthetic alternative to human donor tissue for inducing corneal regeneration: 24-month follow-up of a phase 1 clinical study, *Sci. Transl. Med.,* 2, 46–61.

In the public interest: Nine points to consider in licensing university technologies, Association of University Technology Managers (AUTM), March 2007.

ICH Guidelines (Nov 2005) ICH Steering Committee. Available at www.ich.org.

Lagen om Arbetstagares rätt till uppfinningar ("Law on Employee's Rights to Inventions" [author's translation]), Swedish Law, 1949, 345.

Liedberg, B., Ivarsson, B., and Lundström, I. (1984) Fourier transform infrared reflection absorption spectroscopy (FT-IRAS) of fibrogen adsorbed on metal and metal oxide surfaces, *J. Biochem. Biophys. Method*, 9, 233–243,

Liedberg, B., Nylander, C., and Lundström, I. (1995) Biosensing with surface plasmon resonance – how it all started, *Biosens. Bioelectron.* 10, i–ix.

Nanoscience and nanotechnologies: opportunities and uncertainties. The Royal Society & Royal Academy of Engineering, July 2004.

Reflection paper on nanotechnology-base medicinal products for human use, Committee for Medicinal Products for Human Use (CHMP), EMEA, June 2006.

Thomson, I. S. I. (2005) Nanotechnolgy Benchmarking Project Report. *Australian Academy of Science*, Australia, 2004.

Chapter 11

Public-Private Partnerships — an Example from the Netherlands: The Industrial Partnership Programme

Pieter de Witte

Foundation for Fundamental Research on Matter (FOM),
P.O. Box 3021, 3502 GA, Utrecht, the Netherlands
pieter.de.witte@fom.nl

11.1 Introduction

11.1.1 *General Introduction*

The development of nanotechnology applications from fundamental nanoscience can be a luring perspective for innovative companies and entrepreneurs. Some examples of applications are nanofilters, spintronic devices, and nanobatteries, while new applications in fields like drug delivery, quantum computing, and optoelectronics are being developed or hold a future promise.* One way for companies to enhance their technological innovation capabilities

---

*This chapter is not intended to provide a review on nanotech applications. There are many books available on nanotech applications, mostly focusing on an application area. A flavour of applications can also be found in Chapter 1.

Nanotechnology Commercialization for Managers and Scientists
Edited by Wim Helwegen and Luca Escoffier
Copyright © 2012 Pan Stanford Publishing Pte. Ltd.
ISBN 978-981-4316-22-4 (Hardcover), 978-981-4364-38-6 (eBook)
www.panstanford.com

and develop these new applications is to collaborate with universities and public knowledge institutes.

A specific example of public private collaboration in basic research is the Industrial Partnership Programme (IPP) in the Netherlands. An IPP is intended for longer-term basic research by academics in close contact with industrial researchers in areas with a good potential for innovation and challenging scientific questions. The industrial partner and FOM together finance an IPP.

This chapter is devoted to a practical description of the IPP. In describing the setup of the programme, and explaining crucial factors that make it a successful programme, I intend to provide an insight into how fundamental science can add to innovation, and, ultimately, commercialization in firms.

In this section the IPP is introduced and is described in a context of other public private initiatives and of open innovation. In Section 11.2 the organization of an IPP is described from the embryonic stage to the execution phase. In Section 11.3 I will qualitatively discuss the results obtained so far and elaborate on the role of knowledge transfer, and the network. I will also share some experiences from partners. The last section draws conclusions and provides an outlook for future developments.

The IPP is developed, managed, and executed by the Dutch science foundation in physics, FOM, and could provide a tool to help shape the (open) innovation process from a basic science perspective.

11.1.2 *Public-Private Partnerships in Research*

In their effort to move the frontiers of knowledge, universities are important contributors to technological innovation, either through basic or applied research. Generic links between universities and industry like graduate recruitment, the use of scientific publications or university patents add to the innovation capacity of companies. However, organized university–industry relationships such as research partnerships, contract research, and consulting may also play an important role in driving these innovation processes [1]. Public-private partnerships (PPPs) are an example of such relationships.

Commonly, companies develop one-to-one relationships with universities or specific research groups through dedicated funding of research activities based on the needs of the company. An example of this kind of partnerships is contract research, for example, funding of a PhD position. In this way, firms have access to specific knowledge, with the partner research group preferably near, but often further away from, the company.

PPPs are also realized by establishing a physical research lab at or near the academic premises. These are generally more collaborative and based on longer-term relationships. Examples, among many others, are IBM Zurich, Microsoft Station Q at UCSB, and Nokia at Cambridge. These are examples where a company reaches out to one or several universities that are globally distributed, for their particular expertise or excellence. In this way the firms are able to tap locally from the large body of specific knowledge and human capital and talent present in the university.

Every IPP is a research programme, consisting of multiple projects. The IPP is different from the examples given above, in that within an IPP a company can be connected to several universities distributed in a regional network (in contrast to global distribution).

With the Netherlands being a compact area the universities are all in close proximity, hence it can be considered as a "regional" programme. The participating university groups form a research network, viz., a connected group of laboratories that carry out the IPP projects. FOM serves as a link between the universities and the companies.

11.1.3 *Foundation FOM*

In order for you to appreciate the background and basic ideas of the IPP, it may be helpful to understand the organization FOM. FOM is the Foundation for Fundamental Research on Matter, which was founded in 1946, and which mission has since been to advance the fundamental research in physics in the Netherlands. FOM's activities benefit the common interest, specifically higher education and businesses.

Besides being a funding agency, FOM is a research organization and employs ca. 1000 people. It performs research of internationally high quality at three institutes (FOM Institute for Atomic and Molecular Physics AMOLF, National Institute for Subatomic Physics Nikhef in Amsterdam, and the FOM Institute for Plasma Physics Rijnhuizen in Nieuwegein, which is currently being transformed into the Dutch institute for Fundamental Energy Research) and in research groups at almost all Dutch Universities. International experts assess the research that takes place within FOM against stringent criteria. FOM thematic counsels are scientific committees that give advice to the FOM executive board about scientific proposals for new research programmes. Each year FOM turns out about 100 young doctoral researchers. Most of these remain in the research world at Dutch and foreign universities and in industrial R&D. FOM receives most of its funding from the Netherlands Organisation for Scientific Research (NWO), besides receiving incomes from partnerships with companies and from funds from the national government and European funds. Its total turnover is approximately 90 million euros per year. With scientific quality always being the prominent factor in allocating the research funds, FOM aims to contribute to innovation in firms, and, consequently, the Dutch economy, by conducting basic research.

In 2004, the FOM executive board decided to establish the IPP to increase its contribution to society by also carrying out fundamental research that is directed toward the needs of society, and companies in particular. Part of this trajectory is aimed at creating a change in culture: generated knowledge should find its way more efficiently and effectively to existing industries and entrepreneurs. FOM has other activities, complementary to IPP, to support commercialization and entrepreneurship.

11.1.4 *The Industrial Partnership Programme (IPP)*

An Industrial Partnership Programme [2] is intended for longer-term basic research by FOM staff in close contact with industrial researchers in areas with a good potential for innovation and challenging scientific questions. In other words, joint research that might yield groundbreaking innovations.

Born from the wish to contribute more visibly to the Dutch knowledge based economy without compromising the high scientific standards, the Industrial Partnership Programme started in 2004 with a long-term commitment of 3 M€ per year. It is particularly directed toward multinational firms who have their own R&D laboratories.

An IPP is financed jointly by FOM and the industrial partner(s), the latter contributing at least 50% in cash of the costs. This leads to a leverage on the research funds for IPP. The minimum budget amounts to 1.0 M€ (corresponding to about four PhD projects of 4 years each), which makes the IPP a genuine research programme. This provides for a broad embedding of knowledge also in the academic system. The scientific quality is the ultimate determinant for project funding. The firms "buy" their access to knowledge generated from the programme and also get access to the physics network. Conducting excellent research and the possibility to publish in scientific peer reviewed journals are a point of departure of every IPP.

Since 2004, the accumulated budget has increased from 3 M€ in 2004 to 50 M€ in 2009 (see Fig. 11.1). From this amount, FOM has contributed 18 M€ in cash, the companies have contributed 22 M€ in cash and 10 M€ in kind.

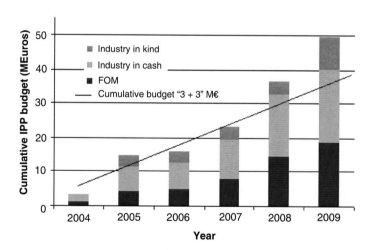

Figure 11.1. IPP total budget in time. Black line is the cumulative budget if the available 3 M€ per year for IPP was used and matched by 3 M€ from industry.

An IPP can be "open" or "closed." In a closed programme, all of the projects are already defined and the research consortium is established when the application is submitted. In the case of an open programme, a programme proposal is written after which a call for research project proposals is organized. More about the different forms of IPP can be found in the next section.

This chapter provides a qualitative description of the IPP. A quantitative study about the results of the IPP in light of its goals (excellent science and innovations) will be conducted at a later stage [3].

11.1.5 *The Advent of Open Innovation and the Rise of the IPP*

The process of deconcentration in "corporate research" is accompanied by a decline in in-house fundamental research. Instead of keeping all the knowledge internal and secret, firms now link to external sources of knowledge, which is also partly brought about by the increasing complexity of technological developments. Industrial innovation is more and more dependent on R&D outsourcing and the ability to participate in strategic alliances, especially for the larger companies in the high tech industry [4] — a process that has intensified with the rise of open innovation [5]. The open innovation paradigm states that firms should use external ideas as well as internal ideas, and both internal and external paths to market, when realizing innovations. The use of deliberate input and output of knowledge in combination with the expansion of markets for external use of innovations should enable firms to achieve high-risk discontinuous innovations more easily. Firms are, therefore, more susceptible to external knowledge and ideas. In short, R&D is treated as an open system [6].

Due to this development, the role of FOM in the R&D ecosystem has changed and the IPP is a response to the need of firms for fundamental knowledge and new ideas. Corporate R&D labs appear to act more like science parks and incubators and seek strategic alliances. An example is the High Tech Campus of Philips in Eindhoven, that underwent a transformation from a hermetically sealed campus to an open terrain where other firms are established and clean rooms can be rented. Philips was one of the early proponents of open innovation in the Netherlands, and, not accidentally, was our first IPP partner. We established a FOM research group on their campus.

11.1.6 *Summary*

An IPP is intended for longer-term basic research by academics in close contact with industrial researchers in areas with a good potential for innovation and challenging scientific questions.

The IPP is an example of a public-private partnership programme in basic science. The form is somewhat different from most existing examples of PPPs in that a company has access to a network of research groups through one point of contact. The rise of the IPP coincides with the appearance of the open innovation paradigm.

11.2 Description of the Industrial Partnership Programme

11.2.1 *Introduction*

Besides the "fundamental" character of the research and the potential for technological innovations that I mentioned in the previous section, other principles of an IPP are the in-cash contribution from the company that amounts to at least 50% of the total research costs, and the minimum size of an IPP being 1 M€. New programmes are awarded on the basis of peer review assessment; there is no guarantee *a priori* that a programme will start, even with the in-cash contribution of the company.

This section describes the structure of the Industrial Partnership Programme in more detail. I will discuss the characteristics of an IPP, the practical aspects related to the development of a new IPP, the collaboration agreement, and the management.

FOM created an IPP office that is the contact desk for industry. Besides the programme management, it is instrumental in setting up new programmes, in the brokerage of research partners, in network events, and in the communication between all partners. These are key elements in the IPP, as will become clear in the next section.

11.2.2 *Characteristics of the Programme*

Below is a set of basic conditions that apply when starting a new IPP, followed by some practical aspects of an IPP, from the embryonic phase to the execution of a programme.

FOM and the industrial partner(s) together determine the precise design of the programme, however, there are six basic conditions that all programmes must meet:

- It concerns fundamental research by personnel employed by FOM in close cooperation with researchers from one or more companies.
- The partners jointly formulate the research objectives.
- Companies finance at least half of the programme budget in cash.
- Each programme budget amounts to at least 1 million euros.
- Every programme proposal is to be assessed for its scientific quality by international independent referees.
- Agreements are made about the intellectual property rights and disclosure of information.

The IPP is organized according to the form-follows-function principle; it can be a multidisciplinary programme, other industries can be sought to enter a broader application field, or funding agencies from other disciplines can be approached for joint funding of the programme.

An important aspect of the IPP is the training of young researchers both through regular PhD courses offered by FOM, and through courses provided by the partner companies.

11.2.3 *Forms of IPP: Open, Closed, or FOM Group at Company Laboratory*

Industrial Partnership Programmes can be open or closed. In a closed programme, all of the projects are already defined when the application is submitted. The consortium is therefore already established. In the case of an open programme, the programme outline is written by a scientific project leader together with the industrial partner. Subsequently, a call for proposals is issued.

Closed IPPs generally address issues related to problems associated with products or processes in a company. These programmes tend to be more focused and have more defined research questions. Generally, they result in filing of more patents than open programmes and are closer to commercialization.

Open programmes are generally aimed at building up a broader knowledge base. They have a more generic reference to existing products or processes and their focus is generally on creating completely novel ideas, hence they tend to be somewhat further from the market than closed IPPs.

A special type of a closed IPP is the FOM research group at an industrial laboratory. FOM employs the research personnel that is stationed at the company, among the industrial researchers. These young FOM researchers remain real scientists, but they are embedded in an industrial environment and are fully involved in the life of the laboratory. The company provides lab and office space, research facilities and bench materials, and a part of the personnel costs. The FOM group takes part in relevant meetings of the company. This group is scientifically supervised by a FOM group leader at an institute or university. The FOM group consists of a group leader and several PhDs and possibly post-docs.

11.2.4 *The Start-Up Phase of an IPP*

11.2.4.1 The embryonic stage of a potential new IPP

We discern three ways in which an IPP can germinate:

1. Researchers already have good contacts with companies, or vice versa, and submit a proposal.
2. The IPP office acts as a matchmaker/broker for researchers with an idea or a company with a problem. The IPP office can organize a brokerage event, for example, a scientific workshop in which academic and industrial participants seek partners (scientific expertise or funding partners) and establish the scientific challenges. The network activities that the IPP office organizes facilitate this matchmaking process.

3. The IPP office acquires companies from outside the network. This latter activity is particularly relevant, since firms often need to find new partners and networks to acquire external knowledge.

If a consortium is formed, parties jointly draw up the research proposal in consultation with the IPP office. The latter supports items such as budget and form of collaboration. Every programme is custom made.

11.2.4.2 Application and review procedure

After the programme proposal is received by the IPP office, the FOM executive board decides whether the application is admissible or not. Criteria are the cash contribution of the private partner, the scientific value, the value of the collaboration, and the potential impact of the IPP on the research landscape.

The scientific assessment is performed according to standard FOM quality procedures.

For a closed IPP, the proposal is subsequently assessed by independent international referees. The applicants can give rebuttal. Based on the referee's comments and the rebuttal the FOM thematic counsels give a funding advice to the executive board.

In case of an open IPP, thematic counsels give advice to the executive board about the scope of the overarching programme proposal, viz. the research theme, fit in the research landscape, anchorage of the knowledge in the research infrastructure (both at the private partner and at FOM). On this programme proposal a call for (pre)proposals for research projects is based. Preproposals are screened by the Programme Committee, wherein also representatives from the company have a seat. Full proposals are assessed by a panel of international independent experts, who have interviews with the applicants.

The final granting decision is made by the FOM executive board.

11.2.4.3 Success rate of IPP applications

The success rate of a closed IPP application is 73%. This means that about a quarter of the applications do not pass the independent scientific assessment, even while the partnering company has already guaranteed its cash contribution.

The success rate for open IPPs is 100% for the overarching programme proposal. For project applications within an open IPP (call for proposals) this number is typically 40–50% for the preproposal stage and 50% for the full proposals, leading to an overall success rate of ca. 25% for the projects in an open IPP.

11.2.4.4 The collaboration agreement

The collaboration agreement is made up simultaneously with the review procedure. The agreement reflects both parties' interests. For a company this implies access to useful knowledge, inventions, and trained people, while for FOM publications, PhD defences and sharing revenues from exploitation are important. There is no guarantee that FOM delivers dedicated results; the programme is built up from the science case and the collaboration is an "effort-obligation."
Some typical agreements are as follows:

- There are no obstacles for researchers to obtain their PhD. A defence can never be delayed. If the company decides to quit its activities related to the IPP, it is obliged to support the completion of the thesis work of the researchers.
- All publications are screened by the company before publication.
- It contains standard conflict resolution procedures.

Usual intellectual property rights (IPRs) agreements are as under:

- The rights in the results are vested in FOM. The partner company gets the first right to apply for a patent in the particular area. FOM maintains the right to apply for a patent if the company decides not to make use of this right.
- If the company makes use of this first right of refusal the rights are transferred and FOM receives an incentive bonus, that is available for the research group that made the invention. All costs incurred by the patent applications are borne by the company.

- A market fee based on revenue sharing arrangements is agreed. This is based on the market value research of the patent and the chance for commercial exploitation. Based on the outcome of this research an agreement about extra allowance is made.
- Research results are always available for research and education purposes.
- In case more companies join in one IPP, rights are mostly established according to their application field, for example, food versus non-food applications.
- Specific agreements vary from sector to sector.

11.2.5 *The Execution Phase of an IPP*

11.2.5.1 Organization and management

Besides funding of the research projects, and the programme management, the IPP office takes an active role in the construction of an IPP, the partner selection, the knowledge transfer, and in maintaining networks. It has dedicated tasks in steering and supporting the application and in the process of agreement by the partners on the content of the proposal and the results and outcome. The IPP office manages IPR negotiations and contractual issues as well.

To facilitate knowledge transfer, several schemes are in place:

- A formal reporting scheme consisting of regular progress meetings with all partners, progress reports, and exchange of researchers.
- Mirror projects and project leaders at the company; every project has a contact person at the company that is available for discussions, inventions, and access to infrastructure. This person also takes care of the in-company embedding of the knowledge.
- Education of IPP researchers through trainings.
- Informal meetings are a relevant aspect of knowledge transfer, see Section 11.3.3.

Before research personnel (PhD students and postdoctoral researchers) is hired by the project leaders, they are presented to the company for approval.

During the execution of the projects, tuning of research focus and picking up new chances can occur after deliberation between project leader and the company.

11.2.5.2 Governance

Every programme has a proper governance structure wherein both FOM and the company are represented. Usually an IPP has a steering committee and a programme committee. The former is a group of responsible representatives from all stakeholders. FOM is represented by its head of the research policy department as delegate from the executive board. The programme committee consists of the scientific managers within a programme. Every IPP has a programme leader who is responsible for the scientific execution and chairs the programme committee. The staff of the IPP office acts as secretaries to these committees.

11.2.5.3 Financial aspects

Companies pay at least 50% of the programme costs in cash, with a minimum of 500 k€. This in cash contribution includes 15% overhead costs. FOM executes the programme and invoices the companies annually. The programme includes budget for meetings, workshops, and knowledge transfer.

11.2.6 *Summary*

In an IPP the industrial partner contributes at least 50% in cash of the total research costs and the execution is done by FOM. The scientific assessment of an IPP application is performed by international independent referees.

IPPs can be open or closed, the choice of the company usually depends on whether it has specific problems associated with its products or processes or whether it wishes to explore completely novel ideas. A special closed form is a FOM group at an industrial research lab.

The collaboration agreement reflects both parties' interests. It balances the freedom to conduct and publish basic research and the confidential disclosure of information to the company and agreements for the valorization of IPR.

The IPP office has an active role in setting up a new IPP, in bringing together parties, and in knowledge transfer.

11.3 Experiences and Results

During the last FOM Strategy meeting in spring 2010 with participants from academia, industry, research agencies, and other stakeholders, it became clear that the IPP had gained in popularity among the researchers. Those same researchers were fairly sceptical during the preceding strategy meeting five years earlier when the IPP concept was introduced. It was agreed that the IPP will be reviewed in 2012 when enough data will be available from completed programmes. Since we have not yet conducted a review, I will provide you here with a qualitative insight into the obtained results, based on the experiences from the IPP office and from the participants. I will discuss in what way the IPP has been successful, and what aspects of the IPP seem to be critical factors in its success.

11.3.1 *Success Comes in Different Shapes and Sizes*

From Fig. 11.1 it is apparent that in the last few years, the IPP is generating more external budget than was put in by FOM. From the growth of the budget and the fact that the cash contribution from industry is higher than the cash contribution from FOM, it may be concluded that the IPP is a successful instrument. This is substantiated by the fact that some companies are involved in several partnerships, and that some companies have started a new IPP after their IPP had ended.

How to measure if your programme is successful? For FOM an IPP is successful if it results in scientific publications, in human capital (trained PhD students and Post-docs), and if the firms acquired relevant new knowledge that facilitated their innovations. Table 11.1

Table 11.1.   IPP nanotechnology related programmes. Numbers are a snapshot of the status in 2010.

Name	Industrial partner	IPP form	Budget (M€)	No. of projects	No. of researchers[1]	No. of publications[2]	No. of patent applications	Duration
Microphotonic light sources	Philips	FOM group at Philips	1.8	3	6	33	8	2005–2010
Extreme multilayer optics	Carl Zeiss SMT	Closed	7.7	9	10	20	13	2005–2010
Microscopy and modification of nanostructures with focused electron and ion beams	FEI electron optics	Closed	2.7	5	10	10	1	2007–2012
Innovative physics for oil and gas — a deep dive into the nano domain	Shell International Exploration and Production	Open	2.2	6	10	0	0	2009–2014

Notes: (1) PhDs and post-docs; (2) scientific peer reviewed publications.

gives an overview of the nano-related IPPs, depicting some of the key parameters.

The first two factors are fairly straightforward to establish. We know that the number and impact of publications is comparable to other FOM research projects. Due to the independent scientific assessment according to regular FOM procedures the scientific quality of the granted IPP projects is high [7]. This could be an asset that draws firms to the IPP. Also, the number of PhD students and Post-docs can be easily monitored. It will be interesting to see in the coming review whether PhD students and post-docs have found a job at the company they collaborated with in an IPP.

The question how knowledge transfer is successfully achieved is harder to answer. Exactly what the impact is of university–industry partnerships on the innovation process appears to be a matter of debate [8].

Knowledge transfer should not be measured only by the number of patents. While we have experienced that for some of our industrial partners the potential to generate new patents is the prime reason to be involved in an IPP, several other partners have different drivers to participate in an IPP, which is also the common perception for industries liaising with public research institutes. Work by Cohen *et al.* [9] shows that knowledge transfer activities from universities to industry are of a much broader spectrum than the activities related merely to the commercialization of intellectual property rights. Access to tacit knowledge that resides in the programme (people, institutes, and infrastructure) is an important reason for companies to engage in collaborations. Less formal activities such as informal contacts, networking, and university–industry mobility of researchers have been mentioned as important drivers in knowledge transfer [10].

Encouraged by these signals, it has been suggested in the UK to use an agreed wide set of knowledge transfer metrics to establish the effect of knowledge transfer into the commercial sphere [11]. Examples of these metrics are the length of the relationship, number of researcher exchanges, number of research training for industries, number of participants in industry–academia network events. In qualitative terms, several of these aspects will be discussed further on.

Who are the scientists involved in an IPP? We can see that scientists who are most successful in obtaining basic research grants at FOM are also most successful in an IPP. Ambos *et al.* have shown that most researchers tend to prefer a traditional academic publishing career over a career that is more open to producing commercial outputs, but also that scientific excellence is significantly associated with the generation of successful commercial outputs [12]. Also most FOM scientists pursue an academic career in fundamental research and scientific excellence. This is a pool of great potential. With the focus of the IPP on basic research, we can draw from the large pool of basic scientists. The IPP "utilizes" what comes naturally to the basic researchers, without expecting them to become all application-driven scientists or entrepreneurs. However, awareness by the researchers for the application perspective is desired to facilitate communication. This sometimes implies a cultural change for some researchers.

One aspect that stands out is the commitment to the programme of the companies, which can probably partly be ascribed to their in-cash contribution of at least 50%. This makes that they make available time and resources, for example, in the form of mirror project leaders, bilateral meetings, or research infrastructure and support.

It is worth mentioning here the secondary results of an IPP. A significant spinoff from the FOM group at Philips is the fact that the scientific group leader obtained external funding for another two PhD and two post-doc positions from FOM and EU programmes (not indicated in Table 11.1). Besides, it has resulted in a PhD thesis of one of the Philips employees. These spinoff activities have more or less doubled the scientific impact of the FOM group.

In the remainder of this paragraph I will focus mainly on this less tangible area of knowledge transfer, networks, and interorganizational relationships.

11.3.2 *The Network*

As I discussed in Section 11.1.2 interorganizational networks are an important aspect in innovation. Network formation is key in the transfer of tacit knowledge through people. Geographic proximity facilitates network formation. "Geographic, cultural, and institutional proximity leads to special access, closer relationships, better information, powerful incentives, and other advantages in productivity and innovation that are difficult to tap from a distance. The more the world economy becomes complex, knowledge based, and dynamic, the more this is true" [13].

The IPP has contributed to the strengthening of the network on a micro level, viz. between companies and individual physicists. Once involved in a collaboration we see those physicists frequently maintain their contacts with that company and start new programmes. Besides, they often extend their connections and collaborations to other companies, and, vice versa, companies extending their links to other research groups through the initial contact.

On a macro level the IPP office focuses its efforts at the R&D ecosystem. Since the Netherlands have nine universities and institute AMOLF involved in nanoscience and nanotechnology research that are all within a few hours travel, this facilitates greatly the network formation. Besides, the Netherlands have a national nanotechnology research network NanoNed and a research infrastructure NanoLab [14]. Through FOM, companies have access to the entire physics and a good deal of the nanotechnology community.

To promote the integration of industry and academia into a physics network, FOM

- may offer relevant industries a seat in the FOM executive board, the board of governors, and the thematic counsels.
- involves industrial researchers in the organization of large scientific conferences.
- organizes problem-solving workshops and site visits between academic and industrial researchers.
- partners with the leading technology institutes in the Netherlands; [15] these public-private research institutes form an additional scientific tie between FOM and industry.
- maintains an alumni network.

These activities are important for the visibility and proper functioning of the "R&D-ecosystem." Besides, it creates trust and helps maintaining our long-term relationships.

We recently observe a better visibility and participation of the company's researchers in the FOM network, for example, by higher participation in our national physics conference and network events and by an increase in bilateral contacts. I ascribe this to both increased external activities of the firms because of the open innovation paradigm, and to our efforts mentioned above.

11.3.3 *Knowledge Transfer*

While knowledge transfer is generally realized through publications and invention disclosures, channels for the transfer of tacit knowledge are formal meetings and reports, exchange of students in laboratories, and informal bilateral contacts [16]. According to Perkmann and Walsh we can distinguish three kinds of relationships between industry and academia that drive the innovation process: technology transfer mechanisms, human mobility, and interorganizational relationships [17].

The formal IPP progress meetings are a stepping stone for further contact, especially the exchange of researchers and the informal bilateral meetings wherein materials were exchanged, research infrastructure was used, and measurements conducted, add to the knowledge exchange.

Aspects that we find instrumental in the exchange of knowledge are as follows:

- Establishing mirror projects and project leaders at the company who are contact persons for the academic researchers and are instrumental in embedding knowledge at the company. In the regular progress meetings, all project leaders and mirror project leaders, and often other researchers from both the company and academia, attend the meeting
- The availability from the companies of resources like relevant in-depth information, expertise, and technologies
- Problem-solving sessions, and scientific cases with realistic data from the companies
- A stay of the PhD students and post-docs for a longer period (months) at a company to apply the knowledge

- Additional training of PhD students and post-docs by spending time in a lab of one of the partner industries

We incorporated several mechanisms into the IPP that should help manage expectations of both parties on the outcome:

- The company is involved in writing of the programme proposal (closed IPP) or in the screening of preproposals (open IPP). In this way it can determine the main lines of research.
- The agreements about goals, confidentiality, and publications must be explicitly discussed during the first meetings of a new programme. This is important to build confidence and trust among the participants.
- Awareness among research staff regarding the objectives, value, and significance of intellectual property protection is important in helping to understand a company's strategy. This means that it offers IP and business courses to its graduate students and post-doc researchers. Besides, it has other programmes to support entrepreneurs that build on spill-over results stemming from the projects.
- During meetings also the companies present themselves and their motivation and their specific research questions.

11.3.4 *IPR*

The IPR strategy is aimed at establishing and maintaining long-term relationships with R&D firms. This implies that we transfer the rights in IP to the companies and that we make the revenues thereof available to the inventors for research purposes.

Currently, FOM does in principle not pursue a licensing strategy by building up a portfolio of patents since the breadth of research themes covered by the IPP would lead to a scattered patent portfolio. Besides, this would be a cost- and resource-intensive activity, which the IPP office at FOM is not equipped for.

FOM always remains the right to apply for a patent if the company decides it will not do so. FOM will do so when IP is of clear strategic importance, for example, with the aim to establish start-ups or as a spin-off from long-term relationships with certain industries, for example, in the field of nanophotonic structures, and EUV multilayer optics. With the gradual emergence of these naturally evolving portfolios, and the expected role of FOM in fundamental energy research, FOM may reconsider its current IP strategy, also in light of the strategic role IP has in open innovation.

As I described in the previous paragraph, a difference between open and closed IPP is the output of patents. A trend that is apparent from the whole range of IPPs (not depicted in Table 11.1) is that closed programmes tend to create more IP than open programmes. We ascribe this to the nearness to the market (see Section 11..2.3). In Table 11.1 you can see that the number of patent applications varies also among the different closed IPPs (see difference between Carl Zeiss and FEI). This hints at the notion that IP is merely one of the reasons for a company to participate in an IPP.

11.3.5 *Experiences from Industrial Partners*

Besides access to knowledge and expertise, reasons to engage in research partnerships for firms are their access to the network, research infrastructure, highly educated human capital, and the one-to-one learning option. Many of the relevant companies (viz. mostly multinationals with own R&D facilities) in the Netherlands, and some firms in neighbouring countries, collaborate in an IPP.

Sometimes a company aims to develop a new area of expertise and to build up new knowledge, for example, a company that makes a move from traditional bulk chemistry products to advanced coating applications and needs to acquire colloid or polymer physics expertise. We observe that several companies traditionally not linked to physics, including chemically oriented companies like Shell, DSM, and AkzoNobel are partnering in our IPP.

I present here some quotes from interviews [18] we held with our industrial partners about the IPP that display the breadth of motivations and opinions. This is not intended as a comprehensive and quantitative overview:

Tata: "The biggest asset of the programme is probably the structural contact that occurs between experts from different perspectives — physics and chemistry, the steel industry, oil recovery, and the chemical industry — who are all working on the same problem [...] The real exchange of knowledge does not occur via reports and books but via people. This is also why the PhD students do not remain in their own laboratories but undertake internships at the participating companies. In that way they develop a feeling for the pivotal problems, and we gather an idea of what their capabilities are."

FEI: "We have assigned four senior scientists to act as contact persons. About once every two weeks they spend a day in the laboratory of the so-called scientific partner. Compared to the formal progress meetings we spend the whole day talking to each other. The exchange of information is more profound; you get a better feeling for what is going on, can exert more influence and you gain more from the partnership."

Shell: "The secret lies in the one-stop shopping....The IPP office gives Shell access to the entire physics community in the Netherlands, because in some way or other all physicists are connected with FOM. None of the hassle of eight different contracts with eight different professors."

These opinions reflect my conclusions from the preceding paragraphs about the importance of the network, the role of the IPP office, and the fact that tacit knowledge is an important driver for companies to participate and that it is transferred through informal contacts.

11.3.6 *Experiences from Academics*

The academic community was somewhat sceptical at the start of the IPP in 2004. Researchers feared a potential loss of academic freedom and less budget for free basic research. Since then, we gradually observed an "image" improvement of the IPP: research projects are considered still free in execution with interesting physics and research budgets have doubled due to the 50% in cash contribution.

From an academic perspective, reasons to participate *a priori* are the opportunity for additional funding and the social relevance of the research. Once researchers are involved in an IPP, several other benefits appear:

- The extension of the network and the potential for new relations, collaborations, and funding
- The access gained to new infrastructure and expertise
- The access to realistic data for their models
- An interesting feeling for issues at stake in companies, an understanding of their strategies, and the awareness for patent opportunities
- Training of young researchers in an industrial environment
- A motivating connection between fundamental research and the use of this knowledge by industry

To get a feel for the experience of the scientific project leaders that are involved in the programmes I have cited some quotes [19]:

"The beauty of this programme is that the focus is on fundamental physics [...] I am happy to see that Shell finds it as important as we (scientists) that the scientific quality of the projects is high."

"We bring together different disciplines of chemistry and physics together in one programme [...] I get inspired to convert my fundamental research results into industrial practice."

11.3.7 *Summary*

Measured by the large cash contribution from companies, that exceeds the contribution from FOM and by the fact that companies start new IPPs, it can be concluded that the IPP is a successful instrument. This is supported by the fact that several firms extend or renew their relationship with FOM. The impact of IPP on innovation must not just be measured by its output in terms of patents and publications, but as well by the involvement of companies, networking, and interaction between researchers of companies and academia. A review of the IPP based on a broad set of metrics should quantitatively show the scientific impact and to what extent and how the IPP contributes to the innovation capabilities of the industrial partners.

The main reasons for researchers to join in an IPP are conducting groundbreaking basic science of social relevance and additional research funding. Companies' motivations are access to excellent research results, new knowledge, networks, and trained people.

The IPP instrument is a symbiosis of these two motivations. Fundamental research projects with high scientific excellence and the possibility to publish and of knowledge transfer activities and invention disclosures.

The commitment of companies in an IPP is high, next to their cash contribution they put in time and resources.

The initial fear of loss of quality and academic freedom that existed within the scientific community gradually disappeared when researchers got involved in an IPP.

The network is an important asset. FOM works on the R&D ecosystem and focuses its attention on the individual level and on the institutional level. It has several schemes to maintain and develop this network.

11.4 Conclusions and Outlook

Universities and research institutes are an important source of knowledge and human capital, and, as such, they have a stake in the innovation process. Ultimately their scientific breakthroughs often are at the basis of new nanotechnology related products.

With the embracing by many firms of open innovation, they increasingly rely on interorganizational relationships like public private collaborations as a tool to enhance their innovation capacity. At the most fundamental level, a public research organization can contribute to the innovation process without compromising its core business of research and teaching. An example of such a public–private collaboration is the Industrial Partnership Programme, which was established to contribute more visibly to the Dutch knowledge-based economy, and is particularly aimed at multinational firms with R&D laboratories.

11.4.1 *Conclusions*

Judged by the growth in budget to 50 M€, of which more than 50% is contributed in cash by industry, and by the fact that companies return to FOM to renew their collaboration, the IPP is a successful instrument. The companies who participate in an IPP show a firm commitment to the programme. The majority of relevant companies in the Netherlands collaborate with FOM in some form of IPP.

Success should not only be measured by the number of patents, scientific publications and PhD defences. Access to tacit knowledge that resides in the programme through informal contacts, networking, and university–industry mobility of researchers are also metrics that should be considered when measuring success. Besides, there are secondary spin-off effects from IPPs like additional funding for researchers, or new contacts for both industry and universities.

The IPP is a collaboration model that is founded on basic science projects that have been granted in competition. This works well: it attracts many scientists to send in proposals, especially the better scientists in their fields, which makes it also scientifically a successful programme. The difference in culture could be a potential obstacle in public–private collaborations. Whereas the protection of knowledge tends to be more important for companies, scientists attach more importance to data collection and publishing in the public domain. We think that the IPP is successful in uniting these two "opposing" interests. Companies understand well that an IPP evolves around basic research and what they can expect from partnering in an IPP. With a clear policy on confidentiality, publications, and intellectual property, the output is a balance between the dissemination of public knowledge by publications and of intellectual property rights and commercialization.

The IPP "utilizes" what comes naturally to the basic researchers. The fact that the academic community was critical at the start of the IPP but gradually became more positive seems to support this view. Although the programme is aimed at drawing from the large pool of basic scientists, we sense a raised awareness toward the value of patents and commercialization from these scientists compared to research programmes without involvement from industries.

The IPP office has a major role in knowledge brokerage, matchmaking, in facilitating knowledge transfer, and in the administration of IPPs. Considerable effort is spent on building up confidence and trust among participants, especially among new partners. Besides, the IPP office also focuses its activities on sustaining and extending the network by organizing conferences, network events, workshops, and by involving firms in the FOM organization.

The Netherlands are a relatively small area, with a high concentration of research activities; a "physics region." This proximity is an attractive dimension for R&D firms since it supports the collaborative character of the IPP.

11.4.2 *Outlook*

Current developments in the (open) innovation landscape [20] may trigger the IPP office to take an even more active role in knowledge brokerage, since firms need to seek new partners and knowledge and their research funding at universities will probably increase. They may also cause FOM to reconsider its current IP strategy since IP has a more strategic role in open innovation.

Concerning the development of new strategic partners I see two options to extend the IPP concept:

- Collaborations with R&D firms in the vicinity of the Netherlands, that are still in practical reach. Currently, we have fruitful collaborations with BASF in Ludwigshafen and Carl Zeiss SMT in Oberkochen.
- With the IPP specifically aimed at the larger firms, that have fairly long innovation horizons, SMEs are kept out of range. We are currently working on schemes to also get the innovative SME's on "board."

Could the IPP concept be a valuable instrument in other countries or for other scientific fields? Besides the criteria mentioned in Section 11.2.2, I think there are three aspects that are vital in setting up similar activities and that were discussed in this chapter:

- The innovation region (geographical proximity as to facilitate collaboration and the transfer of knowledge)
- A dedicated office that develops activities like knowledge brokerage, networking, management
- First and foremost, scientific excellence

Acknowledgements

I kindly thank my colleagues Marcel Bartels and Hendrik van Vuren for their valuable discussions and comments.

References

1. M. Perkmann, K. Walsh (2007), University-Industry relationships and open innovation: Towards a research agenda, **9**(4), 259–280.
2. When I write "the IPP" I refer to the overarching programme; when I write "an IPP" I refer to a particular research programme that we also call IPP.
3. Since the start of the IPP in 2004, thus far 1 programme has finished, the IPP with Philips. Due to its success a new programme with Philips was started in 2010. Several programmes will finish in 2010 and 2011. It is foreseen that in 2012 a review of the IPP will take place. This review will focus on both aspects of the IPP: scientific excellence and innovation capacity at the company.
4. Hagedoorn, J., and Duysters, G. (2002) External sources of innovative capabilities: the preference for strategic alliances or mergers and acquisitions, *Journal of Management Studies*, **39**(2), 167–188.
5. Chesbrough, H. (2003) *Open Innovation: The New Imperative for Creating and Profiting from Technology*, Boston: Harvard Business School Press, ISBN: 1-57851-837-7.
6. Chesbrough, H., Vanhaverbeke, W., and West, J. (2006) *Open Innovation: Researching a New Paradigm*, Oxford University Press, Oxford.
7. The scientific impact of physics and materials science in the Netherlands is high compared to world average. Available at www.sciencewatch.com
8. Poyago-Theotoky, J., Beath, J., and Siegel, D. S. (2002) Universities and fundamental research: reflections on the growth of university-industry partnerships, *Oxford Review of Economic Policy*, **18**(1), 10–21.
9. Cohen, W. M., Nelson, R. R., and Walsh, J. P. (2002) *Links and Impacts: the Influence of Public Research on Industrial R&D*, **48**(1), 1–23.
10. Perkmann, M., and Walsh, K. (2009) *The two faces of collaboration: impacts of university-industry relations on public research*, Industrial and corporate change, **18**(6), 1033–1065.
11. Holi, M. T., Wickramasinghe, R., and van Leeuwen, M. (2008) *Metrics for the Evaluation of Knowledge Transfer Activities at Universities*, Cambridge, Library House.

12. Ambos, T. C., Mäkelä, K., Birkinshaw, J., and D'Este, P. (2008) When does university research get commercialized? Creating ambidexterity in research institutions, *Journal of Management Studies*, **45**(8), 1424–1447.

13. Porter, M. E. (1998) Clusters and the new economics in competition, *Harvard Business Review*, 76, 77–90.

14. Robinson, D. K. R., Rip, A., and Mangematin, V. (2007) Technological agglomeration and the emergence of clusters and networks in nanotechnology, *Research Policy*, **36**(6), 871–879.

15. Guinet, J., Freudenberg, M., and Jeong, B.-S. (2003) Public private partnerships for research and innovation: an evaluation of the Dutch experience, OECD.

16. Cohen, W. M., Nelson, R. R., and Walsh, J. P. (2002) Links and impacts: the influence of public research on Industrial R&D, *Management Science*, **48**(1), 1–23.

17. Perkmann, M., and Walsh, K. (2009) The two faces of collaboration: impacts of university-industry relations on public research, *Industrial and Corporate Change*, **18**(6), 1033–1065.

18. Quotes taken from booklet *The Profits of Cooperation*, Stichting FOM, 2008, p. 21, 31, 41.

19. Quotes translated and taken from FOM quarterly magazine 'FOM Expres'.

20. Gassmann, O., Enkel, E., and Chesbrough, H. (2010) The future of open innovation, *R&D Management*, **40**(3), 213–221.

Chapter 12

University and Employees' Inventions in Europe and the United States

Niklas Bruun[a] and Michael B. Landau[b]

[a] P.O. Box 4, 00014, University of Helsinki, Finland
[b] Georgia State University College of Law, 140 Decatur Street,
Atlanta, Georgia 30303
niklas.bruun@helsinki.fi; mlandau@gsu.edu

12.1 Employee Inventions in Europe

The European countries regulate ownership to intellectual property rights (IPRs) to employee inventions in different ways. Traditionally the European legislation could be divided into two main categories: (a) regulation by the law of contract or practice, and (b) regulation by statutory provisions [1]. Today most countries have introduced at least some kind of statutory provisions, even though the matter might to a large extent be regulated by contracts in practice.

There are no harmonized rules on employees' inventions in Europe with the minor exception of a choice-of-law rule in the European Patent Convention (EPC). EPC Article 60 prescribes that if the inventor is an employee, the right to the European patent shall be determined in accordance with the law of the State in which the

Nanotechnology Commercialization for Managers and Scientists
Edited by Wim Helwegen and Luca Escoffier
Copyright © 2012 Pan Stanford Publishing Pte. Ltd.
ISBN 978-981-4316-22-4 (Hardcover), 978-981-4364-38-6 (eBook)
www.panstanford.com

employee is mainly employed. If this cannot be determined, the law to be applied shall be that of the State in which the employer has his place of business [2].

Within the wide diversities in Europe, the main rule might be described as follows: The employer can either claim the right to inventions created during an employment due to the employment relationship or at least he is entitled to obtain the rights (or usage rights) to such inventions. However, the rules for when an employer may claim the rights to an employee's invention vary considerably among the European countries. A common conceptual notion has been to make a distinction between *dependent inventions*, to which the employer usually can claim rights and *free inventions*, to which the employer usually has no or limited rights. However, the definitions of these categories of inventions differ among countries. The employed inventor, on the other hand, as a rule, has the right to be remunerated for the right to an invention acquired by the employer from the employee. Whereas some countries, such as the Nordic ones, have had a tradition of awarding employees a right to a separate compensation for each invention, others have seen the remuneration as being a part of the employee's salary.

12.1.1 *European Traditions for University Inventions*

In several European countries it has traditionally been held that the general legislation on employees' inventions should not be applied on the academic staff within the universities. Germany, Austria, and the Nordic countries have had a tradition of granting so-called teacher's (or professor's) privileges to university professors and teachers [3]. The legislation in the Nordic countries used to contain an exemption stating that university teachers and researchers were not considered employees in the meaning of the Acts on Employees' Inventions. They could therefore freely dispose over the intellectual property rights stemming from their research and its results. In Germany the legislation contained provisions stating that inventions made by professors, lectures, and assistants were to be considered so called "free inventions," meaning they could not be claimed by the university based on the German Employee Inventions Act.

The teacher's privilege has been considered justified by the status of faculty at universities characterized by the principle of "academic freedom." Universities did not usually supervise and instruct each employee concerning his or her research, but it was at least in principle up to each researcher to decide on the content and focus of his or her research activities. The researcher had the right not only to decide the focus of his research but also the time and means for publication of the results.

However, as will be explained below in recent times the teacher's privilege has been questioned and partly abolished in many countries.

12.1.2 *A Changing Role of Universities Leading to Increased University Ownership*

European research policy of the last decade has aimed at intensifying the cooperation between industry and academia [4]. Universities and public research organizations (PROs) carry larger responsibilities in creating new knowledge and new technologies with direct applicability for society. Accordingly, there is a clear responsibility of universities and PROs in the transferring of such knowledge and technologies into society. In a European Commission Working Paper it is expected that the importance of PRO research will grow even further in the future as a consequence of the following: industrial companies are concentrating on core business and short-term profitability; new products and services increasingly involve cross-technology approaches; it is increasingly difficult for companies to cover all the new fields they wish to invest in; the time lag between new technology being discovered and having new technologies on the market has been reduced. Furthermore it is pointed out that companies are progressively outsourcing a significant share of their basic research to PROs in order to benefit from their expertise, infrastructure, low costs, and background technology [5].

The increased cooperation between industry and universities including the commercialization process has put in motion initiatives to strengthen the universities' position when it comes to rights of ownership to university inventions. Consequently, there has been much debate on whether the university personnel exemption is still appropriate. It has, for instance, been pointed out that university staff in Europe has not been active enough regarding patenting and commercializing their inventions [6].

In different countries the question of how to maximize the commercialization of university inventions has been solved using different means. Some countries have chosen to revise the laws regulating employee inventions including university staff. Others have introduced separate legislation for university inventions. A few have, on the other hand, chosen to maintain more flexible systems of researcher ownership, leaving the door open for contract-based regulation as to details.

12.1.3 *Outline*

The following sections on Europe focus on the legislation regarding university *inventions*, which for the intellectual property part means that focus lies on patent legislation. Since the legislation variety among European countries is quite large, the overview is divided into categories using a few countries as examples. The legislation of ownership to university inventions is divided into the following:

a. University ownership as a general rule

   1. Ownership regulated by labor law
   2. Ownership regulated by patent (or intellectual property) law but negotiable by way of contracts

b. Ownership to university inventions covered by the "professor's privilege"
c. Hybrid systems: ownership regulated by special legislation

Different ways of regulating ownership to university inventions might pose different challenges not only for scientists working in an international research community but also for industry research partners. Some of the features especially relevant to nanotechnology are briefly elaborated after each legislation overview as well as in a final concluding discussion.

12.2 University Ownership as a General Rule

12.2.1 *Ownership Regulated by Labor Law*

12.2.1.1 Germany

In Germany patent ownership is regulated in the Employees' Inventions Act, containing special provisions for academic research. Germany abolished its "professor's privilege" system in the year 2002 and moved toward a system of university ownership [7]. The purpose of this change was to stimulate patenting at universities by giving them the legal means to claim patents for themselves and exploit them commercially [8]. The aim of the new legislation was to promote knowledge and technology transfer at universities and bring out more innovations [9]. The law applies to everyone employed at the university [10], and to research carried out with private funding and in cooperation with industry partners, even commissioned research [11].

According to the German Employee Inventions Act (*Arbeit-nehmererfindungsgesetz, ArbEG*) inventions are divided into job-related inventions and free inventions (Section 4). Job-related inventions are inventions which are created during the employment. They can either directly arise from the duties assigned to the employee or be attributed to the experience gained or work made during the employment (Section 4.2). Job-related inventions must be notified to the employer without delay (Section 5). If the employer wishes to claim the rights to the invention he must notify the employee within four months. During this period of time the employee must keep the invention secret (Section 24). Inventions that are not job-related inventions are considered "free inventions." However, the employer must be notified of their existence, and he must be offered at least a non-exclusive license (Sections 18 and 19 *ArbEG*).

Section 42 *ArbEG* contains special provisions for universities. According to Section 42.1, the inventor has the right to publish his job-related inventions as long as he notifies the employer two months (as a rule) in advance. This means that the general rule of secrecy laid down in the Act is shortened by two months. According to Section 42.2, on the other hand, a university employee also has the right *not* to publish an invention of his, whereby he does not have to notify the university about the invention. In this case the university does not have the right to use the invention, even though it learns of it by other means. Section 42.4 regulates the right to remuneration. According to this rule a university employee is entitled to 30% of the gross-profits from the exploitation of his invention, which is far more than a normal employee receives. Furthermore a university employee has the right to use his invention for the purpose of teaching and further research (Section 42.3).

It can be noted that in Germany the universities have to decide whether to patent or not in a fairly short period of time. However, the rule is regarded as a compromise between the scientists' freedom of publication and research and the universities' interests as an employer.

12.2.1.2 Norway

Whereas two of the other Nordic countries, Denmark and Finland, have introduced an Act on University Inventions, Norway has chosen to maintain the provisions concerning university inventions in the Act on Employee Inventions [12]. The Act was revised in 2003 whereby the teacher's privilege was abolished and new regulations as to university inventions where introduced [13].

The rule according to Section 4 of the Norwegian Act on Employee Inventions is that the employer has the right to claim the rights to inventions made by an employee carrying out his work responsibilities, when the invention falls under the field of business of the employer. According to Section 5 the employee shall notify the employer of an invention without unnecessary waiting. The employer then has a period of four months to decide whether he wishes to claim the rights to the invention, whereby the employee has to be notified in writing (Section 6.1).

According to Section 6.3, teachers and scientific personnel at universities have a right to publish an invention if this was mentioned in the notification of the invention and if this does not violate the rights of a third party. In this case the employer (the university) does not have the right to claim the invention in accordance with Section 4. However, if the inventor has not published the invention within one year from the notification, the university can claim the rights to the invention. Furthermore, an employee who has claimed the right to publish cannot apply for a patent without the consent of the employer.

None withstanding the rules in the Norwegian Act on Employee Inventions the university (or the parties) can choose to negotiate other conditions.

12.2.2 *Ownership is Regulated by Patent (or Intellectual Property)*

12.2.2.1 United Kingdom

In the United Kingdom, rules on employee inventions are included in the Patents Act of 1977. According to Section 39.1 an invention made by an employee shall, as between him and his employer, be taken to belong to his employer if (a) it was made in the course of the normal duties of the employee or in the course of duties falling outside his normal duties, but specifically assigned to him, and the circumstances in either case were such that an invention might reasonably be expected to result from the carrying out of his duties; or (b) the invention was made in the course of the duties of the employee and, at the time of making the invention, because of the nature of his duties and the particular responsibilities arising from the nature of his duties he had a special obligation to further the interests of the employer's undertaking. Any other invention made by an employee shall, as between him and his employer, be taken for those purposes to belong to the employee (Section 39.2). However, the employer and the employee can freely regulate ownership between them by contracts, with the limitation in Section 42, stating that any term in a contract of employment which diminishes the right of the employee is unenforceable against the employee. Here it can be worth noting that Section 39 of the UK Patents Act refers to "inventions" as such, which means that the rule is not limited to patentable inventions [14].

In determining the scope of an employee's normal duties the job description and the employment contract are used. Whether the invention is considered to be made in course of these duties, for instance, depends on the nature of the invention, the relationship between the invention and the field in which the employer operates, the circumstances under which the invention was made, etc. [15]. Monotti & Ricketson hold that as inventions must be secret before their publication in a patent application, an important circumstance must be whether the relationship between the employer and the employee is of a confidential character [16]. The fact that an employee has used the employer's facilities and equipment to make an invention does according to them not necessarily lead to a situation where the employer possesses the rights to the invention.

Even though it at a first glance seems reasonable to conclude that the university based on Section 39 often might be entitled to inventions made by academics, it can be very difficult to show that an invention was made in the course of employment when it comes to academics [17]. Apart from the scope and duties of the employee and the interpretation of the employment contract, the university's ownership also depends on any express terms concerning ownership and any other agreements, statutes or policies dealing with intellectual property issues. For academics it can be difficult to determine when an academic is employed to invent [18]. Furthermore the general principles on employee inventions can be hard to apply for non-commercial activities [19].

British universities oftentimes regulate IP ownership in academic research by way of contracts with the employees or general policy statements. There does not seem to be one single model for IP policies in the United Kingdom, but each university approaches its intellectual property policy in its own way depending on its own objectives [20]. In the United Kingdom, research is often funded by the Research Councils, which might refer to resulting IP in their funding guidelines. For instance, the ESRC states that the ownership of the intellectual property arising from a research project should be clear from the outset. The ESRC assumes that intellectual property rights rest with the RO receiving the ESRC grant, unless otherwise stated to the contrary. Furthermore the guidelines contain rules on the distribution of royalty and income sharing [21].

In addition, the United Kingdom has introduced the so called "Lambert toolkit" for universities and companies that wish to undertake collaborative research projects with each other. The toolkit consists of a set of Model Research Collaboration Agreements and Consortium (multi-party) Agreements including documents for helping the parties to use and understand the agreements. Questions to address before entering into joint research projects are, for instance, questions regarding IP ownership (including the possibilities for the other parties to use the results in further research) and questions regarding the set-ups for secrecy and publication.

12.2.2.2 The Netherlands

In the Netherlands the ownership to employee inventions is regulated in Article 12 of the Dutch Patent Act of 1995. Article 12.1 prescribes that the rights to an invention made by an employee and for which a patent application has been filed, belong to the employee, unless the nature of the employment entails the use of the employee's special knowledge for the purposes of making such inventions. If this is the case the employer is entitled to the patent. Where the invention for which a patent application has been filed has been made by a person who performs services for another in the context of a training course, the person for whom the services are performed is, according to Article 12.2 entitled to the patent unless the invention has no connection with the subject of the services.

The implications of the Dutch legislation have been debated [22]. When an employee makes an invention which falls under the field of business of the employer, the employee is entitled to the invention if the nature of his employment does not require him to use his special knowledge [23]. None withstanding the rule in Article 12, the employer and the employee might, however, regulate the ownership by contracts, stating that the employee does not have any claims on future inventions falling outside his working field, as long as this does not take unfair advantage of the fact that the employee is financially dependant of the employer. These kinds of conditions seem to be quite common in Dutch collective agreements and individual employment contracts [24].

The ownership to university inventions is regulated in Article 12.3, stating the following: "Where the invention has been made by a person carrying out research in the service of a university, college, or research establishment, the university, college, or research establishment shall be entitled to the patent." This means that the starting point for university staff is the opposite of the one regarding other employees. All inventions made by persons carrying out research in the service of the university must be offered to the university — even if they were made, for example, on holidays and in a totally different field of science than the researcher is employed within [25].

12.3 Researcher Ownership Through "Professor's Privilege"

12.3.1 *Italy*

Italy has chosen to move in the other direction than most European countries, abolishing the universities' possibilities to register patents made by university employees. Instead Italy introduced an "academic privilege" by law 383 in 2001, stating that intellectual property rights to inventions made by public employees belong to the employees themselves. The universities, however, have the right to between 30% and 50% of the revenues of patented inventions. The legislation was amended in 2005 (*Decreto Legislativo* 10/2005) whereby a rule was introduced, stating that intellectual property is granted to the inventor's employer if the inventor receives funding from private organizations or a public organization different from their own [26].

12.3.2 *Sweden*

As the only Nordic country, Sweden has chosen not yet to revise its legislation concerning ownership to university inventions. However, the issue has been subject to discussion, and an official report (SOU 2005:95) was presented in 2005 whereby two different approaches were evaluated: (a) giving universities the right to acquire ownership to university inventions (technically maintaining the teacher's privilege in a reduced form), and (b) introducing a system with an obligation to notify inventions, leaving the door open for negotiations as to details. However, both solutions would require substantive legislative amendments, including amendments to the Higher Education Act, and no such changes have yet been made or announced.

12.4 Hybrid Systems: Ownership Regulated and Balanced Through Special Legislation

12.4.1 *Denmark*

Denmark introduced legislation on University Inventions in the year 2000 with the aim of securing that research results achieved through public funding benefit society through commercial exploitation (Section 1). The law is applicable on inventions that can be protected under patent or utility model law (Section 3).

According to Section 8, a research institution may claim the rights to an invention made by an employee (or several employees) as part of his work tasks. If an employee is part of a collaboration between several institutions, the institutions shall agree upon how the rights are to be divided between them. Regarding research projects carried out in cooperation with or funded by a third party, the institution might, according to Section 9, on behalf of itself and the employee agree to partly or wholly wave the rights to the inventions resulting from the project.

An employee who has made an invention shall, in accordance with Section 10 notify the employer thereof without delay. Institutions can regulate the details of the notification procedure, and they can decide that the obligation to notify inventions does not apply in fields of science where inventions are usually not made. The employee is not allowed to publish his invention before he has received a confirmation of the notification from the employer. According to Section 11 the employer has to decide whether to protect the invention within two months from the notification. The prescribed time limits can be prolonged by contract between the employer and the employee.

12.4.1.1 Finland

The Finnish Act on University Inventions came into force in 2007 with the aim of promoting the recognition, protection and exploitation of inventions made at Finnish universities (Section 2).

The Finnish law distinguishes between three kinds of research: (a) open research, (b) contract research, and (c) (inventions made in) other forms of research (Section 3). Open research is (a) research conducted at the university without any contract based involvement or funding from third parties, (b) research conducted using external funding, but without any other conditions than as to publication of the results, and (c) research which would fall under the definition of contract research but for which the parties have agreed that the research is to be open research. Contract research is defined as research conducted for a service fee and research where at least one external body participates as a co-researcher, investor, or participator having obligations regarding the research results or the implementation of the project.

Section 5.1 prescribes a duty for inventors to notify the university about inventions without delay. Hereby the inventor shall present his view on whether the invention is the result of open research, contract research, or some other kind of research. The university shall, no later than two months after the notification, notify the inventor of the actions planned in accordance with the Act, including the university's view on the inventor's statement as to the circumstances under which the invention was made.

According to Section 6 the university can claim the rights to inventions made in open research only if the inventor does not publish the invention or express a wish to exploit it commercially within six months from the notification. However, the inventor must be approached and asked about his intentions in advance. Regarding contract research, on the other hand, Section 7 prescribes, that the university has the right to claim ownership to such inventions within a time period of six months. As to inventions made in other forms of research, the university has a priority right to negotiate about the rights. If the invention is necessary for the activity of the university, the university has the right to get compensation for the right to use the invention (Section 8). It has been pointed out that the line between open and contract research is hard to draw, since researchers can be active in both activities at the same time [27]. If the university possesses a right to claim ownership to an invention according to Section 7, the inventor is not allowed to publish anything that could endanger a patent application, unless otherwise agreed (Section 11).

The starting point according to the Finnish legislation is that the researcher himself is entitled to his inventions, but that the rights can be transferred to the university under certain circumstances, either by law or by the way of contract [28]. It is, however, possible to draw up research contracts with conditions differing from the statutory provisions. Funding organizations such as Tekes usually have their own conditions for the exploitation of research results. The Finnish Act on University Inventions is only applicable on patentable inventions. Finally, it is also important to note that students and researchers with external funding are not covered by the act.

12.4.2 *Comments*

The legal situation regarding employee inventions is very different and not harmonized in Europe. This fact indicates that the importance of contractual arrangements are crucial, especially when research is conducted within the framework of large international cooperation contracts as is often the case within the nanotechnology sector. Generally speaking the European legislation allows for contractual arrangements, there are only very few legislative arrangements that are mandatory within Europe. In the following chapter we will discuss further some of the areas where actors have to be alert when entering into research cooperation.

12.5 Discussion

12.5.1 *Overview*

An overview of the legislation regarding ownership to university inventions in a number of European countries shows that there are some remarkable differences between them. Although many countries have revised their legislation almost within the same time frames, some of them have introduced complete opposite rules than the others. Some of these differences are elaborated and discussed below; however, a detailed comparative analysis is not possible within the scope of this chapter.

12.5.1.1 Inventions subject to the legislation

Whereas the Finnish legislation only covers patentable inventions, the German and the Danish ones include inventions that can be registered as utility models. The UK legislation goes one step further, since Section 39 of the Patents Act is not limited to inventions which can be protected.

12.5.1.2 The tension between publication and secrecy

Whereas some laws contain detailed rules on the researchers' rights to publish, others do not mention this at all. In some countries the researcher's right to publish might be a hinder for university ownership whereas university ownership is a condition for publication, for example, in Germany. According to the German law a researcher has two possibilities: he can either publish his results on the condition that the university can apply for a patent, or he can keep his results secret. This has lead to some criticism because it means that the German legislation does not allow researchers to put their research results in the public domain. For instance, Leistner has seen this as a fundamental problem, since the possibility of putting research results in the public domain for free use is part of the constitutional right of academic freedom set out in Article 5 *Grundgesetz* [29].

12.5.1.3 The definitions of inventions belonging to the employer

Several countries make a distinction between job-related inventions and free inventions. When this distinction is made, the rule seems to be that the employer has at least some rights to the job-related ones, even when the employer is a university or research institution. In the Netherlands, however, the employer has the right to *all* job-related inventions of university employees. Free inventions are, as a rule, the property of the employee, unless they are of some significance to the employer. In Germany also, free inventions must be offered to the university (and it is uncertain which inventions fall into this category since most inventions seem to fall under job-related inventions).

The distinction between free inventions and job-related inventions is not without uncertainties when it comes to university inventions. The fact that academics seldom do research for one single university at a time makes the definition of the boundaries for job-related inventions difficult. Researchers move between universities and research institutions and collaborate with other teams as well as industry partners. Possibly funding can be received from several sources, with several entities possessing ownership claims to different research results.

For instance, Finland has chosen to make a distinction between open research and contract research (and "other" research), which seems be at least a bit easier to identify. Finland has decided not to give university ownership to inventions made within the normal duties of a university-funded researcher (open inventions), but to extend the "professor's privilege" for these kinds of research results to all university employees. On the contrary, the results of contract research belong to the university. In this respect the Italian legislation with its newly introduced academic privilege is not that different from the Finnish one.

12.5.2 *Implications for Nanotechnology*

Research in nanotechnology is characterized by the following:

- Research in the field of nanotechnology is usually conducted in large interdisciplinary research teams, including industry partners. This might complicate ownership issues, notwithstanding which law applies.
- Research needs the infrastructure and facilities of either university or industry laboratories or both. This might have different consequences depending on which legislation applies; however, it mitigates the possibilities that inventions are considered so-called free inventions and belong to the researchers themselves.

- Research in the field of nanotechnology is usually funded through public funding as well as private funding. In some cases private funding can be a prerequisite for university ownership.

No matter which legislation applies — concerning nanotech inventions it is of utmost importance to regulate ownership issues in advance by contracts, including explicit rules on how to handle secrecy issues, and making sure that national rules do not contain binding publications rights for university researchers. Here it is especially important to note the German legislation, as it is as a rule not possible to draw up contracts with content not consistent with the provisions.

12.6 The US Perspective

12.6.1 *University Inventions in the United States*

Inventions by employees and faculty [30] in the United States vary by subject matter, and are governed by different laws. From 1978 onward, common law copyright was abolished for most forms of copyrightable subject matter, and in essence it is governed by exclusively federal law with exclusive federal jurisdiction [31] Patents are subject to exclusive federal jurisdiction as well [32]. Trademarks, the area that involves university faculty the least, is governed by federal law for interstate commerce, and by state law for intrastate commerce.

12.6.1.1 Copyrights

In the United States, copyrightable subject matter is treated very differently from patentable subject matter. The difference is so striking that it warrants discussion. Under the "works made for hire" doctrine [33], if the creator of the subject matter is an employee of the organization or university, as a matter of law, the organization or university is the "author" of the work [34]. Title to the work resides with the employer [35]. For "commissioned works," courts will look to a number of factors to decide whether the party working on the work is an "employee" or is an "independent contractor"

[36]. If the work is a "work made for hire," regardless of which employee did the work, he or she will not receive any credit for it. For example, if John Doe and Sally Smith wrote a software program for Microsoft, the program would be registered in the name of Microsoft Corp. and would carry the copyright notice of © 2010 Microsoft Corp. Registration is still required for US works as a prerequisite for bringing suit; registration is not required for foreign works [37]. Copyright notice was made optional for US works as of March 1, 1989, as a result of the United States joining the Berne Convention. In Patent Law, ownership of the work is not automatic, but is done by an agreement of assignment. Each inventor who makes a "conceptual contribution" to a claim — not necessarily the entire invention — must be listed as an inventor. The inventor(s) are always listed on the patent, even when the patent is assigned. Therefore, if John Doe and Sally Smith invented a new and improved computer keyboard, and assigned it to Microsoft, the patent would be in the names of Doe and Smith, with Microsoft listed as an assignee. Patent Law is discussed in more detail, *infra.*

Being declared the "author" is different from having an owner-ship interest in the work. The distinction is important. If a work is assigned, the transfer may be terminated by the grantor after a period of years, despite an agreement to the contrary [38]. On the other hand, if the employer is the statutory author of a work made for hire, the creator of the work may never get the rights back, and the employer enjoys all of the benefits of authorship.

Despite the seemingly obvious appearance of "work made for hire" status of works created by professors, a "teacher exception" was applied to those works for years [39]. It should be noted that a "teacher exemption" is not found anywhere in the Copyright Act. It is entirely judicially created. In *Weinstein v. University of Illinois* [40] Judge Easterbrook — a former University of Chicago law professor — wrote:

> A university "requires" all of its scholars to write. Its demands – especially the demands of departments deciding whether to award tenure – will be "the motivating factor in the preparation of" many a scholarly work. When the [plaintiff's dean] told [plaintiff] to publish he was not simultaneously claiming a right on the ground that the work had become a "requirement of duty" within the

meaning of [the university policy]. The University concedes in this court that a professor of mathematics who proves a new theorem in the course of. His employment will own the copyright to his article containing that proof. This has been the academic tradition since copyright law began [41].

Judge Posner — also a former University of Chicago law professor — wrote similarly in *Hays v. Sony Corp. or America* [42]. In recent years, however, the tables have been turning. In *Shaul v. Cherry Valley-Springfield Central School Dist.* [43] the court concluded that although the teacher's tests and homework assignments were created at home, the kind of materials were incidental to the teaching job and were works made for hire owned by the school district. A similar result was reached in *Vanderhurst v. Colorado Mt. College Dist* [44].

Over the last several years, owing to the recent development of online courses, computer software, video, multimedia, and other forms of copyrightable subject matter that can be valuable assets to a university and provide a university with a competitive advantage, the "teacher exception" has been interpreted to apply much more narrowly, with universities putting forth more claims to copyright. At least for now, journal articles and books belong to the professor, but that, too, may change [45].

12.6.1.2 Trademarks

The goodwill and reputation of a university is maintained through a program of trademark licensing. The university wants to make sure that the items sold are of good quality and reflect well on the university. The university also wants to generate revenue for the university. Some trademarks are very valuable, such as the reported $9 million per year that the University of Florida is paid by PepsiCo for the use of the mark "Gatorade" [46]. Some universities screen potential licensees. Michigan State posts an "Application for a Trademark License" on its website so that a potential licensee can submit samples, information relating to the company, and, of course, a processing fee. The University can then evaluate and approve or reject a trademark application [47].

Very few trademarks are conceived or developed by university employees such as professors and graduate students, except of course employees of a university's graphic arts department. Infringement is more of an issue.

12.6.1.3 Patents

In the United States, patents are governed by one set of federal laws, 35 U.S.C. 101 *et seq.* With the exception of a few issues of state law that may be related to contract interpretation, but not dealing with standing, there is no diversity of law. On appeal, all cases — regardless of the regional district court in which the case originated — go to the Court of Appeals for the Federal Circuit [48]. Then the case proceeds to the US Supreme Court. Unlike the other areas of Intellectual Property Law, where there can be inter-circuit splits of authority that need to be resolved by either Congress or the US Supreme Court [49], in patents there is only the Federal Circuit law. Of course, when the US Supreme Court disagrees with the Federal Circuit, the case will go up to the Supreme Court for clarification [50].

Universities have always been sights of research and development. In recent years, however, the number of inventions and the licensing activity have increased greatly owing to the amount of money that comes in to a patent owner. These are but a few of the examples of recent commercialized inventions. Stanford University developed and licensed the basic technology behind the Google search engine. Carnegie Mellon University did the same for the Lycos search engine. Florida State University developed and transferred the technology for Taxol, a cancer drug from Bristol-Meyers Squibb [51]. The University of Texas at Arlington developed and patented the software which the maker of the cellphone BlackBerry, Research in Motion, infringed [52]. For one of the largest university patent deals in history, Professors Raymond Schinazi and Dennis Liotta of Emory University licensed AIDS drugs to Gilead Sciences and Royal Pharma for $525 million dollars [53].

These large patent transactions have brought publicity and revenue to many universities. Universities themselves, however, do not have the funding for all of the research. Therefore, the outside sources, whether they be private commercial entities, public commercial entities, or the government are needed to satisfy the fiscal demands of university research. To complicate matters, many different parties, such as professors, students, fellows, the university, the government, funding foundations, and private industry may have different agendas and goals.

The following sections address some of the issues, concerns, and litigated cases related to faculty created inventions.

12.7 University Ownership is the General Rule in the United States As Well

12.7.1 *Duty to Disclose and University Ownership*

The biggest single problem with university faculty is disclosure. There are two main kinds of disclosure. One is disclosing all business that one conducts to the Conflict of Interest Office. Universities have stringent policies on outside work and many faculty either ignore or underreport. There are many different issues involved with conflict disclosure. The first is "How much money is the faculty member making?" Related to that is disclosing that a company is underwriting the research and the resultant paper. These are conflict of interest disclosures and should be made by all faculty. The second type of disclosure relates to disclosing all inventions on which the faculty member is work- ing. The discussion below concentrates on the second type of disclosure.

Most universities have patent policies that state that the university owns any invention that "was developed in the course of or pursuant to a sponsored research agreement" [54] with the university or "was developed with significant use of funds or facilities" administered by the university [55]. The policies are slightly different and are often written by attorneys with varying intellectual property experience [56]. In general, though, a faculty member has a duty to disclose technology and inventions that grow out of his research, and most professors, graduate students, fellows, etc., do so willingly [57]. Despite general support for the policies, there can be disputes over whether an invention was created for the university, or whether it was invented on the employee's time and expense.

In those cases in which a university employee inventor fights over ownership or challenges ownership by the university, most courts have held that the school's Patent Policy controls and the ownership of the invention resides in the university [58]. Most universities have patent policies, but they vary slightly from school to school, and can have a bearing on who ultimately owns the invention. In the vast majority of cases, though, the Patent Policy of the University is in binding. The three cases that immediately follow are different takes on "duty to disclose" cases.

Fenn v. Yale University [59] deals with what can happen when the professor is not forthright with the University. Dr. John B. Fenn was a professor emeritus at Yale [60]. He first publicly disclosed his new invention in June 1988 at the Annual Conference of the American Society of Mass Spectrometry ("ASMS") in June 1988. Dr. Fenn and others recognized the importance of his invention in the field of medicine because it solved a "long-standing scientific problem." His co-inventors believed that the invention was a "scientific breakthrough" and was "revolutionary" [61]. Dr. Fenn did not disclose the invention at that time to Yale, in violation of the Administrative Policy.

Dr. Fenn finally disclosed the invention to Yale's Office of Cooperative Research ("OCR") on April 6, 1989. While Dr. Fenn knew of the importance of the invention, he relayed to Yale that it was of limited commercial value because it was a "use" patent instead of an "apparatus" patent, and that it would be very difficult to prevent infringement. He also said that a patent application must be filed before June 1, 1989, the one-year anniversary of the pubic disclosure of the invention [62]. Dr. Fenn did not disclose at the time that Pfizer had expressed an interest in the commercial viability of the invention. Because of his lukewarm reports, Yale asked Dr. Fenn if the company that he and some other scientists started, Analytica, would be willing to pay for a patent application in return for a license to the invention. Fenn claimed that he did not know whether Analytica had an interest; in reality Analytica was very interested in commercializing the invention.

On May 5, 1989, the OCR called Dr. Fenn to discuss the invention. Fenn told them that he was "up to his ears preparing for several scientific conferences" and that beginning on May 20, 1989, he would be away from Yale for several weeks until after June 1989, the filing deadline. He did not tell OCR that his co-inventor was available and in town. Yale did not file the patent application. Because Dr. Fenn was an expert in the field, Yale relied on his representations that the invention would have little commercial value, if any.

Little did they know that Dr. Fenn had filed a patent application covering the same technology in his name on May 19, 1989. The patent application was financed by Analytica. The patent, US Patent No. 5,130,538, was issued on July 14, 1992. There was already a license agreement in place between Fenn and Analytica. Dr. Fenn did not disclose this either. Yale found out about the invention when a third party contacted Yale about licensing the patent. Yale asked Dr. Fenn to assign the patent, and Dr. Fenn refused. Yale then asked Analytica to enter into a license agreement. Analytica held in escrow the monies due to Dr. Fenn. Analytica initially made payments to Yale, but in 1995, it ceased. The amount of back royalties owed to Yale was over $1.7 million.

Dr. Fenn sued alleging conversion, theft, tortuous interference with business, and violations of the Connecticut Unfair Trade Practices Act [63]. Yale counterclaimed seeking an accounting and assignment of the patent, as well as damages for breach of contract and fiduciary duty, fraud, negligent misrepresentation, conversion, theft, and unjust enrichment. I will discuss only the breach of contract and fraud portions of the opinion.

The court held that "university patent policies such as Yale's have long been recognized as a valid and enforceable part of the employment" [64]. The court continued that "employment agreements [pursuant to which an employee assigns to her employer inventions she developed during the course of employment] are valid and enforceable and...do not violate public policy" [65]. Dr. Fenn claimed that he was only bound by the 1975 Patent Policy, and that Yale had modified, amended, or superseded that policy. The District of Connecticut disagreed, holding that there was language in the 1975 Patent Policy that clearly stated that Yale "could revoke or amend that policy at any time" [66]. Yale did amend its policy in 1984, 1988, and 1989, each time reserving the right to make further modifications. In addition, there was evidence that Dr. Fenn had assented to the policies.

The District Court also concluded that there was sufficient evidence to show that Dr. Fenn engaged in fraudulent misrepresentation and fraudulent nondisclosure. "The essential elements of an action in common law fraud are that (1) a false representation was made as a statement of fact; (2) it was untrue and known to be untrue by the party making it; (3) it was made to induce the party to act upon it; and (4) the other party did so act upon the false representation to his injury" [67]. Fraud by nondisclosure has the first three elements, but "involves the failure to make a full and fair disclosure of known facts" [68]. The Court found that Dr. Fenn made the misrepresentations and omissions in order to induce Yale not file a patent application, and Yale acted upon them. The Court ruled in favor of Yale. The case was subsequently affirmed [69].

In *Chou v. University of Chicago* [70], Joany Chou was graduate student and eventual post-doctoral fellow in the University of Chicago's Department of Molecular Genetics and Cell Biology. She was the graduate assistant of Dr. Roizman, the named sole inventor of US Patent No. 5,328, 688, a drug that was used to fight herpes simplex infections. Dr. Chou did significant work on that invention and told Roizman that the drug should be patented. Dr. Roizman disagreed with her, but filed the Patent anyway in his own name. The University of Chicago has a Patent Policy that required assignment of all patents to the university. The Patent was then assigned by the University of Chicago to Institut Merieux, a French Company that had supported the research. Institut Merieux, in turn, licensed the patented invention to ARCH, which in turn licensed the invention to Aviron. She sued the University of Chicago, Dr. Roizman, and Aviron.

Dr. Chou claimed that the University of Chicago contracts did not include patents [71]. The Faculty Handbook, however, makes reference to the patent statutes as "patent policies" within a section called "Academic Policies."

> Every patentable invention or discovery that results from research or other activities carried out at the University, or with the aid of its facilities or funds administered by it, shall be the property of the

University, and shall be assigned, as determined by the University, to the University, to an organization sponsoring the activities, or to an outside organization deemed capable of administering patents [72].

The above policy stated that "the contents of this handbook do not create a contract or agreement between the University and the individual" [73]. So it seems as though Dr. Chou did not sign away her invention.

The court disagreed. The Federal Circuit claimed that the statement must be read in light of the statement that followed it. "The basic terms and conditions of the employment agreement are set out in the letter of appointment received from the Provost's Office" [74]. Dr. Chou's letter of appointment said that the appointment was subject to "the administrative policies of the University," which included the obligation to assign inventions to the University. Chou was obligated to assign her inventions "even though she never specifically agreed to do so" [75].

Contrast the above with *Board of Trustees of the Leland Stanford Junior University v. Roche Molecular Systems, Inc.* [76]. The University sued Roche, the assignee of a patent for patent infringement by practicing Stanford's patent. The patent in question was invented by Mark Holodniy and others [77]. In 1988, Holodniy joined a laboratory at Stanford as a Research Fellow in the Department of Infectious Disease and signed a "Copyright and Patent Agreement" ("CPA") that obligated Holodniy to assign his inventions to the university [78]. In February 1989, Holodniy began to do research with a company, Cetus, and signed a "Visitor's Confidentiality Agreement ("VCA") with Cetus. In that agreement, Holodniy "will assign and do[es] assign to CETUS, my right, title, and interest in each of the ideas, inventions and improvements" that he devised as a result of his work with Cetus [79]. Holodniy had competing assignments with both Stanford and Cetus. Later, certain Cetus inventions, including the one in question, were later purchased by Roche [80].

The Federal Circuit had the task of construing the two competing assignments. The court looked at the Holodniy/Stanford agreement [81]. The language in the Stanford agreement read, "I *agree to assign or confirm in writing* to Stanford and/or Sponsors that right, title and interest in ... such inventions as required by Contracts or Grants" [82]. The Federal Circuit construed such language as "an agreement to assign," which requires a subsequent document in order to be valid. It is a promise to assign rights in the future, not an immediate transfer of expectancy interests. Therefore, Holodniy agreed to only assign his invention rights to Stanford at a future undisclosed time [83]. Stanford, in order to distinguish itself from other schools, stated in its *Administrative Guide to Inventions, Patents, and Licensing* that "[u]nlike industry and many other universities, Stanford's invention right's policy allows all rights to remain with the inventor if possible" [84].

Holodniy's CETUS agreement stated: "I will assign and *do hereby assign* to CETUS, my right, title, and interest in each of my ideas, inventions and improvements" [85]. In contrast to the Stanford agreement, the language of "do hereby assign" constituted a present assignment of Holodniy's future inventions to Cetus [86]. Cetus gained immediate title, whereby Stanford did not. The situation a conflicting assignments was resolved, though not in Stanford's favor [87].

On June 6, 2011, the United States Supreme Court affirmed the Federal Circuit's decision by a seven-to-two holding [88]. The Supreme Court held that unless there is an agreement to the contrary, an employer — in this case, a university — does not automatically get ownership rights in an invention. The Court held that the inventor must expressly grant his ownership rights to the employer in order for the employer or university to obtain those rights. Because the agreement with Stanford was an ambiguous "agreement to agree" and the agreement with Cetus was a clear-cut valid assignment, the Cetus agreement was binding, notwithstanding the existence of the Bayh-Dole Act.

The *Stanford* decision is the exception, rather than the rule, but it shows that even at a school the caliber of Stanford, the people who write the documents purporting to effectuate title should be very careful. Note that in *Chou*, the inventor did not sign anything and an assignment was deemed to have happened because of the university's policies, and that in *Stanford*, the inventor did sign an agreement, but the agreement had a flaw in it that caused it to not be an assignment.

Once an invention is disclosed, it goes to the Technology Transfer Committee. The Technology Transfer Committee will then evaluate the invention and decide whether (1) it is interested in patenting and/or commercializing the invention, (2) additional information is needed before the university can proceed with its decision, or (3) it is not interested in patenting and/or commercializing the invention. In the event that the university decides that it is not interested in the invention, the rights to the invention will revert back to the faculty member. The faculty member can then decide whether to patent and commercialize or not.

12.7.2 Co-Inventorship

The situation can become quite complicated in the university setting with a professor having several graduate students or fellows working for him. There is a difference between the concept of authorship on an academic paper versus a true inventive contribution, They can be difficult to distinguish, at times Graduate students and technicians are often doing the actual experiment that is designed by the faculty member. Professors can have a tendency to want to claim sole inventorship because of their egos. In those cases in which the patent is valuable, there is also a financial incentive for professors to claim sole inventorship — increased royalties.

Graduate students, fellows, and other faculty have the Patent Act as their friend in the situation of being left off an invention. If nonjoinder of an actual inventor is proved by clear and convincing evidence, a patent can be rendered invalid [89]. That is an extreme measure and is not invoked all that often. But, it is a very good bargaining chip to play. More commonly, an un-named patentee claims that the inventor made a "mistake" by not naming him or her [90]. If a patentee can demonstrate that he or she should be a co-inventor — make a conceptual contribution to a claim — a district court must order correction of the patent under 35 U.S.C. § 256, thus saving it from being rendered invalid [91]. Co-inventorship must be shown by clear and convincing evidence [92]. In *Chou v. University of*

Chicago [93], the graduate student who was omitted from the patent was able to show that she should be listed as a co-inventor.

In situations in which there are multiple parties working on a project of an invention, it is the safest to name almost everyone who has worked on the project, and let the Technology Transfer Committee ultimately decide, after consultation with several people, who is a co-inventor and who is not. It is better to initially be over-inclusive than under-inclusive.

12.7.3 *Universities and Payment of Faculty Inventors*

In modern times, universities are quite generous with the remission of royalties to inventors. Long gone are the days of a bonus of $100 when a product is patented. Remember that there are no "work made for hire" provisions in Patent Law, so everyone on the team who makes a conceptual contribution to a claim in entitled to be named as a co-inventor. The higher royalties paid to faculty inventors in recent years, are the result of the financial benefit of the university [94], especially after the passage of the Bayh-Dole Act [95]. They are also meant to encourage prompt and full disclosure.

Universities have wide ranging royalty schemes, but they all reward the faculty or staff inventor with a substantial royalty [96]. The following are examples of university royalty payment schedules:

12.7.3.1 Cornell University [97]

The royalty is based upon "net royalty expenses" (gross revenues received by university less directly assignable expenses such as patent prosecution and licensing costs) [98]. In the case of a Cornell invention, the university, through CCTEC, will receive all license revenue and, in recognition of the efforts and contributions of the inventor, distribute total net license revenue as follows:

- One-third (33.3%) to the university inventor(s) in recognition of their contribution. In the case of university co-inventors, this distribution will be shared.
- One-third (33.3%) will be divided as follows: (a) 60% to the inventor's research budget, subunit (typically the inventor's department, school, section, or center) and university

unit (typically the inventor's college) in a manner to be determined by the dean of the unit (or, for research centers in the Research Division, the Vice Provost for Research), and (b) 40% to the university for general research support.

- One-third (33.3%) to the university to provide CCTEC with operating funds to cover the cost of service provided to the university with regard to intellectual property matters and particularly to cover direct costs, where license revenue or other cost recovery has not been achieved.

12.7.3.2 University of Michigan [99]

Up to $200,000

- 50% to the inventor(s)
- 17% to the originating unit(s)
- 18% to the originating school, college, division, or other responsible center(s)
- 15% to central administration

$200,000 to $2,000,000

- 30% to the inventor(s)
- 20% to the originating unit(s)
- 25% to the originating school, college, division, or other responsible center(s)
- 25% to the central administration

Over $2,000,000

- 30% to the inventor(s)
- 35% to the originating school, college, division, or other responsible center(s)
- 35% to central administration

12.7.3.3 Yale University [100]

Distribution of Net Royalties. The Net Royalties as defined above shall be divided between the Inventor(s) (as defined under the patent law) and the University as follows:
The first $100,000 of net royalties

- 50% to the inventor(s)
- 50% allocated to the general support of University research

Net royalties between $100,000 and $200,000

- 40% to the inventor(s)
- 60% allocated to the general support of University research

Net royalties exceeding $200,000

- 30% to the inventor(s)
- 70% allocated to the general support of University research

12.7.3.4 Emory University [101]

Gross revenue up to $25,000

- 100% to the contributor (inventor)

Net revenue up to $4 million

- 33% to the contributor
- 33% to the department
- 10% to the school/center
- 25% to the President/general share

Net revenue above $4 million

- 25% to the contributor
- 33% to the department
- 17% to the school/center
- 25% to the President/general share

12.8 Experimental use in the United States

12.8.1 *The Experimental use Exception: What Is Left of It*

Many researchers use patented material to do their own research. After, *Madey v. Duke* [102], the "experimental use" defense is very narrow with respect to universities. Before discussing the *Madey* case, it is helpful to look at the origin of the "experimental use defense," also called the "research exemption." The "experimental use" defense does not have a statutory basis. Its origin appears to be traced back to 1813, when Justice Story stated:

> [I]t could never have been the intention of the legislature to punish a man who constructed such a machine merely for philosophical experiments, or for the purpose of ascertaining the sufficiency of the machine to produce its desired effects [103].

In the following years, it was accepted that "an experiment with a patented article for the sole purpose of gratifying a philosophical taste, or curiosity, or for mere amusement is not an infringement of the rights of the patentee" [104]. Many in academia agreed with the experimental use defense, and "have generally agreed that the unique nature of non-commercial university research required a research exemption" [105].

In 1984, in *Roche v. Bolar* [106] the Federal Circuit found the research exemption to be "truly narrow." Roche's patent on flurazepam hcl [107] (Dalmane) expired on January 17, 1984. Bolar wanted to get into the very lucrative business of having a generic equivalent to Dalmane. Bolar obtained 5 kilograms of flurazapam from a foreign manufacturer in order to do the studies that were necessary for a New Drug Application to the US Food and Drug Administration. The patent was still in term while Bolar was doing experiments on the drug. Roche sued Bolar for patent infringement. Bolar's defense was that it was an experimental use and that public policy favored their research because to wait would give the patent holder another two or three years of protection – the time that it took for FDA approval.

The court followed *Pitcairn v. United States* [108], and held that a "tests, demonstrations, and experiments...[which] are in keeping with the legitimate business of the [alleged infringer]" are infringements for which "experimental use is not a defense" [109]. Because Bolar's use was solely for business purposes and not for "amusement, to satisfy idle curiosity, or for strict philosophical inquiry" it did not qualify for the exemption, and was patent infringement. (*Roche v. Bolar* led to Congress acting very quickly to pass The Drug Price Competition and Patent Term Restoration Act of 1984 [110], popularly known as the Hatch-Waxman Act, which "effectively established a robust generic drug industry in the United States") [111].

In *Madey v. Duke* [112], the Federal Circuit continued its experimental use must be for "amusement, to satisfy idle curiosity, or for strict philosophical inquiry" and rejected Duke's experimental use defense.

> Our precedent clearly does not immunize use that is in any way commercial in nature. Similarly, our precedent does not immunize any conduct that is in keeping with the alleged infringer's legitimate business, regardless of commercial implications. For example, major research universities, such as Duke, often sanction and fund research projects with arguably no commercial application whatsoever. However, these projects unmistakably further the institution's legitimate business objectives, including educating and enlightening students and faculty participating in these studies. These projects also serve, for example, to increase the status of the institution and lure lucrative research grants, students, and faculty [113].

The case was remanded back for a determination of all aspects of the use. Administrators and faculty need to be careful with what they might think is "experimental use." Under *Madey,* even if the use is non-commercial and relates to the universities goals, it might be classified an infringing use, and not an experimental use.

12.9 Government Ownership of Inventions in the United States

12.9.1 *The Bayh-Dole Act*

Universities in recent years have been the places where many valuable inventions are developed. Universities have limited research budgets, so often it is the federal government that provides the funding for the basic of applied research. Prior to 1980 — the year that the Bayh-Dole Act [114] was implemented — the federal government had ownership of the inventions and would grant a non-exclusive license to any party who wanted to practice or commercialize the invention [115]. This arrangement did not provide an incentive for universities to work on federally funded inventions, and as a consequence, few did. Because a large amount of university research was paid for with federal funds, very little intellectual property was developed or offered to third parties. The Bayh-Dole Act changed all that.

> It is the policy and objective of the Congress to use the patent system to promote the utilization of inventions arising from federally supported research or development; to encourage maximum participation of small business firms in federally supported research and development efforts; to promote collaboration between commercial concerns and nonprofit organizations, including universities; to ensure that inventions made by nonprofit organizations and small business firms are used in a manner to promote free competition and enterprise without unduly encumbering future research and discovery; to promote the commercialization and public availability of inventions made in the United States by United States industry and labor; to ensure that the Government obtains sufficient rights in federally supported inventions to meet the needs of the Government and protect the public against nonuse or unreasonable use of inventions; and to minimize the costs of administering policies in this area [116].

For the first time, university recipients of federal funding now have the option of retaining ownership of the inventions and licensing the inventions commercially. The federal government retains (1) a non-exclusive right to practice the invention [117] and (2) "march-in" [118] rights that allow the government to require the owning party to third parties for health and safety issues [119]. Federal rights attach to all federally funded inventions whether or not they have been disclosed to the university. There is a preference to license inventions to US companies [120]. Universities must have blanket, effective, invention disclosure policies so that at a minimum, they can comply with the law and properly administer the government's rights [121]. Because the government can have a non-exclusive right to practice the invention, care should be taken when granting licenses. On cannot have an "exclusive" license if the government is a non-exclusive licensee.

Since the Bayh-Dole Act was passed, universities have developed or "beefed-up" their technology transfer offices to help exploit the inventions that are now owned by the university. The technology transfer process is not the same at all universities, but it does involve the same disclosure procedure as discussed below. A technology committee is made up of personnel with various degrees and specialization. Nonetheless, they evaluate the invention to see if it is commercially viable. Often someone from patent counsel's office will file a provisional patent. The decision to file an actual patent is sometimes done by university patent counsel (if the school has one) or by outside patent counsel.

It cannot be overstated how much of a boon to business the Bayh-Dole Act has been for universities [122].

12.10 Industry May Have a License in Certain Circumstances

12.10.1 The "Shop Right"

The government does not fund all university research. Most of the research is funded by private industry. Private industry does not provide the capital for a project or general research unless they can get remunerated for it. There is no work made for hire doctrine in patent law where the university gets automatic ownership and authorship. But private industry can get a "shop right." "A 'shop right' is generally accepted as being a right that is created at common law when the circumstances demand it, under principles of equity and fairness, entitling an employer to use without charge an invention patented by one or more of his employees without liability for infringement" [123]. It is dictated by principles of equity and fairness, but not all courts agree on the particular set of circumstances for that right [124].

For example, many courts characterize a "shop right" as being a type of implied license [125], and thus the focus is often on whether the employee engaged in any activities, for example, developing the invention on the employer's time at the employer's expense, which demand a finding that he impliedly granted a license to his employer to use the invention. Other courts characterize a "shop right" as a form of equitable estoppel [126]. The focus on those cases is on whether the employee's actions — consent or acquiescence to his employer's use of the invention — demand a finding that he is estopped from asserting a patent right against his employer.

Regardless of how it is labeled, there is a slight possibility that a department, or even a faculty member, of a university that is funded by industry may have to give industry a non-exclusive license to practice the invention for its own purposes. This problem does not occur too often because industry has long and carefully drafted agreements for participating faculty members. Faculty members must disclose the work or consulting that they are doing to avoid a conflict of interest.

12.11 Conclusion

In the United States, generally, the university will own through assignment all patents that are prepared in connection with a researcher's work at the university. There are either assignment documents that are signed by the faculty member, graduate student, fellow, etc., or the assignment is part of the policies that are included by reference in the employment agreement that the faculty member signs. There can be inarticulate drafting which can defeat an assignment, but they are few and far between. All people who made a conceptual contribution to a claim must be mentioned co-inventors on an invention.

The universities that commercialize a faculty created invention are now quite generous with the royalty remittances to the faculty member(s) and the department(s), and some faculty members can become quite rich as a result. The days of "pure academic research" and detachment from the corporate world are getting fewer and fewer, especially in this current economic climate. Interdependence and cooperation are the new rule.

12.12 Nanotechnology: Some Final Reflections

Nanotechnology lies in the forefront of a new technology that is very much driven by large-scale university research. It represents a field where public and private, noncommercial and entrepreneurial interests are intertwined in many complicated ways.

It is important that researchers in this field are motivated and rewarded. In this field it does not seem to be a realistic option that individual researchers act as patent applicants or patent holders. What seems to be important is that patent holders apply a responsible policy for licensing and that early patenting does not lead to obstacles for further research and scientific progress.

Our conclusion, based on the study of the legislative framework in Europe and the United States, is that there are no real obstacles for a reasonable common policy, but that it requires careful contractual management of research projects and of cooperation between universities and researchers from different states, private corporations and other institutions. Within this framework it is reasonable that common platforms of patented knowledge are created that can be openly used by everyone for free or against a relatively low royalty. Inventors should be rewarded by other means than by ownership to patents and public-private ownership should be used in order to guarantee access to the new technology.

Acknowledgements

The authors want to recognize considerable help by Pia Björkwall in the drafting of the European part of the article for which they express their thanks. The responsibility for all mistakes and shortcomings remains on the authors.

References

1. *See*, e.g., Berndt Godenhielm, Copyright and Industrial Property. Chapter 7. Employee Inventions, in *International Encyclopedia of Comparative Law*, vol. XIV.
2. This rule reflects the general approach adopted by the Rome Convention of 1980 on rules concerning the law applicable to contractual obligations, Article 6. The Convention has now been replaced by the Regulation (EC) No. 593/2008 of the European Parliament and of the Council of 17 June 2008 on the law applicable to contractual obligations (Rome I).

3. *See*, e.g., Niklas Bruun, Högskoleforskning och immaterialrätt. *NIR* 2002, 612. *See* German Employee Inventions Act 1957, Section 42.1.

4. *See*, e.g., the Lisbon Strategy 2000.

5. *See* European Commission Working Paper: Management of intellectual property in publicly-funded research organizations: Towards European Guidelines. Available at http://ec.europa.eu/research/era/pdf/iprmanagementguidelines-report.pdf. Accessed 16 July 2010.

6. For a German discussion *see*, e.g., Stephan Barth, Zum 40. Geburtstag des Hochschullehrerprivilegs nach § 42 ArbnEG. *GRUR* 1997, 880 ff. For the Nordic discussion, *see* Wolk, Sanna, Arbetstagares immaterialrättigheter. Rätten till datorprogram, design och uppfinningar mm. i anställningsförhållanden. Norstedts juridik. 2006., 195–214.

7. About the amendments *see*, e.g., Bartenbach & Volz, Erfindungen an Hochschulen – Zur Neufassung des § 42 ArbEG. *GRUR* 2002, 743–758.

8. Falck & Schmaltz, University Inventions: Classification and Remuneration in Germany, the Netherlands, France, the UK, the U.S and Japan, *IIC* 2005, 912.

9. BT-Drs 14/5975, 2.

10. Bartenbach & Volz, Erfindungen an Hochschulen – Zur Neufassung des § 42 ArbEG. *GRUR* 2002, 745.

11. Leistner, Farewell to the "Professor's Privilege" – Ownership of Patents for Academic Inventions in Germany under the Reformed Employees' Inventions Act 2002. *IIC* 2004, 863.

12. Lov av 17. april 1970 nr 21 om retten til oppfinnelser some er gjort av arbeitstakere.

13. Revised by Lov 6 des 2002 nr. 73 (i kraft 1 jan 2003 iflg. res. 6 des 2002 nr. 1348).

14. Bently & Sherman *Intellectual Property Law*, Oxford University Press, 2009, 534–535.

15. Bently & Sherman, *Intellectual Property Law*, Oxford University Press, 2009, 536.

16. Monotti & Ricketson, *Universities and Intellectual Property*, Oxford University Press, 2003, 164.

17. Bently & Sherman, *Intellectual Property Law*, Oxford University Press, 2009, 534.

18. Monotti & Ricketson, *Universities and Intellectual Property*, Oxford University Press, 2003, 245.

19. Monotti & Ricketson, *Universities and Intellectual Property*, Oxford University Press, 2003, 246–248.

20. Monotti & Ricketson, *Universities and Intellectual Property*, Oxford University Press, 2003, 297–298.

21. *See* ESRC Research Funding Guide April 2010, 21. Available at http://www.esrcsocietytoday.ac.uk/ESRCInfoCentre/Images/ESRC%20Research%20Funding%20Guide%20April%202010%20v2_tcm6-9734.pdf. Accessed 6.7.2010. Whoever holds the intellectual property, it should be clear what the distribution of any income might be. For income up to $10,000 the ESRC will assume that this will normally rest with the Grant Holders and the research team, unless otherwise stated to the contrary. For income between $10,000 and $50,000 the ESRC will assume that the major part will accrue to the RO, but the ESRC reserves the right to reclaim up to one third of the total, up to the value of its original grant. For income over $50,000 the ESRC requires that the major part of any income should accrue to the RO, and again reserves the right to reclaim up to one third of the total up to the value of its original grant. In any event, the RO's regulations on the internal sharing of royalty income, as set out in its Statutes, shall take precedence.

22. Quaedvlieg, Denker im Dienstverhältnis. Kernfragen des Arbeitnehmer-Immaterialgüterrechts — Eine Analyse nach niederländischem Recht. GRUR Int. 2002, 907.

23. Quaedvlieg, Denker im Dienstverhältnis. Kernfragen des Arbeitnehmer-Immaterialgüterrechts — Eine Analyse nach niederländischem Recht. GRUR Int. 2002, 907.

24. Quaedvlieg, Denker im Dienstverhältnis. Kernfragen des Arbeitnehmer-Immaterialgüterrechts — Eine Analyse nach niederländischem Recht. GRUR Int. 2002, 907.

25. Quaedvlieg, Denker im Dienstverhältnis. Kernfragen des Arbeitnehmer-Immaterialgüterrechts — Eine Analyse nach niederländischem Recht. GRUR Int. 2002, 907.

26. Baldini, Grimaldi & Sobrero, To patent or not to patent? A survey of Italian inventors on motivations, incentives, and obstacles to university patenting, *Scientometrics*, **70**(2), 2007, 333–354.

27. Bruun & Välimäki, *Korkeakoulukeksinnöt. Uuden lain tavoitteet, tulkinta ja käytäntöön soveltaminen*, Helsinki 2007, 44.

28. Bruun & Välimäki, *Korkeakoulukeksinnöt. Uuden lain tavoitteet, tulkinta ja käytäntöön soveltaminen*, Helsinki 2007, 125.

29. Leistner, Farewell to the "Professor's Privilege" – Ownership of Patents for Academic Inventions in Germany under the Reformed Employees' Inventions Act 2002, *IIC* 2004, 868–869.

30. We are using the term "faculty" to refer to all university employees.

31. 17 U.S.C. § 301. There is some debate as to whether non-copyrightable subject matter, like unfixed works, is covered by common law copyright or not.

32. *See*, Bonito Boats, Inc. v. Thunder Craft Boats, Inc., 489 U.S. 141 (1989) (State law that governs patentable material is pre-empted.)

33. The "works made for hire" doctrine is codified in 17 U.S.C. § 107. A "work made for hire" is —
    (1) a work prepared by an employee within the scope of his or her employment; or
    (2) a work specially ordered or commissioned for use as a contribution to a collective work, as a part of a motion picture or other audiovisual work, as a translation, as a supplementary work, as a compilation, as an instructional text, as a test, as answer material for a test, or as an atlas, if the parties expressly agree in a written instrument signed by them that the work shall be considered a work made for hire. For the purpose of the foregoing sentence, a "supplementary work" is a work prepared for publication as a secondary adjunct to a work by another author for the purpose of introducing, concluding, illustrating, explaining, revising, commenting upon, or assisting in the use of the other work, such as forewords, afterwords, pictorial illustrations, maps, charts, tables, editorial notes, musical arrangements, answer material for tests, bibliographies, appendixes, and indexes, and an "instructional text" is a literary, pictorial, or graphic work prepared for publication and with the purpose of use in systematic instructional activities.

34. Despite its importance, the term "author" is not a defined term in the U.S. Copyright Statutes. This is particularly interesting for the definition of a "compilation" and a "derivative work" in § 101, and § 102 – "Copyrightable Subject Matter" include the terms "original work of authorship." In 1990, in Feist Publications, Inc. v. Rural Telephone Service Co., Inc., 499 U.S. 340, 353 (1991), the Supreme Court held that "author" is the creator of original material. Before that, an "author" could be someone who collected data through an expense of labor, time, and capital.

35. *See*, e.g., Community for Creative Non-Violence v. Reid, 490 U.S. 730, 109 S.Ct. 2166 (1989).

36. *See*, e.g., Community for Creative Non-Violence v. Reid, 490 U.S. 730, 109 S.Ct. 2166 (1989). In determining whether a hired party is an employee under the general common law of agency, we consider the hiring party's right to control the manner and means by which the product is accomplished. Among the other factors relevant to this inquiry are the skill required, the source of the instrumentalities and tools, the location of the work, the duration of the relationship between the parties, whether the hiring party has the right to assign additional projects to the hired party, the extent of the hired party's discretion over when and how long to work, the method of payment, the hired party's role in hiring and paying assistants, whether the work is part of the regular business of the hiring party, whether the hiring party is in business, the provision of employee benefits, and the tax treatment of the hired party. *See* Restatement § 220(2) (setting forth a nonexhaustive list of factors relevant to determining whether a hired party is an employee). No one of these factors is determinative. *See also,* Michael B. Landau, *"Works Made For Hire" After Community for Creative Non-Violence v. Reid: The Need For Statutory Reform and The Importance Of Contract,* 9 CARDOZO ARTS & ENT. L.J. 107 (1990); Marci A. Hamilton, *Commissioned Works as Works Made for Hire Under the 1976 Copyright Act: Misinterpretation and Injustice,* 135 U. PA. L. REV. 1281, 1283 (1987).

37. Registration does not apply to foreign works. 17 U.S.C. § 411.
    Section § 411(a) provides as follows:
    Registration and civil infringement actions
    (a) Except for an action brought for a violation of the rights of the author under section 106A(a), and subject to the provisions of subsection (b), no civil action for infringement of the copyright in *any United States work* shall be instituted until preregistration or registration of the copyright claim has been made in accordance with this title. (emphasis added.)

38. Owing to the termination provisions that were added to the Copyright Act of 1976, and amended in 1998, for all works that were transferred before January 1, 1978, all transfers of a work may be terminated after the 56th year of the copyright term or again after the 75th year of the copyright term if they were not terminated exercised at the 56th year despite an agreement to the contrary. 17 U.S.C. § 304(c)( ). Works transferred after 1 January 1978 may be terminated during a five year window that begins thirty-five years after the grant was executed, or if the grant involves the right of publication, between thirty-five and forty years after the grant was executed. 17 U.S.C. § 203.

39. The two cases in which the "teacher exception" have been applied were written by Judge Richard Posner and Judge Frank Easterbrook, federal judges who used to be law school professors at the University of Chicago. To say the least, each one had a strong "self-interest" in coming our with a teacher exception.

40. 811 F.2d 1091 (7th Cir. 1987).

41. 811 F.2d 1091, 1094 (7thCir. 1987); *see also* Williams v. Weisser, 78 Cal. Rptr. 542 (Cal. App. 1969) (lectures delivered in class are the professor's common law copyright property).

42. 847 F.2d 412 (7th Cir. 1988).

43. 363 F.3d 177 (2d Cir. 2004).

44. 16 F.Supp.2d 1297 (D. Colo. 1998); *cf. Pavlica v. Behr*, 397 F.Supp.2d 519 (S.D.N.Y. 2005) (there was a triable issue of fact regarding ownership of the course curriculum and accompanying manual); Gilpin v. Siebert, 419 F. Supp. 1288 (D. Or. 2006).

45. Because there is no set number of journal articles and books that must be written, the teacher exception applies to them. Course materials are entirely another matter.

46. Robert, W., Gomulkiewicz, *et al.,* Licensing intellectual property, at 466–467 (Aspen 2008).

47. *See* http://licensing.msu.edu/licapp.pdf.

48. The Court of Appeals for the Federal Circuit was formed on 1982 by combining the Court of Customs and Patent Appeals ("CCPA") with the Court of Claims. Pre-1982 cases of the CCPA are binding precedent for the Federal Circuit.

49. Michael Landau & Donald E.Biederman, The Case For A Specialized Copyright Court: Eliminating The Jurisdictional Advantage, 21 HASTINGS COMM/ENT L. J. 717 (1999).

50. Although there are no inter-circuit splits of authority that need Supreme Court resolution, in recent years, the U.S. Supreme Court has taken a disproportionate number of patent appeals. It appears as though the Supreme Court has a different view of the law from that of the Court of Appeals for the Federal Circuit, and that the Supreme Court wants the last word. *See*, e.g., Microsoft Corp. v. i4i Ltd. Partnership, 131 S.Ct. 2238 (2011)(stating that the defenses to patent infringement, even those involving prior art that was not before the Patent & Trademark Office, must be proved by clear and convincing evidence); Board of Trustees of Leland Stanford Junior University v. Roche Molecular Systems, Inc., 131 S.Ct. 2188 (2011)(holding that

there must be a valid assignment in order to effectuate a transfer of ownership to a company or university in Patent Law); Global Tech Appliances, Inc. v. SEB, S.A., 131 S.Ct. 2060 (2011)(holding that inducing infringement of a patent requires knowledge that the induced acts indeed are patent infringement); Bilski v. Kappos 130 S.Ct. 3218 (2010)(redefining the test for patentable subject matter); Quanta Computer, Inc. v. LG Electronics, Inc., 553 U.S. 617 (2008)(authorized sale of a patented article exhausts patent holder's rights and prevents patent holder from controlling postsale use of the item); KSR Intern. Co. v. Teleflex, Inc., 550 U.S. 398 (2007)(obviousness); Microsoft Corp. v. AT&T Corp., 550 U.S. 437 (2007)(what constitutes a component part); eBay Inc. v. MercExchange, L.L.C., 547 U.S. 388 (2006)(setting proper equitable standards for injunctions); Festo Corp. v. Shoketsu Kinzoku Kogyo Kabushiki Co., Ltd, 535 U.S. 722 (2002)(doctrine of equivalents); Florida Prepaid Postsecondary Educ. Expense Bd. v. College Savings Bank, 527 U.S. 627 (1999)(state sovereign immunity and the constitutionality of the Patent Remedy Clarification Act); and Markman v. Westview Instruments, Inc., 517 U.S. 370 (1996)(claim construction is a matter of law and letting it go to the jury is reversible error).

51. Robert W. Gomulkiewicz, *et al.,* Licensing Intellectual Property, at 465 (Aspen 2008).

52. *See* UT System Wins $1.8M Settlement from the BlackBerry Maker, Austin Business Journal, August 1, 2005. Available at http://austin.bizjournals.com/austin/stories/2005/08/01/daily6.html.

53. Rick Mullen, *Emory Gets $525 Million for AIDS Drug,* Chemical & Engineering News, Volume 83, Number 30, July 25, 2005, p. 11, available at: http://pubs.acs.org/cen/news/83/i30/8330Emory.html.

54. *See* Guide to the Ownership, Distribution and Commercial Development of M.I.T. Technology, June 2010, 2.1(a)(1) available at: web.mit.edu/tlo/www/ downloads/pdf/guide.pdf.

55. *See* Guide to the Ownership, Distribution and Commercial Development of M.I.T. Technology, June 2010, ¶ 2.1(a)(2) available at: web.mit.edu/tlowww/downloads/pdf/guide.pdf.

56. For example, in M.I.T.'s Guide to the Ownership, Distribution and Commercial Development of M.I.T. Technology, June 2010, there is an obvious error in 3.3 Copyrights; Asserting and Registering. It reads: "In order to maintain a copyright for the period prescribed under the copyright statute, notice of copyright must be affixed to the copyrightable material. Failure to affix the proper notice will cause the

copyright to be lost aster a certain period of time has elapsed from the first publication of the work." (P. 11) That has not been the law since March of 1989, when the United States became a signatory to the Berne Convention.

57. Cornell University requires inventors to assign to the university or its designee all rights and titles of their inventions and related property rights that result from activity conducted in the course of an appointment with the university and/or using university resources, including those provided through an externally funded grant, contract, or other type of award or gift to the university. *See* Cornell Patent Policy, available at: http://studentlife.gradschool.cornell.edu/index.php/guide-to-graduate-study/patent-policy.

58. Under U.S. Patent Law, all appeals, whether they are appeals of a decision not to allow a patent in the Patent and Trademark Office or whether they are appeals from litigation in the regional district courts, end up in the Federal Circuit. From there the cases are further appealed to the Supreme Court. In copyrights and trademarks, litigation moves from the regional district court to the regional circuit court of appeals. Then the case moves to the Supreme Court.

59. 283 F.Supp.2d 615 (D. Conn. 2003), *motion denied*, 393 F.Supp.2d 133 (D. Conn. 2004), *modified*, 2005 WL 327138 (D. Conn 2005), *aff'd*, 184 Fed.Appx. 21, 213 Ed. Law Rep. 156 (2d Cir. 2006).

60. Under the then existing mandatory retirement policy, Dr. Fenn retired from his position as a full professor, but continued his work at Yale for another seven years in the title of "Professor Emeritus" and "Senior Research Scientist." 283 F.Supp.2d at 620.

61. 283 F.Supp.2d at 625.

62. 35 U.S.C. 102 – Conditions for patentability; novelty and loss of right to patent.
    A person shall be entitled to a patent unless — (b) the invention was patented or described in a printed publication in this or a foreign country or in public use or on sale in this country, more than one year prior to the date of the application for patent in the United States.

63. Conn. Gen. Stat. § 42–110a *et seq.*

64. 283 F.Supp.2d at 628–629, *citing*, Chou v. University of Chicago, 254 F.3d 1347, 1356–1357 (Fed. Cir. 2001); University of West Virginia Bd. of Trustees v. VanVoorhies, 84 F.Supp.2d 759, 769–771 (N.D.W.Va 2000).

65. 283 F.Supp.2d at 628-629, *quoting,* Goldwasser v. Smith Corona Corp., 817 F.Supp. 263 (D. Conn. 1993), aff'd, 26 F.3d 137 (Fed. Cir. 1994).

66. 283 F.Supp.2d at 629.

67. 283 F.Supp.2d at 632, *quoting,* Barbara Weismen, Trustee v. Kaspar, 233 Conn. 531, 539 (1995).

68. 283 F.Supp.2d at 632–633, *quoting,* Gelinas v. Gelinas, 10 Conn. App. 167, 173 (Conn. App. 1987), cert. denied, 204 Conn. 802 (1987).

69. Fenn v. Yale University, 184 Fed.Appx. 21, 213 Ed. Law Rep. 156 (2d Cir. 2006).

70. 254 F.3d 1347 (Fed. Cir. 2001).

71. Dr. Chou also claimed that she was omitted from being named on the patent, and moved for a certificate of correction under 35 U.S.C. § 256. That part of the case is discussed, *infra,* in CO-INVENTORSHIP.

72. 254 F.3d at 1357.

73. 254 F.3d at 1357.

74. 254 F.3d at 1357.

75. 254 F.3d at 1357. The court noted that Chou had indeed assigned other inventions on which she was a named co-inventor; *see also, Regents of the University of New Mexico v. Knight,* 321 F.3d 111 (Fed. Cir. 2003)(Both Scallen and Knight were bound by the UNM Patent Policy. Each year Scallen and UNM entered into a written faculty contract, which, at the times relevant here, incorporated the 1983 UNM Patent Policy contained in the Faculty Handbook. The Patent Policy's terms and conditions broadly apply to all "staff members," which includes "any faculty member, student, or any other person associated with the teaching or research staffs of the University." That certainly includes Scallen, who was a faculty member. Although Knight, as a faculty staff member, had no employment contract with UNM, he too was bound by the Patent Policy. Under New Mexico law, a written personnel policy may form an implied employment contract. The 1983 UNM Patent Policy thus created an implied contract between Knight and UNM that governed the relationship between Knight and UNM. Thus, we agree with the district court's conclusion that both Scallen and Knight were bound by the Patent Policy); University of West Virginia Bd. of Trustees v. Van-Voorhies, 84 F.Supp.2d 759, 769–770 (N.D.W.Va 2000).

76. 583 F.3d 832 (Fed. Cir. 2009), aff'd, 131 S.Ct. 2188 (2011).

77. The other inventors, Thomas Merigan and Thomas Katzenstein signed a "Materials Transfer Agreement" that permitted Stanford to use materials developed by Cetus and gave Cetus licenses to technology

that Stanford developed as a result of its working with Cetus's materials. 583 F.3d at 837.

78. 583 F.3d at 837.

79. 583 F.3d at 837.

80. 583 F.3d at 837.

81. The question of who owns patent rights and on what terms is usually a question of state law and is resolved in state courts. *See* Jim Arnold Corp. v. Hydrotech Systems, 109 F.3d 1567, 1572 (Fed. Cir. 1997); *see also* MyMail, Ltd. V. America Online, Inc. 476 F.3d 1372, 1376 (Fed. Cir. 2007). There is, however, an exception to the rule. "Although state law governs the interpretation of contracts generally, the question of whether a patent assignment clause creates an automatic assignment or merely an obligation to assign is intimately bound up with the question of standing in patent law. We accordingly have treated it as a matter of federal law." DDB Techs., LLC v. MLB Advanced Media L.P., 517 F.3d 1284, 1290 (Fed. Cir. 2008).

82. 583 F.3d at 841. Emphasis in original.

83. 583 F.3d at 841.

84. 583 F.3d at 841.

85. 583 F.3d at 842.

86. 583 F.3d at 842. (emphasis in original).

87. The problem of conflicting assignments is more common than one might think. Proper due diligence, and the addition of a contract clause stating, "I hereby assign and do assign all inventions and technology related to my work at [University]. I further agree not to execute any agreement or to engage in any activity which will be in conflict with my assigning all inventions and technology to the University."

88. *See* Board of Trustees of the Leland Stanford Junior University v. Roche Molecular Systems, Inc., 131 S.Ct. 2188 (2011). Note that the case in Patent Law is different from the case in Copyright Law where an employee is automatically the "author" of all works created by his employees within the scope of their employment.

89. 35 U.S.C.A. § 102(f).

90. *See* Chou v. University of Chicago, 254 f.3d 1347 (Fed. Cir. 2001); *see also* Ethicon, Inc. v. United States Surgical Corp., 135 F.3d 1456 (Fed. Cir. 1998); Pannu v. Iolab Corp., 155 F.3d 1344 (Fed. Cir. 1998).

91. 35 U.S.C.A. § 256 provides as follows:

    35 U.S.C. 256 Correction of named inventor.

    Whenever through error a person is named in an issued patent as the inventor, or through error an inventor is not named in an issued patent and such error arose without any deceptive intention on his part, the Director may, on application of all the parties and assignees, with proof of the facts and such other requirements as may be imposed, issue a certificate correcting such error. The error of omitting inventors or naming persons who are not inventors shall not invalidate the patent in which such error occurred if it can be corrected as provided in this section. The court before which such matter is called in question may order correction of the patent on notice and hearing of all parties concerned and the Director shall issue a certificate accordingly.

92. *See* Vanderbilt Univ. v. ICOS Corp., 601 F.3d 1297 (Fed. Cir. 2010)(Vanderbilt failed to show by clear and convincing that three additional professors were co-inventors.); *cf.,* Univ. of Pittsburgh of Commonwealth System of Higher Educ. V. Hedrick, 573 F.3d 1290 (Fed. Cir. 2009)(independent researchers were not co-inventors as they completed research after the university's researcher had completed conception of patent).

93. 254 F.3d 1347 (Fed. Cir. 2001).

94. An example of the royalty payment to inventor faculty increasing can be found in Fenn v. Yale University, *supra.* In the 1966 Patent Policy, the faculty member's royalty was 15%. In the 1989 Patent Policy it ranged from 30% to 50% depending on the sales of the invention.

95. The Patent and Trademark Law Amendments Act of 1980, Public Law No. 96–517, 94 Stat. 3014 (codified as 35 U.S.C. § § 200–212 and 37 CFR 401).

96. BethLynn Maxwell *et al., Overview of Licensing Technology from Universities,* 762 PLI/Pat 507 (Sept 2003)("Milestone payments are common way for rewarding developers for sharing the risk of development.")

97. Cornell University, Inventions and Related Property Rights, Policy 1.5 (last updated June 4, 2010) at 11.

98. Robert W. Gomulkiewicz, *et al,* Licensing Intellectual Property, at 467 (Aspen 2008).

99. Robert W. Gomulkiewicz *et al.,* Licensing Intellectual Property, at 468 (Aspen 2008).

100. Royalty Figures – Yale Patent Policy, available at: http://www.yale.edu/ocr/pfg/policies/patents.html.

101. Royalty Figures –§ 7.6.05, Distribution of Cumulative Net Revenue, available at: http://www.policies.emory.edu/7.6.

102. 307 F.3d 1351 (Fed. Cir. 2002).

103. Whittmore v. Cutter, 1 Gall. 429, 29 F.Cas. 1120, 1 Robb. Pat. Cas. 28 (No. 17600)(C.C. Mass. 1813).

104. Poppenhausen v. Falke, 4 Blatchf. 493, 19 F.Cas. 1048, 2 Fish.Pat.Cas. 181, No. 11,279 (C.C.S.D.N.Y. 1861); *see also,* Baxter Diagnostics, Inc. v. AVL Scientific Corp., 924 F.Supp. 994, (C.D.Cal. 1996);Pfizer, Inc. v. International Rectifier Corp., , 217 U.S.P.Q. 157, 161 (C.D.Cal. 1982).

105. *See, generally,* Maureen E. Boyle, *Leaving Room For Research: The Historical Treatment Of The Common Law Research Exemption In Congress And The Courts, And Its Relationship To Biotech Law And Policy* 12 YALE JOURNAL OF LAW AND TECHNOLOGY 269 (Spring, 2009-2010); Shamnad Basheer, *The "Experimental Use" Exception Through a Developmental Lens,* 50 IDEA 831 (2010); Maureen A. O'Rourke, *Toward a Doctrine of Fair Use in Patent Law,* 100 COL. L. REV. 1177 (2000)(arguing that the narrowing of the experimental use doctrine is wrong); *Cf.,* Anthony J. Caruso, *The Experimental Use Exception: An Experimentalist's View,* 14 ALBANY J. L. SCI & TECH. 217 (2003)(patent infringement suits are rarely brought against non-commercial defendants regardless of how broadly it is defined).

106. 733 F.2d 858, 863 (Fed. Cir. 1984), *cert denied,* 469 U.S. 856 (1984).

107. U.S. Patent No. 3,299,053 ('053 patent) titled, "Novel 1 and/or 4 substi-tuted alkyl 5-aromatic-3H-1, 4-benzodiazepines and benzodiazepine-2-ones." One of the chemical compounds claimed in the patent is flurazepam hydrochloride, known to the public as "Dalmane."

108. 547 F.2d 1106, 1125–1126 (Ct.Cl. 1976), *cert denied,* 434 U.S. 1051 (1978).

109. 733 F.2d at 863.

110. Public Law No. 84–417, 98 Stat. 1585 (1984).

111. Martin J. Adelman, *et al,* Cases and Materials on Patent Law (3d ed. 2009), at 798.

112. 307 F.3d 1351 (Fed Cir. 2002), *cert. denied,* 539 U.S. 958 (2003). It should be noted that the exemption for experimentation "reasonably related" to the submission to the Food and Drug Administration ("FDA") for approval also applies to experimentation on drugs that are

not ultimately submitted for FDA approval. *See* Merck KGAA v. Integra Lifesciences I, Ltd, 545 U.S. 193, 125 S.Ct. 2372 (2006).

113. 307 F.3d at 1362.

114. The Patent and Trademark Law Amendments Act of 1980, Public Law No. 96-517, 94 Stat. 3014 (codified as 35 U.S.C. § § 200–212 and 37 CFR 401).

115. BethLynn Maxwell, *et al., Overview of Licensing Technology from Universities,* 762 PLI/Pat 507, 514 (2004).

116. 35 U.S.C. § 200.

117. 35 § 202(c)(4).

118. 35 U.S.C. § 203.

119. 35 U.S.C. § 203 (a)(2).

120. 35 U.S.C. § 204.

121. 35 U.S.C. § 206; *See* Disclosure Requirements and University Ownership, *supra.*

122. Kenneth Sutherlin Dueker, *Biobusiness on Campus: Commercialization of University Developed Biomedical Technologies,* 52 FOOD DRUG L. J. 453, 457 (1997).

123. McElmurry v. Arkansas Power & Light Co., 995 F.2d 1576 (Fed. Cir. 1993); *see also* P. Rosenberg, *Patent Law Fundamentals,* § 11.04, 11–20 (1991).

124. Although a creature of common law, most courts rely upon the leading Supreme Court decisions *McClurg v. Kingsland,* 42 U.S. (1 How.) 202, 11 L.Ed. 102 (1843), *Gill v. United States,* 160 U.S. 426 (1896), and *United States v. Dubilier Condenser Corp.,* 289 U.S. 178, 53 S.Ct. 554, 77 L.Ed. 1114 (1933), to analyze "shop rights" issues.

125. *See, e.g.,* McClurg v. Kingsland, 42 U.S. 202 (1843); *see also,* Solomons v. United States, 137 U.S. 342, (1890); Mechmetals Corp. v. Telex Computer Products, Inc. 709 F.2d 1287, 219 USPQ 20 (Fed. Cir. 1983).

126. In *Gill v. United States,* 160 U.S. 426 (1896), an influential estoppel case, the Supreme Court stated in pertinent part:

if the inventions of a patentee be made in the course of his employment, and he knowingly assents to the use of such inventions by his employer, he cannot claim compensation therefore, especially if his experiments have been conducted or his machines have been made at the expense of such employer.

Chapter 13

Environment, Health, and Safety Within the Nanotechnology Industry

Kaarle Hämeri

Division of Atmospheric Sciences, Department of Physics, P.O. Box 48 (Erik Palménin aukio 1), FI-00014 University of Helsinki, and Finnish Institute of Occupational Health, Topeliuksenkatu 41 a A, FI-00250 Helsinki, Finland
kaarle.hameri@helsinki.fi

13.1 Introduction

Nanoscience consists of a wide range of topics, methods, and applications that have in common concepts and physical laws that prevail only in nanometer scale. The research in nanoscience is typically interdisciplinary in nature and covers a wide range of traditional scientific fields. Nanoscience deals with objects and phenomena that appear as the characteristic dimension of the system below about 100 nm. At this size range the physics that govern macroscopic world does not fully apply but new phenomena become important. These phenomena include the increasing role of surface atoms (more surface for the same mass of the material), the unclear concept of surface for clusters consisting of just a few atoms, the curvature effects of small objects, and the increasing importance of quantum mechanics for nanometer-sized objects.

Nanotechnology Commercialization for Managers and Scientists
Edited by Wim Helwegen and Luca Escoffier
Copyright © 2012 Pan Stanford Publishing Pte. Ltd.
ISBN 978-981-4316-22-4 (Hardcover), 978-981-4364-38-6 (eBook)
www.panstanford.com

The common unifying concepts and physical laws that prevail in the nanoscale can be collectively termed "nanoscience." The properties of nanomaterials differ significantly from those of the bulk of the same material, such as high tensile strength, low weight, high electrical and thermal conductivity, and unique electronic properties

Nanoscience has been defined somewhat differently depending on the purpose and context. National Science Foundation US National Nanotechnology Initiative gives the following definition:

1. Research and technology development at the atomic, molecular, and macromolecular levels, in the scale of approximately 1–100 nm range
2. Creation and use of structures, devices, and systems that have novel properties and functions because of their small and/or intermediate size
3. Ability to control and manipulate on the atomic scale

The EU defines the field as follows: "Nanosciences and nanotechnologies are new approaches to research and development that concern the study of phenomena and manipulation of materials at atomic, molecular and macromolecular scales, where properties differ significantly from those at a large scale" [1].

Both of these definitions consist of identification of a size-scale as well as appearance of fundamentally different phenomena taking place within this size range. In simple terms, we may define nanoscience as the study of phenomena and manipulation of materials at atomic, molecular, and macromolecular scales, where properties differ significantly from those at a larger scale. Therefore a wide range of topics appears under the concept of nanoscience. The diversity is further exaggerated by the hype covering the topic of nanoscience. As there is more funding focused on the topic of nanoscience and technology, sometimes also conventional science tend to appear under nano-topic.

In contrast to nanoscience, the term nanotechnology refers to the development of a mature know-how for the production of nano-objects and the exploitation of knowledge progress on specific nano-objects to make concrete applications. Nanotechnology is closely related to nanoscience but obey distinct drivers in so far as nanotechnology tries to respond to particular needs while nanoscience is primarily turned toward the discovery and study of novel phenomena and the creation of new concepts to describe them. Nanotechnology is the way discoveries made at the nanoscale are put to work [2].

The definition of nanotechnology has proven to be a difficult task. The most common nanotechnology definitions currently available include, for example, US NNA, EU 7th Framework, Japan Science and Technology Plan, and ISO definitions. As an example, US National Nanotechnology Initiative defines nanotechnology as follows: "Nanotechnology is the understanding and control of matter at dimensions of roughly 1 to 100 nanometers, where unique phenomena enable novel applications. Encompassing nanoscale science, engineering and technology, nanotechnology involves imaging, measuring, modeling, and manipulating matter at this length scale" [3]. The threshold of 100 nm has been frequently used in defining nanotechnology. The threshold of 100 nm is, however, only indicative of a point along the continuum where classical rules of physics start to give way to new nanotechnology phenomena.

The development of nanotechnology is approached from two main principles, top down and bottom up. According to the evolutionary top-down principle, microtechnology is gradually scaled down into nanotechnology (under 100 nm). This top-down principle approach manipulates materials down to the nanoscale through elaborations of various techniques such as lithography, cutting, or milling techniques. According to the bottom-up principle, entirely new structures (including materials) and manufacturing processes are created by applying top science (so-called self-assembly, as in biology). The bottom-up approach creates new materials at the nanoscale through, for example, nanoparticle synthesis (from gas to particles) and liquid-phase processes. The top-down approach has already created plenty of new technology companies and new industrially applicable products and production processes. In the long term, bottom-up technology is also expected to change production methods in many areas.

Many applications of nanotechnology involve the use of nanomaterials. In fact, numerous nanoparticles are already on the market, in products such as paints, sunscreens, cosmetics, medicines, self-cleaning glass, industrial lubricants, advanced tyres, semiconductors, and food. This has created concerns over the safety of engineered nanoparticles where exposure to humans and/or the environment occurs intentionally or accidentally.

The view about the nature and the potential risks of nanoparticles may be summarized as follows:

- There are potential risks to human health and the environment from the manufacture and use of nanoparticles.
- There is a need for more research on what might be the potential risks and how to deal with them.
- The lack of data makes it difficult for manufacturers, suppliers, and users to have effective risk management processes and to comply with their regulatory duties.
- All of the stakeholders (regulators, companies) need to start to address these potential risks now [4].

13.2 Exposure to Nanoparticles

The risk of health effects, which may arise as a result of exposure to nanomaterials, is generally considered to be a function of the toxicity of the material and the dose (amount) which accumulates in the specific biological area of interest. However, it is difficult to quantify the dose. Therefore the quantification and management of the risks is usually done by studying the exposure and using it as a proxy for the dose (property that is linked to dose). The exposure is understood as the presence of pollution and human being in a same place simultaneously (Fig. 13.1). Note that exposure does not need to result to dose. Knowledge and control of exposure is critical in risk assessment and management.

Critical questions in relation to exposure are how much, how long, and how many people are exposed. Exposure is usually measured in terms of the pollutant concentration and duration of people present. Control of exposure effectively removes the risks from the pollutant. Without exposure there is no risk.

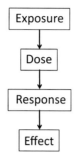

Figure 13.1. Understanding exposure.

The main routes by which people can be exposed to nanomaterials are inhalation, ingestion, and dermal penetration.

Inhalation is considered to be the primary route by which nanoparticles suspended in air can enter the bodies of people. The inhaled particles will deposit in all regions of the respiratory tract. The deposition depends strongly on the particle size.

Ingestion exposure to nanomaterial may arise from hand to mouth contact. The exposure may also be caused by swallowing mucous, which contain deposited particles cleared from the lungs and the respiratory tract. This route of exposure is often much less important than that of inhalation.

Dermal exposure is of increasing concern (e.g., Brouwer *et al.*, 2005). People may be exposed via the skin by handling or touching materials or surfaces coated with nanomaterials. However, this route of exposure is considered also much less important than inhalation.

13.2.1 *Exposure Scenarios*

There are different scenarios for exposure through which humans could become exposed to engineered nanomaterials. In occupational settings the exposure can occur at various situations. One of the most crucial groups of people are those working on R&D at universities, research institutes, or companies. However, the amount of materials produced and used as well as the total number of people at the risk of exposure are relatively small. As the material is produced for commercial purposes, the risk increases significantly.

During the commercial production, the amount of material is much larger and also the number of people involved at various stages of the material handling is higher. The activities at which the exposure may take place include, for example, recovery, packaging, and storage of the nanomaterials.

At later stage of the life cycle of the nanomaterials, the exposure may take place during processing of the nanomaterials (e.g., cutting or finishing), use of the products containing nanomaterials, or finally disposal of the products. For all nanomaterials, there are many exposure scenarios which may or may not occur depending on the details of manufacture, use, and disposal of that material. In all the scenarios, the population exposed, the levels of exposure, the duration of exposure, and the type of the material to which people are exposed are all different.

13.2.2 Exposure Metrics

Workplace exposure limits for most of the hazardous materials and chemicals are based on mass concentration, typically expressed as mg/m^3 of air. This is relatively straightforward to measure by collecting the material into a filter and weighing the collected amount. However, due to very small diameter and therefore also small mass, nanoparticles contain very little mass even in relatively high number concentrations.

The main metrics that can currently be used for measuring nanoparticles are (1) mass concentration (mg/m^3); (2) number concentration (units m^{-3}); and (3) surface area concentration (units m^2/m^3). The surface area concentration is indicated to be linked with inflammation caused by nanoparticles. The number concentration is shown to be linked with respiratory and cardiovascular diseases (Seaton *et al.*, 1995). As no single metrics have been shown to characterize the health risk to nanoparticles, it is recommended that all three metrics are determined simultaneously, if possible.

13.2.3 Exposure due to Release to the Environment

It is possible that emissions from the processes in which nanomaterials are produced, could potentially lead to increased air concentrations of these nanomaterials. Wide environmental exposure could occur as a result of these releases. In such a case, the possibility of human exposure is through release to the environment. However, currently available knowledge of these processes is insufficient to allow quantitative predictions of the environmental fate of nanomaterials.

13.3 Environmental Fate

Nanomaterial release into the environment may occur both intentionally and non-intentionally at any point of the life cycle of nanomaterial production, use, and disposal. For example, wastewater treatment streams, landfill leachates, and waste incineration products are all likely to contain nanomaterials from nano-enabled products that reach the environment following either disposal at the end of their useful life, accidental spills during production or later transport of nanomaterials, or release from wear and tear of products containing nanomaterials. In order to develop a full understanding of the potential risks of nanomaterials, further examination of their environmental transport and fate within air, soil, and water is necessary.

Most of our current knowledge of transport of nanomaterials within air, soil, and water compartments is based on knowledge in aerosol and colloid science. This scientific background provides fundamental information from which one can develop further understanding of nanoparticles' fate and transport.

13.3.1 *Nanomaterials in Air*

The key processes governing nanoparticle behavior in air are diffusion, agglomeration, deposition, and re-suspension of deposited aerosol material. As nanoparticles are very small in size the inertial and gravitational forces are relatively inefficient to be able to influence the nanoparticle motion. Nanoparticles behave thus in a manner of that of gas molecules.

Particle diffusion is a process in which particles suspended in air are constantly and randomly bombarded from all sides by gas molecules, causing them to move randomly. Particle diffusion coefficient is inversely proportional to particle diameter (small particles have high diffusion coefficient). Particles with a high diffusion coefficient (such as nanoparticles) therefore have high mobility and will mix rapidly in aerosol systems. Upon release into the environment, atmospheric diffusion will lead nanomaterials to move rapidly from a high concentration to a lower one, and thus resulting in rapid dispersion and potential for particles to travel a great distance from source [5].

Agglomeration is the process by which particles stick together to form larger particles. Agglomeration rate depends primarily on particle mobility and number concentration, both of which increase as particle size decreases. Thus, nanoparticles tend to agglomerate rapidly, even at a low mass concentration. Aged nanoparticle aerosols can usually not be recognized by a specific particle size range, and tend to have a size range very similar to ubiquitous background aerosol as determined by size distribution measurements. This means detecting nanoparticle concentrations within air is very difficult, unless measurements are taken close to a point source emission.

Deposition is a process by which nanoparticles may be removed from air. Deposition of particles is dependent on the gravitational settling velocity, a characteristic which is proportional to the diameter of the particle. Nanoparticles in air will therefore deposit at a very slow rate and nanoparticles may be transported over great distances in air before being deposited.

Potential re-suspension of nanoparticles into the atmosphere is a complex process. Re-suspension depends on several factors such as particle size, shape, charge, and the energy applied to the potential re-suspension situation. Particles are held together in agglomerates and to the surface on which they are deposited by short-range attractions called van der Waals forces. It is thus questionable as to how likely re-suspension would be, once particles are deposited.

In summary, there is currently still lack of data in relation to environmental fate and behavior of nanomaterials. This creates major gap in developing a holistic view on the fate and transport of nanomaterials within the environment.

13.4 Health Effects and Human Toxicity

The characteristics of nanomaterials vary greatly, for example, in terms of their composition, morphology, and size. It is therefore of interest to evaluate the toxic potential of the nanomaterials under investigation, and to identify whether the same underlying mechanisms drive each of their toxicities, and also to determine whether any generalizations can be made regarding nanomaterials as a whole. In addition, it is important to determine material or particle-specific properties that are particularly relevant in driving nanomaterial toxicity, allowing identification of key characteristics that can influence safety. In an attempt to achieve this, available information regarding the exposure conditions and characteristics of nanomaterials need to be investigated. In addition, there are discrepancies regarding the dose metrics used when expressing the concentration of particles exposed to cells or animals; specifically whether dose should be characterized in terms of particle mass, surface area, or number concentration. This is important, as it has been repeatedly shown that the toxicity of particles is related to their size, so that as particle size decreases, toxicity generally increases. This is thought to be driven by their surface area. However, nanomaterials are a diverse group of materials, and it has become evident that other particle dimensions may also be important in driving their toxicity.

Exposure to nanomaterials is expected to primarily occur through inhalation, ingestion, or dermal routes. Accordingly, a particular focus on the consequences of exposure of the lungs, skin, or gastrointestinal tract has been employed, with inclusion of both *in vitro* and *in vivo* models for each route. However, nanomaterials can distribute from their exposure site, within the blood or even nerves. This means that nanomaterial toxicity may be exerted at a number of targets, including the liver, brain, spleen, and kidneys.

Studies that assess the toxicity of a wide variety of nanomaterials include the utilization of both *in vivo* (usually within mice and rats) and *in vitro* models (using cell lines and primary cells). Rodent models have been used extensively to study a number of disease states, including particle-induced lung disease (including quartz and asbestos-induced disease), and so there is already an understanding of the limitations of such models, in terms of extrapolating hazard assessment to humans. Current expertise does not perfectly allow for inclusion of species differences, but the existing database of particle toxicology combined with the use of such models provide a useful series of protocols to allow benchmarking of new nanomaterials to the relative potential toxicity of other substances of known hazard.

Cell lines are frequently used to investigate the effects of potentially toxic substances. The types of cells available are many and represent a wide range of organ and cell types, including tumor derived and transformed cells. The cell response is often representative of the *in vivo* response, but careful comparisons and controls are required to ensure relevance. One of the biggest problems with *in vitro* models is the use of excessive concentrations to assess toxicity or mechanisms of toxicity.

In vitro studies allow the assessment of toxicity within a controlled system, where a number of different concentrations and time-points can be evaluated. In comparison, *in vivo* study design requires consideration of strategies to reduce the use of animals, resulting in a low number of doses and time-points within an individual study. In order to allow the safe development of nanotechnology while limiting the use of *in vivo* models in the future, it is clear that the development and validation of the predictive capability of *in vitro* models of nanomaterial toxicity is necessary.

Toxicity studies of engineered nanoparticles are currently still rather few. Most of the existing data concerns toxicity of fullerenes, carbon nanotubes, metal, and metal oxide particles. The predominant process causing the pathological effects of particles in the lungs and the cardiovascular system is inflammation, involved in atherothrombosis, asthma, chronic obstructive lung disease, pulmonary fibrosis, and cancer. Therefore, the ability of particles to start or influence on inflammation can be seen as a key property of the nanomaterial. Several studies have examined the pro-inflammatory effects of manufactured nanoparticles, on the basis that their ability to cause inflammation is a major predictor of potential hazard in such particles [6]. The first important finding was that nanoparticles have a more pronounced effect on inflammation, cell damage, and cell stimulation than an equal mass of particles of the same material of greater size [7]. This appears to be true for materials such as carbon black, titanium dioxide, various metals, and polystyrene. Surface area is the property driving the pro-inflammatory effects and this is evident from both *in vitro* and *in vivo* studies. The surface area dose-related inflammatory response is partly related to transition metals but is also found with low-toxicity materials and the cellular mechanism is not well understood. Even a low toxicity surface has the ability to generate free radicals and oxidative stress in cells that has nothing to do with transition metals, as there is no soluble toxic component that could mediate the effect. The low toxicity surface free radical effect is evident in their ability to generate oxidants in cell-free chemical systems. In cells, high-surface-area doses appear to initiate inflammation through a number of pathways but oxidative stress-responsive gene transcription is one of the most important. A wide range of nanoparticles has been found to have the ability to cause pro-inflammatory effects and also mechanisms other than the well-known ones described above may also be significant.

13.5 Risk Assessment

Nanoparticles have a wide variety of applications, which range from industrial production efficiency, energy production, environment remediation, consumer products to medicine, and many others. Besides the promising benefits there are concerns about possible risks for humans and more recently also for the environment.

Very few publications up to now have been dealing with risk assessment methodologies for nanomaterials. Some analyses have been carried out to evaluate the applicability of existing approaches to nanomaterials, in particular the assessment methodologies as described in the Guidance Documents of the EU's chemicals legislation (REACH). The main questions concern nanomaterial characterization, detection methods, nanomaterial fate and transport, standardized *in vivo* and *in vitro* testing development, and the identification of mechanisms of action.

On the basis of the identified information, the risk assessments are carried out following both a quantitative and a qualitative approach. For human health, the quantitative approach requires establishing exposure values for the various routes of exposure (inhalation, dermal, and ingestion) for consumers and workers and the establishment of a derived no-effect level (DNEL), typically based on extrapolation of animal data to the human situation by using appropriate assessment factors. For the environmental assessment, the quantitative approach requires the determination of the predicted exposure concentration (PEC) and the predicted no-effect concentration (PNEC) for each environmental compartment. PEC and PNEC are then compared to identify any risk for environmental compartments. For both human health and environment, the application of assessment factors is based on the REACH (Registration, Evaluation, and Authorization of Chemicals) guidance.

Even if the basic concepts for risk assessment have been developed, a quantitative risk assessment of nanoparticles is not yet possible. The traditional risk assessment process outlined by REACH involves the use of toxicology data to obtain the DNEL of exposure. Risk is assessed by comparing the DNEL with the exposure levels obtained from the different scenarios in the exposure assessment.

13.6 Regulatory Issues

In Europe, regulations regarding nanoparticles are based at present on existing laws and regulations applicable to chemicals. All nanoparticles in chemical substances must meet the requirements of the REACH. Although REACH does not explicitly refer to nanomaterials, they are included by the definition of a "substance." The main objective of the directive is to ensure a high level of relevant protection of human health and the environment.

The general requirements in relation to occupational safety and health of workers at workplaces are presented in the EU Directive 89/391/EEC. The aim of this framework directive is to ensure a high level of protection of workers at work, including those exposed to nanomaterials. The directives adopted relate to risks related to exposure to carcinogens or mutagens at work, risks related to chemical agents at work, the use of work equipment by workers at work, the use of personal protective equipment at the workplace, and safety and health protection of workers potentially at risk from explosive atmospheres.

The Council Directive 98/24/EC on the protection of the health and safety of workers from the risks related to chemical agents at work presents minimum requirements for the protection of workers from risks to their safety and health arising from the effects of chemical agents that are present at the work-place or as a result of any work activity involving chemical agents.

Product legislation lays down requirements regarding specific products, such as medicinal products, plant protection products, cosmetics, food, and feed additives. Consumer products that are not a subject of specific legislation have to meet the requirements of the General Product Safety Directive.

The European Directive 98/8/EC on Biocidal Products provides a framework of rules that apply to the marketing of biocidal (including nanomaterials) substances and products. A biocidal product is any substance, which is used to control or kill harmful organisms, such as bacteria, fungi, moulds, and yeasts.

Environmental regulation relevant in the nanotechnology and nanomaterials context relates in particular to integrated pollution prevention and control (IPPC), the control of major accident hazards involving dangerous substances (Seveso II), the water framework directive, and a number of waste directives.

13.6.1 *Future Perspective*

The Community Strategy for health and safety at work has been published by the EC for the period of 2007–2012. Nanotechnology is addressed there as an important topic in context of the identification of new risks. Furthermore a communication from the European commission toward a European strategy for nanotechnology was published. It has stated that activities connected with research and development as well as technological progress need to be accompanied by scientific studies and assessment of possible health or environmental risks associated with nanotechnology. In relation to health and safety issues the commission highlights the need to

- identify and address safety concerns at the earliest possible stage
- reinforce support for the integration of health, environmental, risk and other related aspects into R&D activities together with specific studies
- support the generation of data on toxicology and ecotoxicology (including dose response data) and evaluate potential human and environmental exposure
- identify and address safety concerns at the earliest possible stage

In 2005, the European Commission adopted the Communication "Nanosciences and Nanotechnologies: An Action Plan for Europe, 2005–2009." This Action Plan defines actions for implementation of a safe, integrated, and responsible approach for nanosciences and nanotechnologies. A risk assessment related to human health, the environment, consumer, and workers should be responsibly integrated into all stages of the life cycle of the technology, starting at the point of conception and including R&D, manufacturing, distribution, use, and disposal or recycling.

13.7 Standardization Activities

The European Committee for Standardization (CEN) and the International Organization for Standardization (ISO) have two Technical Committees, CEN/TC 137 "Assessment of workplace exposure to chemical and biological agents" and ISO/TC 146/SC-2 "Air Quality, Workplace Atmospheres," working in the field of assessment of workplace exposure to different agents. Recently, they both have started to develop documents on the topic of the exposure to nanomaterials.

The only published standard related to nanomaterials exposure is the ISO/TR 27628 (2007) "Workplace atmospheres: ultrafine, nanoparticle and nano-structured aerosols. Inhalation exposure characterization and assessment" prepared by the Technical

Committee ISO/TC 146/SC22. This standard contains guidelines on characterizing occupational nanoaerosol exposures and represents the current state-of-the-art, with an emphasis on nanoscale particles. Furthermore, background information is provided on the mechanisms of nanoaerosol formation and transportation within an occupational environment and on industrial processes associated with nanoaerosol exposure. Exposure metrics appropriate to nanoaerosols are considered and specific methods of characterizing exposures with reference to these metrics are presented. Specific information is provided on methods for both bulk aerosol characterization and single particle analysis.

Both committees are working in parallel to develop the project "Workplace atmospheres. Characterization of ultrafine aerosols/nanoaerosols. Determining the size distribution and number concentration using mobility particle sizers/differential mobility analysers." In addition, two other committees created in 2005, deal specifically with nanotechnologies: ISO/TC 229 and CEN/TC 352.

Notes

1. COM2005 243 final. *Nanosciences and Nanotechnologies: An Action Plan for Future Europe, 2005–2009.*
2. NNI 2007.
3. NNI 2000.
4. Project report: "Engineered Nanoparticles: Review of Health and Environmental Safety."
5. Aitken *et al.*, 2004.
6. Seaton *et al.*, 2009.
7. Donaldson *et al.*, 2000.

References

Aitken, R. J., Creely, K. S., and Tran, C. L. (2004) *Nanoparticles: An Occupational Hygiene Review*, HSE Books, Sudbury, UK.

Brouwer, D. H., Gijsbers, J. H. J., Lurvink, M. W. M. (2004) Personal exposure to ultrafine particles in the workplace: exploring sampling techniques and strategies, *Ann. Occup. Hyg.,* **48**(5):439–453.

Donaldson, K., Stone, V., Gilmour, P. S., Brown, D. M., and MacNee, W. (2000) Ultrafine particles: mechanisms of lung injury. *Phil. Trans. R. Soc. Lond. A,* **358**, 2741–2749, HSE 2005.

NNI 2000 (National Nanotechnology Initiative). Available at www.nano.gov

NNI 2007. Available at www.nano.gov

Project report: *Engineered Nanoparticles: Review of Health and Environmental Safety.*

RR274.COM2005 243 final. *Nanosciences and Nanotechnologies: An Action Plan for Future Europe, 2005–2009.*

Seaton, A., MacNee, W., Donaldson, K., and Godden, D. (1995) Particulate air pollution and acute health effects. *Lancet,* 345, 176–178.

Chapter 14

Regulation of Nanomaterials in the EU

Bärbel R. Dorbeck-Jung

University of Twente, Faculty of Management and Governance,
P.O. Box 217, 7500 AE Enschede, the Netherlands
b.r.dorbeck-jung@utwente.nl

14.1 Introduction

This chapter examines the regulation of nanomaterials in the European Union (EU). It gives an overview of the existing European regulation that applies to nanomaterials. It deals with two cross-cutting regulatory issues that are highly debated in the EU: the definitions of nanomaterials and the applicability of the EU chemicals regulation. The chapter explores the regulation of nanoproducts in certain areas to which the EU regulatory bodies have paid special attention. These are the areas of medicine, cosmetics, and food. Other important issues like safety, environment, and patents are covered by other chapters of the handbook. Nanomaterials are regulated at the EU level, but also at the level of the 27 member states. In this chapter the focus lies on the EU level. Significant initiatives of member states will be mentioned.

Nanotechnology Commercialization for Managers and Scientists
Edited by Wim Helwegen and Luca Escoffier
Copyright © 2012 Pan Stanford Publishing Pte. Ltd.
ISBN 978-981-4316-22-4 (Hardcover), 978-981-4364-38-6 (eBook)
www.panstanford.com

14.2 Regulatory Structure and Policy

To understand the nano-specific regulatory activities in the EU we need to look at its regulatory structure, as well as its policy on nanosciences and nanotechnologies. According to the European Treaties, the regulatory bodies of the EU have certain legal competences in the field of economic and social regulation (among which are customs, competition, internal market, consumer, and health). EU regulation is governed by the subsidiarity principle, which is intended to regulate the lawfulness of the exercise of Community competence. A core meaning of this principle is that in areas which do not fall within its exclusive competence, the Community shall only take action if, and in so far as, the objective of the proposed action cannot be sufficiently achieved by the member states and can therefore be better achieved by the Community. Nanoscientific and nanotechnological (product) development is regarded as one of the issues of central policy which is related to a series of competences in the areas of competition, creation of the internal market, consumer, and worker safety.

Formally, EU legislation is established by the European Parliament and the European Council. In practice the directorate generals (DGs) of the European Commission (EC), regulatory agencies, and advisory bodies prepare and implement EU legislation. With regard to nanomaterials the DGs Health and Consumers, Enterprise and Industry, Environment, and Research and Innovation are involved. In the past five years DG Health and Consumers has taken an active stance in the debate on safety issues of nanomaterials. It has organized, for instance, annual high level dialogues on safety issues.[1] DG Enterprise is mainly involved in the funding of the development of nanosciences and nanotechnologies and in the legislation of chemicals including nanomaterials. DG Research and Innovation supported the Code of Conduct for Responsible Nanosciences and Nanotechnologies Research, which the Commission launched in 2008. Regulatory agencies that are active in the field of nanomaterials are the European Medicines Agency (EMA), the

---

[1]More information on the 4th dialogue see http://ec.europa.eu/health/ nanotechnology/events/ev_20110329_en.htm (accessed on 27-04-2011).

European Food Safety Authority (EFSA), the European Chemicals Agency (ECHA), the European Environment Agency (EEA) and the European Agency for Safety and Health at Work (EU-OSHA).

When preparing policy and regulatory proposals on consumer safety, public health and environmental effects of nanomaterials the European Commission relies mainly on the work of three Scientific Committees. These are the Scientific Committee on Emerging and Newly Identified Health Risks (SCENIHR), the Scientific Committee on Consumer Safety (SCCS), and the Scientific Committee on Health and Environmental Risks (SCHER). When it comes to the regulation of exposure limits for work with nanomaterials the Scientific Committee for Occupational Exposure Limits (SCOEL) can be involved. Members of all these scientific committees are external experts. In the field of nanotechnologies the opinions of SCENIHR on methods of risk assessment (2005, 2007), risk assessment of products of nanotechnologies (2009), and scientific aspects of the existing and proposed definitions relating to products of nanoscience and nanotechnologies (2010) are of particular importance.[2] Furthermore, the SCCS's 2007 Opinion on the Safety of Nanomaterials in Cosmetic Products has to be mentioned.[3] In addition to these Committees, the European Group on Ethics in Science and Emerging Technologies (EGE) provides the Commission with advice on ethical and social aspects. Interestingly, in 2007 this body issued an opinion on nanomedicine that included regulatory issues.[4] In the same area the European Science Foundation published a report in 2005.[5] This report proposed a definition of nanomedicine that is frequently referred to. Another relevant advisory body is the Advisory Committee on Safety and Health at Work (ACSH). Furthermore, the European Committee for Standardization (CEN) deals with nanotechnological issues (CEN/TC 352).

---

[2]See the DG Health and Consumers website http://ec.europa.eu/health/ph_risk/committees/04_scenihr/scenihr_opinions_en.htm#nano (accessed on 15-04-2011).
[3]See http://ec.europa.eu/health/ph_risk/committees/04_sccp/docs/sccp_o_123.pdf (accessed on 15-04-2011)
[4]European Group of Ethics in Science and New Technologies (EPE) (2007) *Opinion on the Ethical Aspects of Nanomedicine*, European Commission, Brussels.
[5]European Science Foundation (ESF) 2005, *Forward Look on Nanomedicine*, see http:// www.esf.org/publications/214/Nanomedicine.pdf (accessed 27-04-2011).

At the national level of the member states we find similar structures of competent bodies, scientific, advisory, and standardization
committees. Some member states currently are launching action
plans on nanoscience and nanotechnologies which are implementing the EU Action Plan 2005–2009.[6] In its Recommendation on
the Action Plan the European Commission clarifies that products
based on nanoscience and nanotechnologies are already in use
and analysts expect markets to grow by hundreds of billions of
euros during this decade. The Commission stresses that Europe
musttransform its world-class Research and Development (R&D)
in nanoscience and nanotechnologies into useful wealth-generating
products in-line with the actions for growth and jobs, as outlined
in the "Lisbon Strategy" of the Union. Cornerstones of the policy
are increasing funding of R&D; improving the dialogue between researchers, public, private decision-makers, and other stakeholders;
building R&D infrastructure and "poles of excellence"; industrial
innovation; and responsible development, including integration of
ethic, social, and legal aspects, as well as international cooperation.

With regard to regulation the implementation document of the
Action Plan includes a couple of interconnected actions. In the
meantime the Commission reviewed relevant EU legislation with
a view to determine the applicability of the existing regulations
to the potential risks of nanomaterials. The review concluded that
although nanomaterials are not mentioned specifically in Union
legislation, current legislation covers in principle the potential
health, safety, and environmental risks in relation to nanomaterials.[7]
On 24 April 2009, the European Parliament adopted a resolution on
regulatory aspects of nanomaterials.[8] At that time the Parliament
voted to include nano-specific provisions in cosmetics and food

---

[6]This Action Plan was issued on 6 June 2005. See ftp://ftp.cordis.europa.eu/pub/
nanotechnology/docs/nano_action_plan2005_en.pdf (accessed 15-04-2011)

[7]See the Communication "Regulatory aspects of nanomaterials" and the accompanying Commission staff document "Summary of legislation in relation to health,
safety and environment aspects of nanomaterials, regulatory research needs
and related measures. http://www.kowi.de/Portaldata/2/Resources/fp7/coop/eu-
nano-policy-2004-08-en.pdf (accessed on 15-04-2011). On this website all key
documents of the EU policy for nanoscience and nanotechnologies are listed.

[8]Resolution European Parliament on Regulatory Aspects of Nanomaterials
(PA_T6(2009) 0328).

legislation. The Resolution contains regulatory constraints for the manufacturers, importers, and distributors who have or plan to incorporate engineered nanomaterials in products and place them on the EU market. According to influential commentators, the most important aspect of this Resolution is that it testifies to the increased readiness within the regulatory community to adopt targeted, nano-specific regulation.[9]

14.3 Overview of Hard and Soft Regulation

For the aim of this handbook I use a broad approach to regulation. Commonly, regulation is understood as the control of conduct or social affairs through rule-making, implementation, oversight, and enforcement.[10] In regulatory practice of science and business, rules matter when they are effective. Effective rules can be based on legislation ("hard regulation"), but also on other rules that although not legally binding have important effects in regulatory practice ("soft regulation") [11]. In the field of nanoscience and nanotech-nologies, examples of soft regulation are the EC Recommendation on the nano-specific Action Plan, the 2008 EC Recommendation on the Code of Conduct for Responsible Nanosciences and Nan-otechnologies Research, the 2006 United Kingdom's Voluntary Self Reporting Scheme for engineered nanoscale materials, and the 2010 Ruling of the Dutch Minister of Social Affairs and Labour on the use of Nanoreference Values in occupational safety risk management. When nanoscientists or the nano-industry conduct activities of research, product development and marketing in the EU, a huge body of legislation and soft regulation applies. Almost all legislative pieces of this regulatory body are general in the sense of being applicable to all kinds of research, development and marketing

---

[9]Van Calster, G., and Diana, M. B. A good foundation? Regulatory oversight of nanotechnologies using cosmetics as a case study, in *International Handbook on Regulating Nanotechnologies* (ed. Hodge, G., Bowman, D., and Maynard, A.), pp. 268–290.

[10]Black, J. (2002) This understanding refers to the influential definition of Julia Black. Critical reflection on regulation, *Australian Journal of Legal Philosophy*, **2**, 71–35.

activities, as well as to certain general product categories. In the EU, nano-specific regulation relies on soft instruments such as guidelines, recommendations, communications, codes of conduct, benchmarks, and standards. When we search for the rules that apply to a certain activity with nanomaterials, the whole governance arrangement, both legislation and soft regulation, is taken into account.

14.3.1 *Hard Regulation (Legislation)*

In the EU a huge and complex body of *general legislation* applies to nanomaterials. Within this legislative field we can distinguish between the domains of product safety, occupational health and safety, environment, privacy and civil liberties, intellectual property, and liability.[11] The legislation that applies in daily practice depends on the kind of activity in which nanomaterials are used, as well as on the area of the nanotechnological applications. For R&D, product marketing and post-marketing in relation to nanomaterials, European legislation on occupational safety, health, and the environment play a role. In addition, the production and use of nanomaterials can be governed by product-specific legislation. Attention should also be paid to legislation in which (non nano-specific) sizes of solid substances are included. The determination of size is specifically mentioned for solid substances/preparations when used as feed additives for animal nutrition.[12] To date, the only nano-specific EU legal provisions can be found in Regulation EC/1223/2009 on Cosmetic Products. Nano-specific legislation is currently being discussed and prepared in the implementation of the REACH (Registration, Evaluation, Authorization, and Restriction of Chemicals) and in the legislation for the regulation of novel foods. More information about the legislative processes is provided below.

---

[11]A similar distinction has been made by Bowman and Hodge: Bowman, D., and G. Hodge, Nanotechnology: Mapping the wild regulatory frontier, *Futures*, **38**, 1060–673.

[12]Such is the case for Commission Regulation (EC) No 429/2008 of 25 April 2008 on detailed rules for the implementation of Regulation (EC) No 1831/2003 of the European Parliament and of the Council as regards the preparation and the presentation of applications and the assessment and the authorization of feed additives.

There are two main reasons for the lack of nano-specific EU legislation. Based on the 2008 regulatory review and the experience with nanoproducts that already have been marketed a common opinion is that the existing legislation is robust enough to cope with the challenges of nanoscience and nanotechnologies. Another reason refers to the huge uncertainties regarding the effects of nanotechnologies. Traditional legislation is based on evidence. When evidence on risks and benefits cannot be provided legislators have to wait. However, since the EU is committed to responsible development of nanosciences and nanotechnologies regulatory bodies cannot afford a *laissez-faire* attitude. Confronted with the uncertainties on the effects of nanotechnological development the regulatory activities of the EU have focused on testing whether legislation is robust enough to cope with nanoscientific and nanoproduct development, as well as on detecting regulatory gaps and collecting information about the properties and risks of nanomaterials. Stimulating and enabling responsible research, development, and public engagement are the cornerstones of EU regulatory policy.

14.3.2 *Soft Regulation*

In the EU nanomaterials are governed jointly by general legislation and nano-specific soft regulation. In our understanding soft regulation refers to non-legally binding rules that are established by private and public organizations. Usually soft regulation is laid down to implement and enforce existing legislation (e.g., interpretation, enforcement, and oversight guidance). Soft regulation serves also to prepare specific legislation. It is set up for specific purposes like standardization and responsible development for which legislation does not seem to be an appropriate tool. With regard to soft regulation that has been laid down to implement legislation, a huge body of general guidelines applies to activities with nanomaterials. Nanopharmaceuticals, for example, are governed by the guidelines that the European Medicines Agency and national competent bodies have issued for clinical trials, product approval, manufacturing processes and post-marketing controls. In 2007 nano-specific guidance has been launched for the use of nanomaterials in medical devices.[13] Member states, in collaboration with the industry have issued guidelines to implement labor conditions legislation. For the same reason guidance and benchmarks for risk management at the work place have been established. Furthermore, general soft instruments for the implementation and preparation of legislation on nanomaterials are the rules of the Precautionary Principle and the As-low-as-reasonable-achievable (ALARA) Principle. According to the 2000 Communication of the EC on the Precautionary Principle, scientific uncertainty about technological risks is no reason for regulatory inaction if there might be immense adverse effects.[14] The ALARA Principle requires to minimize the exposure to nanomaterials at the work place and the release of nanoparticles into the environment as low as reasonably achievable. Other nano-specific soft tools for the implementation and preparation of legislation are the reporting schemes some Member States have set up. These schemes serve to collect information on the properties and risks of activities with nanomaterials. Furthermore, the European Commission is drafting a recommendation to prepare legal definitions of nanomaterials.

Other examples of nano-specific soft regulation that do not focus primarily on legislative support are the EC's Action Plans on Nanosciences and Nanotechnologies, the 2008 Code of Conduct for Responsible Nanosciences and Nanotechnologies Research and the regulatory activities of some member states on labeling and registration of nanoproducts. The main aim of the EC Code of Conduct is to promote integrated, safe, and responsible nanosciences and nanotechnologies research in Europe for the benefit of the society as a whole.[15] Another aim is to coordinate regulatory activities of the member states. The Code provides member states with an instrument to undertake further initiatives to ensure safe,

---

[13] Working Group on New and Emerging Technologies in Medical Devices (N&ET Working Group) 2007. Report on Nanotechnology to the Medical Devices Expert Group. See http://ec.europa.eu/consumers/sectors/medical-devices/scientific-technical-assessment/working-group/index_en.htm (accessed on 27-04-2011)

[14] European Commission (2000). Communication on the precautionary principle, COM(2000), 1, 2.2.2000).

[15] See website quoted at note 7.

ethical, and sustainable development when implementing their national nanotechnological strategy. The Commission recommends that member states consider the Code's general principles and guidelines to be an integral part of institutional quality assurance concerning funding schemes, as well as adopting them for the auditing, monitoring, and evaluation processes of public bodies and using them in their cooperation with third countries and to encourage national dialogue on nanotechnologies. Another recommendation is that the member states encourage the adoption of the Code amongst all relevant stakeholders and that they take the necessary steps to ensure that stakeholder contribute to safe, ethical, and sustainable development. The Code is based on a set of seven guiding principles, namely, comprehensibility, sustainability, precaution, inclusiveness, excellence, innovation, and accountability. It lays down guidelines on actions to be taken. Guidance is given on how to achieve good governance, due respect for precaution, as well as wide dissemination and good monitoring of the Code. To date, the Code has not yet been implemented in the member states.[16] The Netherlands is the first member state that has introduced a mandatory contractual obligation to comply with the Code in its national funding schemes for nanosciences and nanotechnologies R&D. Researchers and companies that apply for funding have to declare that they implement the Code.

14.4 Cross-Cutting Regulatory Issues

14.4.1 *Definition of Nanomaterials*

With the expected increase in the applications of nanotechnology, an urgent need is felt in the EU to identify what can be considered as a nanomaterial by clear and unequivocal descriptions. This need to identify a nanomaterial is caused by the uncertainty surrounding the safety evaluation and risk assessment of nanomaterials. In

---

[16]NanoCode (2010) *Synthesis Report on Codes of Conduct, Voluntary Measures and Practices Towards A Responsible Development of N&N*, see http://www.nanocode.eu (accessed on 19-04-2011).

its Resolution of 24 April 2009, the European Parliament called for the introduction of a comprehensive science-based definition of nanomaterials in Union legislation as part of nano-specific amendments to relevant horizontal and sectoral legislation.[17] The Parliament further called on the Commission to promote the adoption of a harmonized definition of nanomaterial at the international level and to adapt the relevant European legislative framework accordingly. The Commission invited the SCENIHR to provide scientific input on elements to consider when developing a definition of the term "nanomaterial" for regulatory purposes. The SCENIHR opinion "Scientific basis for the definition of the term 'nanomaterial'" was issued for public consultation on 6 July 2010. The Committee concluded that size is universally applicable to nanomaterials and is a key element to a definition. A defined size range would facilitate a uniform interpretation. The lower limit was proposed at 1 nanometer (nm). An upper limit of 100 nm is commonly used by general consensus but there is no scientific evidence to qualify the appropriateness of this value. The Committee thought that the use of a single upper limit value might be too limiting for the classification of nanomaterials and a differentiated approach might be more appropriate. For regulatory purposes, the number size distribution should also be considered using the mean size and its standard deviation to refine the definition. Furthermore, SCENIHR identified certain specific cases where the application of the definition can be facilitated by using the volume-specific surface area as proxy for the internal or surface structure. On the basis of the SCENIHR Opinion of 8 December 2010 the DG Environment drew a Recommendation on a definition of the term "nanomaterial."[18]

---

[17]See note 7.

[18]Interesting are Article 2 (1. Nanomaterial: means a material8 that meets at least one of the following criteria: – consists of particles, with one or more external dimensions in the size range 1 nm–100 nm for more than 1% of their number size distribution; – has internal or surface structures in one or more dimensions in the size range 1 nm; – 100 nm; – has a specific surface area by volume greater than 60 m^2/cm^3, excluding materials; consisting of particles with a size lower than 1 nm. 2. Particle: means a minute piece of matter with defined physical boundaries (ISO 146446:2007) and Article 3 (1. The Commission will carry out a public consultation by 2012 and if appropriate review the adequacy of Article 2 taking into account experience gained, scientific knowledge and the technological development.). Drafted in April 2011 (see,

The European Commission hopes to issue a regulatory definition of nanomaterials before the summer of 2011. However, several issues still need to be resolved before it can make a "political" decision on the issue, according to DG Environment's Henrik Laursen, speaking at the Fourth Annual Nano Safety for Success Dialogue on 29 March 2011 in Brussels.[19] At this Conference it was mentioned that that DG Health and DG Enterprise both gave a negative reaction to the draft definition. The main point of disagreement was DG Enterprise's call for mass, rather than particle numbers, to be used as the measuring unit. At the Nano Safety Dialogue it was obvious that the Commission is coming under growing pressure to produce a standard EU definition for nanotechnology and materials.

14.4.2 New Chemicals Regulation (REACH)

Chemicals regulation plays an important role in the governance of nanomaterials. Regulation 1907/2006 on the Registration, Evaluation, Authorisation and Restriction of Chemicals (REACH) and Regulation 1272/2008 on Classification, Labeling and Packaging (CLP) are the most important pieces of this body of legislation. REACH provides an over-arching legislation applicable to the manufacture, placing on the market and use of substances on their own, in preparations or in articles. It does not contain specific provisions for nanosubstances. Nanomaterials are covered by the REACH Regulation, even though there is no explicit reference to nanomaterials. "No data no market" is a core principle of the Regulation. This means that if a manufacturer, importer or producer cannot show either that a substance is safe, or which precautions need to be taken to ensure its safe use, that substance cannot be lawfully marketed in the EU. The general obligations in REACH, such as registration of substances manufactured at 1 tonne or more and providing information in the supply chain, apply as for any other substance. The first registration deadline under REACH

---

http://ec.europa.eu/environment/consultations/pdf/recommendation_nano.pdf., accessed on 27-04-2011).

[19] http://chemicalwatch.com/7163/eu-commission-directorates-argue-over-nano-definition (accessed on 15-04-2011)

(30 November 2010) applies to substances manufactured or imported at 1000 tonnes or more per year. The registrations of nanomaterials in this tonnage band will help to generate more information useful for the assessment of risks. The European Chemicals Agency (ECHA) receives the registrations. The Agency plays a central role in the collection, evaluation, and dissemination of information on substances and preparations, including nanomaterials.

According to the CLP Regulation nanomaterials that fulfill the criteria for classification as hazardous substances and mixtures must be classified and labeled. Many of the related provisions, including safety data sheets and classification and labeling apply already today, independently of the tonnage in which the substances are manufactured or imported. Substances, including nanomaterials, meeting the classification criteria as hazardous must be notified to ECHA by 3 January 2011. Currently, the Commission is preparing advice on how to manage nanomaterials in accordance with REACH and the CLP Regulation.[20] The preparatory papers provide an overview of how the provisions of REACH apply to nanomaterials and how nanomaterials are classified in accordance with REACH and particularly the CLP Regulation. According to analysts, REACH is not sufficiently implementable to nanomaterials.[21] In this context an important problem is the focus on volume (tonne threshold) that fails to account for particle size and surface area as key determinants of toxicity.[22]

14.5 Regulation of Nanoproducts in Specific Areas

14.5.1 Nanomaterials in Medical Products

Nanomedicine is a field that is developing rapidly in the EU. It can be defined as the application of nanotechnologies in view of diagnosing, treating, or preventing diseases and of preserving

---

[20]See http://ec.europa.eu/enterprise/sectors/chemicals/documents/reach/index_en.htm (accessed on 26-04-2011).

[21]Lee, R. G., and Vaughan, S. (2010) REACHing down: Nanomaterials and chemical safety in the European Union, *Law, Innovation and Technology*, **2**(2), 193–217.

[22]Stokes, E. (2009) Regulating nanotechnologies: Sizing up the options, *Legal Studies*, **29**, 281.

and improving human health.[23] Nanomedicine includes the areas of diagnostics, therapeutics, and regenerative medicine. Since no specific legislation has been established in the EU with regard to these products the current regulatory system of medical products as it stands is applicable. This system contains general and specific medicinal products regulation, specific substances regulation, and medical devices regulation.[24] In the approval procedures the main evaluation principle is that the risks must be outweighed by the benefits of the product (benefit-risk balance). Discussions on regulatory challenges of medical products utilizing nanomaterials have taken place mainly in the field of therapeutic applications. In this field, the European Medicines Agency (EMA) is an active key player. To date, the Agency has approved 18 nanopharmaceuticals without specific "nano"-related evaluation problems. This is why the EMA concludes that the existing evaluation system is robust enough to accommodate nanodrugs. However, the Agency is well aware of the regulatory challenges coming nanomedical applications are posing. To cope with these challenges the EMA's Scientific Committee (CHMP) launched a reflection paper in 2006 on nanotechnology-based medicinal products for human use.[25] Later the EMA established an ad hoc CHMP Expert Group which organized a workshop in 2009. The EMA's Innovation Task Force provides a forum for early exploration of regulatory implications. The Small and Medium Size Enterprises Office offers regulatory support. Since 2006 the EMA actively participates in the dialogue platform of regulators from the EU, US, Canada, and Australia. This platform serves to share experience and understanding across regulatory frameworks including food, medical devices, cosmetics,

---

[23]This definition is used European Medicines Agency, Committee for medicinal products for human use, Reflection paper on nanotechnology-based medicinal products for human use, EMEA/CHMP/79769/2006). It refers to the work of the European Science Foundation (see note 5).

[24]An overview of general legislation that applies to nanopharmaceuticals is provided by B.R. Dorbeck-Jung and N. Chowdhury (2011) Is the European medical products authorisation regulation equipped to cope with the challenges of nanomedicines? Law & Policy, 33(2), 276–303.

[25]This conclusion was made in the EMA's EMEA/CHMP/79769/2006 and at its 2010 Workshop on Nanomedicines. See Workshop Report on http://www.ema.europa.eu/docs/en_GB/document_library/Report/2010/10/WC500098380.pdf (accessed on 28-04-2011).

and pharmaceuticals. In September 2010 the first international workshop on nanomedicines was held with the EMA. One conclusion of the workshop was that the pooling of knowledge and expertise at a global level and across disciplines is required to cope with the challenges of nanomedicines. Regulatory problems were identified with regard to the definition of nanomedicines, the regulatory classification of combination products, nano-specific safety and quality standards, risk assessment methods, and the environmental impact analysis.[26]

In the field of nanomedicine experts agree that combination products pose major regulatory challenges.[27] Highly discussed are nano-enabled drug delivery systems that integrate medical devices and pharmaceuticals. In this context the question arises which regulatory regime applies. Since an approval according to the medicinal products regulation is accompanied by more regulatory obligations than a notification according to the medical devices regime the classification of the combined product does matter in regulatory practice. Medical devices using nanomaterials have been marketed in Europe.[28] Like the EMA, the European Working Group on New and Emerging Technologies in Medical Devices (N&ET WG) concluded in 2007 that the medical device legislation is suitable to deal with medical devices utilizing nanotechnologies.[29] In the view of the Working Group, the main reason is that the legal provisions on risk management are appropriate to address all kinds of risks. However, the Group recommended to develop regulatory guidance that serves to collect information on the experience with medical devices manufactured utilizing nanotechnology (including risk assessment), as well as to provide specific requirements for

---

[26] See note 26.

[27] Gaspar, R. (2010) Therapeutic products: regulating drugs and medical devices, in *International Handbook on Regulating Nanotechnologies* (ed. Hodge, G., Bowman, D., and Maynard, A.), 291–320.
Gaspar, Rogério. 2007. "Regulatory Issues Surrounding Nanomedicines: Setting the Scene for the Next Generation of Nanopharmaceuticals," *Nanomedicine* 2 (2): 143–147.

[28] C. Altenstetter, Medical device regulation and nanotechnologies: Determining the role of patient safety concern in policymaking, *Law & Policy*, 33 (2) 2011, 227–255.

[29] See note 14.

post-marketing control. Currently, the N&ET WG is finalizing a guidance document on these issues.

14.5.2 *Nanomaterials in Cosmetic Products*

Cosmetic products which incorporate engineered nanomaterials are on the market. An established application field of metallic oxide nanoparticles in consumer products is sunscreens with physical UV filters. Due to the nano-scaled dimensions of the UV-absorbing particles the creams are colorless and offer protection from incident radiation over a broad spectrum without irritating the skin. According to the EU Cosmetics Directive 76/768/EEC and its amendments, the manufacturer or importer of a cosmetic product has to assess the safety of the product prior to placing it on the market and document this. This has to be done taking "into consideration the general toxicological profile of the ingredients, their chemical structure, and their level of exposure."[30] The safety evaluation usually is reviewed by the competent authorities of the member states in in-market controls. Apart from these general rules, certain groups of substances, including UV filters, have to be permitted by the European Commission prior to their use in cosmetic products placed on the EU-market. Such permission is preceded by an opinion of the Scientific Committee for Consumer Products (SCCP). In its 2007 report the SCCP concluded that there are large data gaps in risk assessment methodologies with respect to nanoparticles in cosmetic products.

In March 2009, the European Parliament proposed amendments to the recast of the Cosmetics Directive including nano-specific provisions. The final text of the Cosmetics Regulation was adopted by the Parliament and the Council in November 2009 (Regulation EC/1223/2009). It will take some time until this new instrument will enter into force. One of the key provisions of the final text is the definition of the "nanomaterial" in Article 2(1)(k) as "an insoluble or

---

[30]Art. 7a (1) (d) Cosmetics Directive 76/768/EEC http://ec.europa.eu/consumers/ sectors/cosmetics/cosmetic-products/nanomaterials/index_en.htm#h2-what-are-the-rules-in-europe-for-the-use-of-insoluble-nanoparticles-in-cosmetics? (accessed 20-04-2011).

biopersistent and intentionally manufactured material with one or more external dimensions, or an internal structure, on the scale from 1 to 100 nm." According to the Regulation, the responsible person for placing a new cosmetic product containing nanomaterials on the EU market is required to notify the Commission of the presence of the product and to supply certain safety information to the Commission six months before its entry onto the market. Any products containing nanomaterials must indicate the presence of them in the list of ingredients. Article 16(10)(b) obliges the Commission to set up a publically accessible catalogue of all nanomaterials used in cosmetic products placed on the market including the reasonably foreseeable exposure conditions.

14.5.3 *Nanomaterials in Food Products*

It is claimed that nanotechnologies offer a variety of possibilities for applications in the food and feed area — in production/processing technology, to improve food contact materials, to monitor food quality and freshness, improved traceability and product security, modification of taste, texture, sensation, consistency and fat content, and for enhanced nutrient absorption. In the food area nano-scaled micelles that transport water-insoluble nutrients, vitamins, minerals, colorants, and fatty acids are already used. The advantages of nanomaterials are said to be the easy absorption in the bowel and solubility in higher concentrations in the product. Potential applications of nanomaterials in the food area are mineral and vitamin supplements, gas-tight plastic bottles, antibacterial kitchenware, and food storage containers.

In this area regulatory activities started with a scientific opinion of the Scientific Committee on Consumer Safety (SCCS) which had been requested by the European Food Safety (EFSA). The opinion "The Potential Risks Arising from Nanoscience and Nan-otechnologies on Food and Feed Safety" was adopted on 10 February 2009.[31] The SCCS emphasized that the risk assessment processes are still under development with respect to the characterization

---

[31] *The EFSA Journal* (2009) 958, 3–39

and analysis of engineered nanomaterials (ENMs) in food and feed, the optimization of toxicity testing methods for ENMs, and the interpretation of the resulting data. The Committee recommended to take actions to develop methods to detect and measure ENMs in food/feed and biological tissues, to survey the use of ENMs in the food/feed area, to assess the exposure in consumers and livestock, and to generate information on the toxicity of different ENMs. In March 2009, the European Parliament adopted the first reading of a proposal for a regulation of the European Parliament and of the Council on novel foods that included a definition of engineered nanomaterials amending earlier legislation. A year later members of the European Parliament's Environment, Public Health, and Food Safety Committee suggested to control the use of nanotechnologies in foods for humans.[32] The proposed measure entirely excluded entry onto EU markets of any food derived from cloned animals as well as food produced by nanotechnology processes, unless such food has undergone a specific risk assessment of its possible impact on health. The measure also required all ingredients containing nanomaterials to be clearly labeled by listing the names of the ingredients followed by the word "nano" in brackets. At the end of March 2011 the European Council refused to approve the labeling of all food products from cloned offspring as demanded by the European Parliament. According to commentators, this means that the current novel foods Regulation, adopted in 1997, will remain in force, and no new measures relating to nanomaterials will be taken.[33]

14.6 Conclusion

The exploration of the EU regulation that applies to nanomaterials shows a large body of existing general and nano-specific soft regulation. The various activities of European regulatory bodies and

---

[32] http://www.europarl.europa.eu/sides/getDoc.do?language=EN&type=IM-PRESS&reference=20100507STO74257 (accessed 20-04-2011).

[33] http://chemicalwatch.com/7085/nano-food-labelling-falls-foul-of-eu-failure-to-agree-novel-foods-regulation (accessed 20-04-2011).

advisory committees have led to the identification of regulatory gaps. Regulatory bodies are employing activities to prepare nano-specific guidance for the implementation and enforcement of existing legislation, as well as for the preparation of nano-specific legislation. I agree with influential analysts that an increased restlessness among regulators and governments vis-à-vis the technology's uncertainty can be observed[34] While the risk assessment experts emphasize that it will take many more years to get insights into the effects of nanomaterials European industry and consumers urge regulators to come with nano-specific legislation. Considering this tension a final conclusion is that the next years will be crucial for regulatory development in relation to nanomaterials. The coming years will show whether regulatory bodies and regulatory science are capable to stimulate responsible development of nanoscience and nanotechnologies.

---

[34]See Van Calster and Bowman, note 9, at p. 277.

Chapter 15

Nanomaterial Regulation in the United States

Michael E. Heintz

Maryland Energy Administration, 60 West St., Suite 300, Annapolis, MD 21401, USA
mheintz13@gmail.com

15.1 Nanomaterial Regulation in the United States

Nanomaterial regulation in the United States is still very much in the developmental stages. To date, there is no one national regulation for nanomaterials. Instead, each lawmaking authority has individual responsibility for chemicals or uses falling under their statutory mandates. This means that individual federal agencies, states, and local jurisdictions are in the process of developing nanomaterial regulation for their specific areas of concern. While most federal agencies, and several states, have begun the process of evaluating nanomaterial regulation, nothing has developed into a permanent set of controls or guidelines. This includes environmental protection, worker safety, cosmetics, insurance coverage, and related concerns. The following discusses the regulatory efforts concerning nanomaterials in the United States. Readers are urged to check cited

Nanotechnology Commercialization for Managers and Scientists
Edited by Wim Helwegen and Luca Escoffier
Copyright © 2012 Pan Stanford Publishing Pte. Ltd.
ISBN 978-981-4316-22-4 (Hardcover), 978-981-4364-38-6 (eBook)
www.panstanford.com

sources and newly formed resources on a regular basis as the area is fast moving and likely to change in the near future.

15.2 Federal Regulation of Nanomaterials

As stated above, the United States has no central policy or regulation of nanomaterials. Instead, each individual agency is left to determine for itself if it will regulate nanomaterials, as well as the process for such regulation. In an effort to coordinate federal agency activity, the National Nanotechnology Initiative (NNI) was created. The NNI is the "U.S. government interagency cross-cut program that coordinates federal research and development activities in nanoscale science, engineering, and technology and related efforts among various participating agencies."[1] The NNI positions itself as a "clearinghouse" of information between all of the federal agencies addressing questions regarding nanomaterials. While the NNI has no authority to direct individual agency action or policy, it attempts to coordinate those actions in order to maintain a level of consistency across the federal government. In all, twenty-five federal agencies participate in NNI activities.[2] Additionally, the NNI has a 2010 budget of $1.6 billion in order to accomplish its mission.[3] The NNI is a subset of the National Science and Technology Council, a cabinet level office within the President's administration, and is administered with assistance from the National Nanotechnology Coordination Office.[4] Notably, the NNI's and related offices' efforts focus on nanotechnology research

---

[1] Subcommittee on Nanoscale Science, Engineering, and Technology, Committee on Technology, National Science and Technology Counsel (2009), *National Nanotechnology Initiative, Research and Development Leading to a Revolution In Technology and Industry, Supplement to the President's FY 2010 Budget*, Executive Office of the President, Office of Science and Technology Policy, Washington, D.C., available at http://nano.gov/NNI_2010_budget_supplement.pdf at 1 (accessed on June 26, 2010) (hereinafter "*NNI Budget*").

[2] National Nanotechnology Initiative, *About the NNI-Home*, available athttp://www.nano.gov/html/about/home_about.html (accessed on June 26, 2010).

[3] *NNI Budget* at 7.

[4] National Nanotechnology Initiative, *About the NNI-Home*, available at http://www.nano.gov/html/about/home_about.html (accessed on June 26, 2010).

and development, not regulation. However, as the only federal office focusing on cross-agency efforts regarding nanomaterials, its focus can be wide reaching. That said, the NNI has refrained from entering into the regulatory discussion, instead allowing each individual agency to address regulation questions themselves. Most recently, the President's Council of Advisors on Science and Technology recommended changes to the NNI and its operations. This third evaluation of the NNI determined that the organization has had a "catalytic and substantial impact on the growth of US nanotechnology innovation and should be continued."[5] However, the report did make a number of recommendations for improvements, including increased coordination between the NNI and participating agencies, increased focus on environmental, health, and safety issues as well as standards setting, increased focus on commercialization of nanotechnology-based products, continued focus on education and societal efforts, and development of "signature initiatives" by lead agencies.[6] Overall, the report recommended the continuation of the NNI and praised its work while citing room for improvement. Funding for the NNI would be reauthorized by passage of the America COMPETES Act, H.R. 6115, in the Senate.[7]

While Congress has taken up the issue of nanomaterials, the legislation has stalled in committee. The Nanotechnology Advancement and New Opportunities (NANO) Act, H.R. 820, was introduced in February, 2009 and quickly assigned to committee for debate.[8] However, since that time, no further movement has occurred with regard to this legislation. In addition to the funding and incentive

---

[5] Office of Science and Technology Policy Executive Office of the President (2010), "Independent Review Finds Federal Nanotechnology Initiative Highly Effective; Recommends Changes to Ensure Ongoing U.S. Dominance," Washington, D.C., available at http://www.whitehouse.gov/sites/default/files/microsites/ostp/nano-release.pdf (last visited June 26, 2010).

[6] President's Council of Advisors on Science and Technology (2010), *Report to the President and Congress on the Third Assessment of the National Nanotechnology Initiative*, Washington, D.C., at *x*.

[7] America COMPETES Reauthorization Act of 2010, H.R.5116, available at http://www.thomas.gov/cgi-bin/query/z?c111:H.R.5116.EH: (accessed on June 26, 2010).

[8] Nanotechnology Advancement and New Opportunities Act, H.R. 820, available at http://thomas.loc.gov/cgi-bin/bdquery/z?d111:HR00820:@@@L&summ2=m& (accessed on June 26, 2010).

portions of the legislation, the bill, if enacted, would require "the establishment of nanotechnology research grant programs by the. ...Administrator of the Environmental Protection Agency (EPA) to address technologies for remediation of pollution and other environmental protection technologies; [and the] Secretary of Health and Human Services (HHS) to address health related applications of nanotechnology."[9] However, it seems unlikely that this piece of legislation will be enacted, let alone move out of committee.

In early 2010, a second nanotechnology-based piece of legislation was introduced — the Nanotechnology Safety Act of 2010, S. 2942.[10] This bill, introduced in the Senate, proposes to amend the Food, Drug, and Cosmetics Act (see below) to "establish within the Food and Drug Administration a program for the scientific investigation of nanoscale materials included or intended for inclusion in FDA-regulated products, to address the potential toxicology of such materials, the effects of such materials on biological systems, and interaction of such materials with biological systems."[11] The bill proposes allocating $25,000,000 to the Food and Drug Administration to accomplish this goal and otherwise collect, synthesize, and report on nanotechnology impacts in the food and drug sectors. However, the bill does not propose to establish any regulation or further legislative control of nanomaterials. Instead, the bill is a formalized research program for the agency in order to understand the potential risks, if any, associated with nanomaterials in food and drug products. The bill was referred to the Committee on Health, Education, Labor, and Pensions, but no further action is currently scheduled.

Most recently, S. 3117, the Promote Nanotechnology in Schools Act of 2010 was introduced in the Senate on March 15, 2010 and referred to the Committee on Health, Education, Labor, and Pensions.[12] The bills would require the Director of the National

---

[9] *Id.*

[10] Nanotechnology Safety Act of 2010, S. 2942, available at http://thomas.loc.gov/cgi-bin/query/z?c111:S.2942: (accessed on June 26, 2010).

[11] *Id.*

[12] Promote Nanotechnology in Schools Act, S. 3117, available at http://thomas.loc.gov/cgi-bin/bdquery/D?d111:1:./temp/~bdZtWs:@@@L&summ2=m &|/ home/LegislativeData.php| (last visited June 26, 2010).

Science Foundation "to establish a nanotechnology in the schools program awarding matching grants to certain eligible institutions to: (1) purchase nanotechnology equipment and software; (2) service and upgrade such equipment; and (3) provide nanotechnology education to students and teachers."[13] Eligible recipients include high schools, colleges, and informal science and technology centers.

Until Congress decides to act on the pending legislation, or introduces new measures which are then passed, the individual federal agencies are left to direct their own nanomaterial programs and potential regulation within the scope of their areas of concern.

15.2.1 *Environmental Regulation of Nanomaterials*

The United States Environmental Protection Agency (EPA) has moved the closest toward nanomaterial regulation. In addition, in 2006 the American Bar Association's Section on Environment, Energy, and Resources conducted a study of the six major environmental statutes to determine applicability to nanomaterials.[14] In Phase I of the study, the Section concluded that each of the Clean Water Act, Clean Air Act, Comprehensive Environmental Response, Compensation, and Liability Act (aka "Superfund), Toxic Substances Control Act, Resource Conservation and Recovery Act, and the Federal Insecticide, Fungicide, and Rodenticide Act each had the authority to regulate nanomaterials with little or no revision. Phase II is examining the Food Quality Production Act, Endangered Species Act, and National Environmental Policy Act, and is ongoing.[15]

EPA has also undertaken its own review of its authority concerning nanomaterials, and has focused its efforts on the Toxic Substances Control Act (TSCA) as the primary means of data gathering and regulation. While the agency has yet to officially regulate nanomaterials, indications are that the Office of Pollution Prevention and Toxics (OPPT) is moving in that direction and will

---

[13] *Id.*

[14] American Bar Association, Section of Environment, Energy, and Resources, *Section Nanotechnology Project*, available at http://www.abanet.org/environ/nanotech/ (accessed on June 26, 2010).

[15] *Id.*

do so soon. TSCA,[16] the country's chemical registration statute, "provides EPA with authority to require reporting, record-keeping and testing requirements, and restrictions relating to chemical substances and/or mixtures."[17] Principally, TSCA requires the registration and tracking of chemicals produced for use in, or imported to, the United States. Chemical manufacturers may not produce, use, or sell a new chemical product until it is approved for listing by EPA. Because all "new chemicals" must first be registered, EPA has had difficulty determining how best to treat nanomaterials. Because the chemical composition of most nanomaterials is the same as the macro-sized counterparts, EPA has been reticent to regulate nanomaterials as "new chemicals." EPA has been working on ways to keep informed about nanomaterials in advance of regulation.

Most recently, the EPA released a new definition of "nanomaterial." A formal Federal Register announcement was expected in June 2010 for the reportability of nanomaterials under FIFRA. In addition, a presentation by William Jordan defined a nanomaterial as "an ingredient that contains particles that have been intentionally produced to have at least one dimension that measures between approximately 1 and 100 nanometers."[18] According to Mr. Jordan's presentation, the Federal Register notice will "Announce a new interpretation of FIFRA/regulations and proposes a new policy."[19] However, it is expected that the EPA will determine that pesticides containing nanomaterials will require reporting under FIFRA section 6(a)(2) and will apply equally to new and pending registrations.[20] This notice is also likely to include a determination that an existing pesticide or ingredient is "new" if appearing in

---

[16] 15 U.S.C. §2601 *et seq.*

[17] U.S. Environmental Protection Agency, *Summary of the Toxic Substances Control Act*, available at http://www.epa.gov/lawsregs/laws/tsca.html (accessed on June 26, 2010).

[18] Jordan, William (April 29, 2010), *Nanotechnology and Pesticides,* presented at the Pesticide Program Dialogue Committee, at page 5, available at http://www.nanotechproject.org/process/assets/files/8309/epa_newpolicy_nanomaterials.pdf (accessed on June 6, 2010).

[19] *Id.* at page 15.

[20] *Id.* at page 16.

nanoscale form.[21] Consequently, existing pesticides such as silver may need re-registration when used in nanoscale form.

15.2.1.1 Nanomaterials stewardship program

EPA's first action with regard to nanomaterials was a data gathering effort labeled the Nanomaterials Stewardship Program (NMSP). Similar to the United Kingdom's Voluntary Reporting Scheme for Manufactured Nanomaterials (VRS), the NMSP was designed as a voluntary reporting system for nanomaterial manufacturers and users. In this way, companies could submit monitoring data on specific nanomaterials to the EPA so the agency could develop a database addressing environmental risks and concerns. Started in January 2008, and administered by OPPT, the NMSP was to "provide a firmer scientific foundation for regulatory decisions by encouraging submission and development of information including risk management practices for nanoscale materials."[22] The NMSP was designed with two complimentary programs. Both portions of the NMSP were voluntary, and like the VRS, received only minimal to moderate response.

The Basic Program allowed participants to submit available data without creating additional data or performing unique studies. Basic Program participants were asked to submit data reports by December 2008. By December 8, 2008, twenty-nine companies submitted data for 123 nanoscale materials. Additionally, seven more companies had committed to submit data, but had not done so by the due date.[23] Conversely, the In-Depth Program asked participants to develop data over time, and in conjunction with EPA, in an effort to accumulate information not readily available by "business as usual" activities. By December 8, 2008, only four companies committed to participating in the In-Depth Program.

---

[21] *Id.* at page 18.
[22] U.S. Environmental Protection Agency, Office of Pollution Prevention and Toxics, *Nanoscale Materials Stewardship Program*, available at http://www.epa.gov/oppt/nano/stewardship.html (accessed on June 26, 2010).
[23] U.S. Environmental Protection Agency Office of Pollution Prevention and Toxics (2009), *Nanoscale Materials Steward Program Interim Report*, Washington, D.C., (hereinafter "*Interim Report*"), available at http://www.epa.gov/oppt/nano/nmsp-interim-report-final.pdf (accessed on June 26, 2010), at 3.

The NMSP was scheduled to end in January 2010, with a final report published soon after. However, the interim report considers the NMSP a success, despite the data gaps that remain.[24] For example, OPPT determined that approximately 90% of the nanomaterials used in commerce were not reported under the NMSP and many submissions did not contain "exposure or hazard related data."[25] Notwithstanding the data gaps, OPPT feels that the NMSP "contributed to a considerably stronger and better informed understanding of the issues and commercial status of nanoscale materials."[26] EPA is now using the information gathered through the NMSP to develop and implement effective nanomaterial regulations, beginning with carbon nanotubes.[27]

15.2.1.2 Carbon nanotube (CNT) regulations

On October 31, 2008, EPA issued a notice in the Federal Register stating that carbon nanotubes are a distinct chemical structure to other carbon-based chemicals in the chemicals inventory.[28] "EPA generally considers CNTs to be chemical substances distinct from graphite or other allotropes of carbon listed on the TSCA Inventory. Many CNTs may therefore be new chemicals under TSCA section 5."[29] The consequence of this notice is that manufacturers and importers of carbon nanotubes are now required to submit "pre-manufacture notices" (PMN) in connection with carbon nanotubes. Under TSCA, chemicals not registered on the inventory cannot be used in commerce, that is, manufactured or imported, without submitting a PMN at least 90 days before the start of manufacturing. Because users of carbon nanotubes believed them to be significantly similar to other forms of carbon already listed on the chemicals inventory, such as graphene, product production was halted in many instances in order to register the nanotubes and maintain compliance with TSCA. In addition, because of the number of

---

[24] *Interim Report* at 3.
[25] *Interim Report* at 27.
[26] *Interim Report* at 26.
[27] *Interim Report* at 26.
[28] 73 F.R. 64946–64947 (Oct. 31, 2008).
[29] 73 FR 64946 (Oct. 31, 2008).

different types of carbon nanotubes, registration of one did not necessarily allow for manufacture of a different version. EPA stated, "If a particular CNT is not on the TSCA Inventory, anyone who intends to manufacture or import that CNT is required to submit a PMN (or applicable exemption) under TSCA section 5 at least 90 days before commencing manufacture."[30] This continues to cause supply chain and manufacturing issues for those products using unregistered carbon nanotubes.

15.2.1.3 Significant new use rule (SNUR)

Most recently, and most importantly, EPA began the process of promulgating a Significant New Use Rule (SNUR) concerning carbon nanotubes.[31] According to the EPA, this SNUR will "require persons who intend to manufacture, import, or process new nanoscale materials based on chemical substances listed on the TSCA Inventory to submit a Significant New Use Notice (SNUN)."[32] The SNUNs would be required at least 90 days before manufacturing of the nanomaterial is to begin. Additionally, the SNUNs would provide EPA "with a basic set of information on nanoscale materials, such as chemical identification, material characterization, physical/chemical properties, commercial uses, production volume, exposure and fate data, and toxicity data."[33] The SNUR, on the other hand, will identify current uses of nanomaterials based on prior submissions to the EPA, including those from the Nanomaterials Stewardship Program. Comments on the SNUR were due to EPA in early December 2008, but no further information has been released. However, a new SNUR is likely imminent.

Related to the SNUR issue, in November 2008, the Thomas Swan & Company, Ltd. announced it filed the first PMN and accompanying consent decree permitting it to manufacture carbon nanotubes in the United States. Under the consent decree, Thomas Swan's

---

[30] 74 FR 64946, 64947 (Oct. 31, 2008).
[31] 74 Fed. Reg. 57430-36 (Nov. 6, 2009).
[32] U.S. Environmental Protection Agency Office of Pollution Prevention and Toxics, *Control of Nanoscale Materials under the Toxic Substances Control Act*, available at http://www.epa.gov/oppt/nano/#existingmaterials (accessed on June 26, 2010).
[33] *Id.*

New Jersey subsidiary, Swan Chemicals, Inc., is permitted to supply "small quantities" of its multiwalled nanotube for use in the United States. More recently, in March 2010, Swan Chemicals followed the terms of its consent decree and issued a Notice of Commencement of Manufacture and Import (NOCMI) for the same multiwalled nanotube. The NOCMI is required in order for Swan Chemicals to produce the nanotubes in commercial quantities. Because Swan Chemicals believes it is now providing its nanotubes in large enough quantities, the NOCMI was the next appropriate step for the company. However, with the NOCMI comes the requirement for Swan Chemicals to adhere to all terms of the consent decree for the safe and environmentally responsible production and handling of the multiwalled nanotubes. A sanitized version of the consent order is publically available.[34] Expect other companies to follow suit with additional PMNs for additional nanomaterials.

15.2.2 *Worker Safety and Nanomaterials*

Like EPA, the United States' worker protection agencies have been focusing on the health and safety impacts of nanomaterials. Although more focused on worker protection and occupational health impacts, agencies such as the Occupational Health and Safety Administration (OSHA) and the National Institute for Occupational Safety and Health have become more active in the field of nanotechnology.

15.2.2.1 Nanotechnology and the occupational health and safety administration

Part of the Department of Labor, OSHA is tasked with the protection of workers in all contexts. OSHA's mandate extends to not only the factory floor and shipping centers, but also to office and nontraditional working environments. The heart of OSHA's authority comes

---

[34]*In re [Redacted]*, Premanufacture Notice No. P-08-0177, available at http://nanotech.law.asu.edu/Documents/2010/01/EPA-HQ-OPPT-2008-0252-0022%20MWCNT%20AOC_436_9286.pdf (accessed on June 26, 2010).

from the "General Duties Clause" of the Occupational Safety and Health Act, Section 5(a)(1), which requires employers to "furnish to each of his employees employment and a place of employment which are [*sic*] free from recognized hazards that are causing or are likely to cause death or serious physical harm to his employees."[35] Additionally, Section 5(a)(2) requires employers to "comply with occupational safety and health standards" promulgated by OSHA.[36] Because nanomaterials are used increasingly in a wide range of commercial products, OSHA has begun the process of studying the worker health impacts.

OSHA determined that "Employees who use nanomaterials in research or production processes may be exposed to nanoparticles through inhalation, dermal contact, or ingestion, depending upon how employees use and handle them."[37] For that reason, OSHA has regulatory authority concerning nanomaterial use in worker protection contexts. However, OSHA has determined that the risks associated with nanomaterials are not yet fully understood. "Although the potential health effects of such exposure are not fully understood at this time, scientific studies indicate that at least some of these materials are biologically active, may readily penetrate intact human skin, and have produced toxicologic reactions in the lungs of exposed experimental animals."[38] Accordingly, OSHA has not yet set nanoscale specific regulations. OSHA has identified the specific areas of authority may become applicable to nanomaterial production and use in the future, including regulatory standards related to occupational injury and illness, personal protective equipment, eye and face protection, respiratory protection, hand protection, hazard communication, sanitation, and occupational exposure in laboratory settings.[39] Furthermore, OSHA leaves open

---

[35] 29 U.S.C. §654.

[36] *Id.*

[37] U.S. Department of Labor, Occupational Safety and Health Administration, *Health Effects and Workplace Assessments and Controls,* available at http://www. osha.gov/dsg/nanotechnology/nanotech_healtheffects.html (accessed on June 26, 2010).

[38] *Id.*

[39] U.S. Department of Labor, Occupational Safety and Health Administration, *Nanotechnology OSHA Standards,* available at http://www.osha. gov/dsg/nanotechnology/nanotech_standards.html (accessed on June 26, 2010).

the possibility that certain chemical-specific health standards it has set may need revision to address nanoscale counterparts. For example, OSHA set occupational exposure limits for specific chemicals such as cadmium and chromium.[40] Because the impacts from nanoscale counterparts may be completely different from the hazards associated with the current uses, OSHA may require additional protections associated with specific materials. This gives OSHA a second route for nanomaterial regulation: the General Duties Clause, but also chemical specific safety and health standards. However, before acting to regulate nanomaterials, OSHA would prefer to understand the risks associated with occupational exposures. For that reason, OSHA is relying upon the National Institute for Occupational Safety and Health to evaluate and determine occupational health risks associated with nanomaterial exposures.

15.2.2.2 National Institute for occupational safety and health

The National Institute for Occupational Safety and Health (NIOSH) is tasked with gathering "new knowledge in the field of occupational safety and health and to transfer that knowledge into practice for the betterment of workers."[41] While not a regulatory agency per se, NIOSH supports the regulatory mission of health and human services agencies by conducting "scientific research, develop[ing] guidance and authoritative recommendations, disseminat[ing] information, and respond[ing] to requests for workplace health hazard evaluations."[42] NIOSH is the research and investigative arm of occupational safety regulators, including OSHA and others. For that reason, NIOSH is at the forefront of research concerning occupational health exposure and impacts related to nanotechnology.[43]

---

[40] *Id.*; 29 C.F.R. 1910.1027; 29 C.F.R. 1910.1026.

[41] Centers for Disease Control and Prevention, National Institute for Occupational Safety and Health, *About NIOSH*, available at http://www.cdc.gov/niosh/about.html (accessed on June 26, 2010).

[42] *Id.*

[43] Centers for Disease Control and Prevention, National Institute for Occupational Safety and Health, *NIOSH Safety and Heath Topic: Nanotechnology*, available at http://www.cdc.gov/niosh/topics/nanotech/ (accessed on June 26, 2010).

NIOSH's research focuses on three questions regarding nanomaterial exposure: (1) What are the pathways of nanomaterial exposure for workers in the field? (2) What are the interactions between nanomaterials and the human system? and (3) What are the impacts of nanomaterials on humans?[44] From there, NIOSH is beginning to develop an understanding of nanomaterial impacts and is now releasing study results and recommendations associated with worker protection.

NIOSH's plan regarding nanomaterial research and study has been released and revised multiple times over the last several years. Its most recent version, "Strategic Plan for NIOSH Nanotechnology Research and Guidance: Filling the Knowledge Gaps," was issued in November 2009. In short, the Strategic Plan states that NIOSH will focus on six goals during its 2009–2012 research term:

1. Conduct toxicological research on nanoparticles likely to be commercially available.
2. Conduct research to identify long-term health effects of carbon nanotubes (CNT).
3. Develop recommendations for controlling occupational exposure to fine and ultrafine titanium dioxide (TiO_2) including development of recommended exposure limits (RELs). Conduct research on improving sampling and analytical methods, determining the extent of workplace exposures, and controlling airborne exposures below the REL. Identify what medical surveillance is appropriate. Consider to what extent the observed relationship between TiO_2 particle size and toxicity can be generalized to other metal oxides.
4. Develop recommendations for controlling occupational exposures to purified and unpurified single-walled carbon nanotubes (SWCNT) and multiwalled carbon nanotubes (MWCNT) including development of RELs. Conduct research to address gaps in information on sampling, analysis, exposure assessment, instrumentation, and controls. Identify what medical surveillance or epidemiological studies are appropriate.

---

[44] *Id.*

5. Conduct research on how to identify categories of nanoparticles that can be distinguished based on similar physico-chemical properties. Conduct research to develop RELs and ultimately recommended exposure standards for these categories.

6. Conduct research on explosion potential of various nanoparticles.[45]

To that end, NIOSH released several publications for the purpose of education and recommending processes for worker safety.[46] This list serves as a good reference for comprehensive treatment of nanomaterials and workplace safety issues. While NIOSH does not set enforceable exposure limits or worker protection practices, it is actively recommending prudent measures to accomplish both. Most notably, NIOSH published its *Approaches to Safe Nanotechnology: Managing the Health and Safety Concerning Associated with Engineered Nanoparticles*, in March 2009.[47] NIOSH's *Approaches* include the following recommendations: minimizing worker exposure to nanomaterials, limiting airborne exposure in much the same way as for aerosols, and implementation of a workplace risk management program that includes hazard evaluation, engineering control evaluation, exposure assessments, educational programs, and personal protective equipment evaluation. Additionally, NIOSH recommends evaluating and testing engineering controls and personal protective equipment for specific exposure and release scenarios, such as the effectiveness of HEPA filters to control airborne exposure.[48] As NIOSH continues to evaluate worker exposure to nanomaterials and the accompanying impacts, it will likely release to those results as they become available. NIOSH seems committed to understanding the unique workplace concerns that nanomaterials raise.

---

[45] Centers for Disease Control and Prevention, National Institute for Occupational Safety and Health, (2009) *Strategic Plan for NIOSH Nanotechnology Research and Guidance: Filling the Knowledge Gaps*, Washington, D.C.

[46] Centers for Disease Control and Prevention, National Institute for Occupational Safety and Health, *NIOSH Safety and Heath Topic: Nanotechnology, Occupational Safety and Health Practitioners*, available at http://www.cdc.gov/niosh/topics/nanotech/professionals.html (accessed on June 26, 2010).

[47] Centers for Disease Control and Prevention, National Institute for Occupational Safety and Health (2009), *Approaches to Safe Nanotechnology: Managing the Health and Safety Concerns Associated with Engineered Nanomaterials*, Washington, D.C.

[48] *Id.* at vii-viii.

15.2.3 *Food and Drug Regulation of Nanomaterials*

The Food and Drug Administration (FDA) maintains primary oversight of and control for "protecting the public health by assuring the safety, effectiveness, and security of human and veterinary drugs, vaccines and other biological products, medical devices, our nation's food supply, cosmetics, dietary supplements, and products that give off radiation."[49] With the increasing use of nanomaterials in food packaging and cosmetics, to name just two uses, the FDA is concerning itself more with the impacts and potential regulation of nanotechnology.

In 2006, the FDA formed its Nanotechnology Task Force, charged "with determining regulatory approaches that encourage the continued development of innovative, safe, and effective FDA-regulated products that use nanotechnology materials."[50] In July 2007, the Task Force released its report concerning the use and regulation of nanomaterials in food, drug, and cosmetic products.[51] As stated in the Report, "A general finding of the report is that nanoscale materials present regulatory challenges similar to those posed by products using other emerging technologies."[52] The Report recommended continued education and research concerning nanomaterials in regulated products, and determined that the FDA has the general authority under current legislative mandates to regulate nanomaterials under its Pre-Market Authorization authority. The FDA does not see nanomaterials as posing any unique concerns beyond other new technologies the Agency is asked to address and consider.[53] For that reason, it is unlikely that FDA will treat nanomaterials significantly different from other new forms of technology it encounters. The Task Force did, however,

[49] U.S. Department of Health and Human Services, Food and Drug Administration, *About the FDA: FDA Fundamentals*, available at http://www.fda.gov/AboutFDA/Basics/ucm192695.htm (accessed on June 26, 2010).
[50] U.S. Department of Health and Human Services Food and Drug Administration, *Science and Research, Nanotechnology Task Force: About the Task Force* http://www.fda.gov/ScienceResearch/SpecialTopics/Nanotechnology/NanotechnologyTaskForce/default.htm (accessed on June 26, 2010).
[51] U.S. Department of Health and Human Services, Food and Drug Administration (2007), *Nanotechnology Task Force Report 2007*, Washington, D.C.
[52] *Id.* at Executive Summary.
[53] *Id.* at 20.

make several recommendations regarding policy and regulatory issues. Many of the Task Force's recommendations surrounded data requests, labeling, and guidance to help industry and the agency simultaneously understand the concerns and opportunities regarding nanomaterials in food and drug applications.[54]

In the time since the Task Force report, the FDA has continued to evaluate and study the issue of nanomaterials in food, drugs, and cosmetic products. While no guidance or other regulatory statements have been released, the FDA established a "NanoTechnology Interest Group" (NTIG). The NTIG established multiple working groups that consist of multidisciplinary interests. "The goal of these FDA working groups is to share information about nanotechnology and to provide a level of coordination of review for the various product types."[55] The working groups are to identify not only the regulatory hurdles that are presented with nanomaterials in regulated products, but also to recommend and discuss solutions to those challenges. FDA intends to release additional information as it becomes available.

Most recently, FDA Center for Drug Evaluation and Research (CDER) Office of Pharmaceutical Science (OPS) released its Manual of Policies and Procedures (MAPP) for nanomaterials. The *Reporting Format for Nanotechnology-Related Information in CMC Review* issued on June 3, 2010, "provides chemistry, manufacturing, and controls (CMC) reviewers within the [OPS] with the framework by which relevant information about nanomaterial-containing drugs will now be captured in CMC reviews of current and future CDER drug application submissions. This information will be entered into a nanotechnology database under construction and ultimately be used to develop policy regarding these products."[56] The MAPP

---

[54] *Id.* at 20–36.

[55] U.S. Food and Drug Administration, Science and Research, *Nanotechnology Frequently Asked Questions*, available at http://www.fda.gov/Science Research/SpecialTopics/Nanotechnology/FrequentlyAskedQuestions/default.htm (accessed on June 26, 2010), at #13.

[56] Center for Drug Evaluation and Research, Office of Pharmaceutical Science (2010), *Reporting Format for Nanotechnology-Related Information in CMC Review*, available at http://www.fda.gov/downloads/AboutFDA/CentersOffices/CDER/Manual ofPoliciesProcedures/UCM214304.pdf (accessed on June 26, 2010).

provides for a list of information that CMC reviewers should document in the CDER's nanomaterial database for use in future work.

15.3 Insurance Concerns with Nanomaterials

In recent years, insurance companies have begun addressing nano-materials in varying levels. While not formal regulation, insurance companies are now starting to police their clients in an effort to define risk and loss scenarios incorporating nanomaterials. Several of the global companies, such as Lloyds of London and Swiss re, have released statements or short papers on nanomaterial-related risk and coverage. Given the increasing use of nanomaterials in commercial products, the unknowns concerning risk to humans and the environment, coupled with the lack of regulatory controls, insurance companies are understandably cautious concerning nanomaterial-related coverage.

Lloyd's of London was one of the first major insurance carriers to release statements concerning nanomaterials. In late 2007, Lloyd's recognized that nanomaterials in commercial products may create a set of risks not seen previously.[57] Consequently, Lloyd's set out to both begin to understand the risks posed by nanomaterials and educate potential clients as to the risks and possible impacts for insurance coverage.[58] Lloyd's identified the following types of insurance that may be needed for nanomaterials:

- *Professional indemnity/medical malpractice* — protects against being sued for giving incorrect advice or negligence

---

[57] Lloyd's of London (2007), *Nanotechnology-What are the risks?* Available at http://www.lloyds.com/News-and-Insight/News-and-Features/Archive/2007/11/Nanotechnology_what_are_the_risks_23112007 (last visited June 26, 2010); *see also* Baxter, David, *Nanotechnology: An Insurers' Perspective*, available at http://www.safenano.org/Uploads/Features/SAFENANO_Nanotechnology Insurance.pdf (accessed on June 26, 2010).

[58] Lloyd's of London, *Nanotechnology: The Big Questions*, available at http://www.lloyds.com/NR/rdonlyres/A3674547-BF42-42B2-ACF7-9DC17A574BD9/0/Nanotechnology_the_big_questions.pdf.

- *Directors and officers* — protects business leaders against being sued by shareholders for unsatisfactory management or negligence leading to loss of company value
- *General liability* — covers other liabilities the company may be exposed to
- *Employers' liability* — provides compensation to employees for personal injuries as a result of employment
- *Product liability* — covers cost of recall and claims made by consumers[59]

At the same time, Lloyd's recognized the potential safety and remediation potential associated with nanomaterials. In this way, nanomaterials could lessen the risk of loss faced by insurance companies. These concerns resulted in Lloyd's collaborating with the Lighthill Risk Network to host a half-day seminar concerning the risks and questions with nanotechnology.[60] Lloyd's has not announced any formal decisions concerning insurance coverage and nanomaterials.

Similarly, the global insurance provider, Swiss Re, has begun studying the risks and questions associated with nanomaterials. Like Lloyd's, Swiss Re is continuing to study and monitor the development of both the technology and regulation associated with nanomaterials.[61] To that end, Swiss Re released *Nanotechnology, Small Matter Many Unknowns*."[62] A portion of the report attempts to compare the risks posed by carbon nanotubes to the risks

---

[59] Baxter, D. *Nanotechnology: An Insurers' Perspective*, available at http://www.safenano.org/Uploads/Features/SAFENANO_NanotechnologyInsurance.pdf (accessed on June 26, 2010).

[60] The Lighthill Risk Network (2007), *Nanotechnology Seminar*, available at http:// www.lighthillrisknetwork.org/Nanotechnology/nanotechnology_seminar. html (accessed on June 26, 2010); see also, Lighthill Risk Network (2007), *Risks and Opportunities of Nanotechnology*, available at http://www.lighthillris knetwork.org/events/Risks%20and%20Opportunities%20of%20Nanotechnology. pdf (accessed on June 26, 2010).

[61] Swiss Re (2004), *Swiss Re investigates the opportunities and risks of nanotechnology from an insurance perspective*, available at http://www.swissre.com/media/media_ information/swiss_re_investigates_the_opportunities_and_risks_of_nanotechnology _from_an_insurance_perspective_html (accessed on June 26, 2010).

[62] Swiss, Re (2004), *Nanotechnology, Small Matter Many Unknowns*, available at http://media.swissre.com/documents/nanotechnology_small_matter_many_ unk- nowns_en.pdf (accessed on June 26, 2010).

encountered with asbestos.[63] However, Swiss Re does not draw a final conclusion about the comparison. The report concludes with a discussion concerning the risks posed to insurance companies, including related issues such as risk communication, and a review of the precautionary principle, which seems to be the emerging method for addressing nanomaterial risks.[64]

Finally, although no insurance company has taken a final stance with regard to nanomaterial risk and coverage, one company did make a brief statement excluding nanomaterials from coverage. In 2008, the Continental Western Insurance Company released a statement excluding nanomaterials from coverage.[65] While there was some confusion over the ongoing validity of the exclusion, some thought that Continental recalled the notice, given the presence of the exclusion on Continental's website, it can be presumed the exclusion stands. Continental's exclusion opened with, "This insurance does not apply to: 'Bodily injury', 'property damage', or 'personal and advertising injury' related to the actual, alleged, or threatened presence of or exposure to 'nanotubes' or 'nanotech-nology' in any form, or to harmful substances emanating from 'nanotubes' or 'nanotechnology'."[66] The exclusion went on to define both "nanotubes" and "nanotechnology," and exclude coverage from injury related almost any form of exposure to nanomaterials.[67] To date, this appears to be the only blanket coverage exclusion released by an insurance company.

Conversely, one insurance company recently released the first nanomaterial policy. Lexington Insurance Company introduced, LexNanoShield as the first "integrated liability coverage and an array of risk management services to help those insureds man-ufacturing, distributing, or using nanoparticles or nanomaterials

---

[63] *Id.* at 42.

[64] *Id.* at 46–49.

[65] Continental Western Group, *Notice to Policyholders,* available at http://cwgins.com/mike/documents/PN61610708.pdf (accessed on June 26, 2010); *see also* Continental Western Group, *Nanotubes and Nanotechnology Exclusion,* available at http://cwgins.com/mike/documents/CW33690608Nanotubes Exclusion.pdf (accessed on June 26, 2010).

[66] *Id.*

[67] *Id.*

assess and manage their products liability exposures."[68] Aimed at small and medium manufacturers and distributors of nano-enhanced products, Lexington Insurance provides coverage for general liability, products liability, pollution liability, and product recall liability.[69] While only released in March 2010, some of the novel attributes of the coverage include prepaid time with legal counsel, prepaid time with technical consultants, and hazard mitigation consults.[70] While it is not clear how many customers have taken advantage of this coverage, if any, it appears this is the first attempt at comprehensive loss prevention associated specifically with nanotechnology products.

Despite the actions taken by insurance companies regarding nanomaterials, few have taken concrete action with regard to the technology. However, as more information is gained, both in terms of risks and benefits, expect insurance companies to take a more active role with regard to nanomaterials. Especially in an atmosphere of no regulation, insurance companies will likely begin to take proactive steps to protect companies and stockholders from unforeseen losses resulting from the increased use and proliferation of nanomaterials.

15.4 State and Local Regulation of Nanomaterials

Like the federal counterparts, state and local governments have been hesitant to enact laws or ordinances aimed specifically at nanomaterials. While several states have taken legislative action concerning nanotechnology, the efforts have focused on defining the scope and limit of the technology for either future regulatory efforts or development funding programs. To date, Michigan, Arkansas, and Oklahoma have all passed statutes that define nanotechnology in those states. However, all of the definitions are slightly different resulting in no uniform definition between the states.

---

[68]Lexington Insurance Chartis, *LexNanoShield(SM)*, available at http:// www. lexingtoninsurance.com/documents/lexHSLexNanoShield.pdf (accessed on June 26, 2010).

[69]*Id.*

[70]*Id.*

Arkansas currently defines "nanotechnology" as "the materials and systems whose structures and components exhibit novel and significantly improved physical, chemical, and biological properties, phenomena, and processes due to their nanoscale size."[71] Michigan defines "nanotechnology" as "materials, devices, or systems at the atomic, molecular, or macromolecular level, with a scale measured in nanometers" for "strategic funding" purposes.[72] Finally, Oklahoma defines "nanotechnology" as "technology development at the molecular range (1 nm to 100 nm) to create and use structures, devices, and systems that have novel properties because of their small size" for science and technology research and development purposes.[73]

Each definition is slightly different, which could cause the same material to receive different regulatory treatment in the different states. Especially as more states move to statutorily control nanomaterials, these slight differences could result in forum shopping and business location issues for manufacturers, distributors, and nanomaterial users.

Most recently, the State of California released a draft policy addressing nanomaterials.[74] "A Nanotechnology Policy Framework: Policy Recommendations for Addressing Potential Health Risks from Nanomaterials in California" was released in draft form for comments in April 2010, and contains a review of the issues raised by nanomaterials as well as recommendations for several areas including sources, exposure, health effects, product safety, and public engagement. The report is focused on the health concerns and impacts associated with the growing nanomaterial-based industry in California. The report takes specific aim at the risk assessment issues concerning nanomaterials and is meant to inform risk managers and other decision makers concerning the use and exposure to nanomaterials. The report also includes a summary of

---

[71] A.C.A §15-4-2103(5).
[72] M.C.L.A. 206.30-125.2088a.
[73] 74 Okl. St. Ann. §5060.4(14).
[74] Wise, Amber R., Jackie Schwartz, and Tracey J. Woodruff (2010), *A Nanotechnology Policy Framework for California: Policy recommendations for Addressing Potential Health Risks From Nanomaterials,* University of California San Francisco Program on Reproductive Health and Environment.

toxicological and health impact data studies to date as well as short review of the state of regulation. Comments on the draft report are due in early May 2010, and a final report will be following a public meeting to discuss the report.[75]

In looking at the local treatment of nanomaterials, very few local governments have moved to regulate nanotechnology. Only Berkeley, California, and to a lesser extent, Cambridge, Massachusetts have taken formal action concerning nanotechnology. Berkeley was the first to require nanomaterial reporting by manufacturers or users within Berkeley city limits in 2006.[76] These sections require those facilities that manufacture or use nanomaterials to list such materials in the required hazardous materials disclosures made to the city for emergency response purposes.

Then, in 2008, the City of Cambridge, Massachusetts followed suit by adopting a report concerning the use and concerns about nanomaterials to the City Manager.[77] While not a municipal code like Berkeley, the Cambridge report was adopted by the city council and includes recommendations for tracking and inventorying nanomaterials within the city limits. The report did not recommend the adoption of nanomaterial ordinances at that time.[78] The city is continuing to monitor nanomaterial developments in the city over time.

Neither city is expected to release additional regulatory statements on nanomaterials, but both are continuing to evaluate new information as it becomes available. In addition, it is likely other local regulators, be it state or municipal will continue to evaluate and possibly look to their own forms of regulation in the near term. It will be crucial for those working in the industry to keep abreast of local regulation and study of nanomaterials.

---

[75] See additional information and updates at University of California San Francisco, *A Nanotechnology Policy Framework: Policy Recommendations for Addressing Potential Health Risks from Nanomaterials in California*, available at http://prhe.ucsf.edu/prhe/nanomaterialsreport.html (accessed on June 26, 2010).

[76] Berkeley Municipal Code Section 15.12.040(I); 15.12.050(7).

[77] City of Cambridge, Massachusetts (2008), *Recommendations for a Municipal Health & Safety Policy for Nanomaterials: A Report to the Cambridge City Manager*, Cambridge, MA.

[78] *Id.* at 10.

15.5 Conclusion

All levels of government in the United States are researching or considering some form of nanomaterial regulation. All federal, state, and local agencies are moving deliberately concerning specific requirements or prohibitions, all are making progress toward a regulatory decision. Further, most governmental actors seem to be waiting for a counterpart to make a significant decision in order to learn from other's experiences. To that end, US EPA seems to be the leading candidate for short-term nanomaterial regulation through TSCA. However, it is impossible to predict when the next significant regulatory step will be taken by a government in the United States. For that, diligence and stakeholder involvement will be needed to stay up-to-date on the latest developments.

Bibliography

1. 15 U.S.C. §2601 *et seq.*
2. 29 U.S.C. §654.
3. 29 C.F.R. §1910.1027.
4. 29 C.F.R. §1910.1026.
5. 73 Fed. Reg. 64946–64947 (Oct. 31, 2008).
6. 74 Fed. Reg. 57430–36 (Nov. 6, 2009).
7. 74 Okl. St. Ann. §5060.4(14).
8. A.C.A §15-4-2103(5).
9. America COMPETES Reauthorization Act of 2010, H.R.5116, available at: http://www.thomas.gov/cgi-bin/query/z?c111:H.R.5116.EH: (last visited June 26, 2010).
10. American Bar Association, Section of Environment, Energy, and Resources, *Section Nanotechnology Project*, available at: http://www.abanet.org/environ/nanotech/ (last visited June 26, 2010).
11. Baxter, David, *Nanotechnology: An Insurers' Perspective*, available at: http://www.safenano.org/Uploads/Features/SAFENANO_Nanotechnology Insurance.pdf (last visited June 26, 2010).
12. Berkeley Municipal Code Section 15.12.040(I); 15.12.050(7).

13. Center for Drug Evaluation and Research, Office of Pharmaceutical Science (2010), *Reporting Format for Nanotechnology-Related Information in CMC Review*, available at: http://www. fda.gov/downloads/AboutFDA/CentersOffices/CDER/ManualofPolicies Procedures/UCM214304.pdf (last visited June 26, 2010).

14. Centers for Disease Control and Prevention, National Institute for Occupational Safety and Health, *About NIOSH*, available at: http://www.cdc.gov/niosh/about.html (last visited June 26, 2010).

15. Centers for Disease Control and Prevention, National Institute for Occupational Safety and Health, *NIOSH Safety and Heath Topic: Nanotechnology*, available at: http://www.cdc.gov/niosh/topics/nanotech/ (last visited June 26, 2010).

16. Centers for Disease Control and Prevention, National Institute for Occupational Safety and Health, (2009) *Strategic Plan for NIOSH Nanotechnology Research and Guidance: Filling the Knowledge Gaps*, Washington, D.C.

17. Centers for Disease Control and Prevention, National Institute for Occupational Safety and Health, *NIOSH Safety and Heath Topic: Nanotechnology, Occupational Safety and Health Practitioners*, available at: http://www.cdc.gov/niosh/topics/nanotech/professionals.html (last visited June 26, 2010).

18. Centers for Disease Control and Prevention, National Institute for Occupational Safety and Health (2009), *Approaches to Safe Nanotechnology: Managing the Health and Safety Concerns Associated with Engineered Nanomaterials*, Washington, D.C.

19. City of Cambridge, Massachusetts (2008), *Recommendations for a Municipal Health & Safety Policy for Nanomaterials: A Report to the Cambridge City Manager*, Cambridge, MA.

20. Continental Western Group, *Notice to Policyholders*, available at: http://cwgins.com/mike/documents/PN61610708.pdf (last visited June 26, 2010).

21. Continental Western Group, *Nanotubes and Nanotechnology Exclusion*, available at: http://cwgins.com/mike/documents/CW33690608 NanotubesExclusion.pdf (last visited June 26, 2010).

22. *In re [Redacted]*, Premanufacture Notice No. P-08-0177, available at: http://nanotech.law.asu.edu/Documents/2010/01/EPA-HQ-OPPT-2008-0252-0022%20MWCNT%20AOC_436_9286.pdf (last visited June 26, 2010).

23. Jordan, William (April 29, 2010), *Nanotechnology and Pesticides,* presented at the Pesticide Program Dialogue Committee, at page 5 (available at: http://www.nanotechproject.org/process/assets/files/8309/epa_newpolicy_nanomaterials.pdf) (last visited June 6, 2010).

24. Lexington Insurance Chartis, *LexNanoShield(SM),* available at: http://www.lexingtoninsurance.com/documents/lexHSLexNanoShield.pdf (last visited June 26, 2010).

25. The Lighthill Risk Network (2007), *Nanotechnology Seminar,* available at: http://www.lighthillrisknetwork.org/Nanotechnology/nano-technology_seminar.html (last visited June 26, 2010).

26. The Lighthill Risk Network (2007), *Risks and Opportunities of Nanotechnology,* available at: http://www.lighthillrisknetwork.org/events/Risks%20and%20Opportunities%20of%20Nanotechnology.pdf (last visited June 26, 2010).

27. Lloyd's f London, *Nanotechnology: The Big Questions,* available at: http://www.lloyds.com/NR/rdonlyres/A3674547-BF46-42B2-ACF7-9DC17A574BD9/0/Nanotechnology_the_big_questions.pdf

28. Lloyd's of London (2007), *Nanotechnology-What are the risks?* Available at: http://www.lloyds.com/News-and-Insight/News-and-Features/Archive/2007/11/Nanotechnology_what_are_the_risks_23112007 (last visited June 26, 2010).

29. M.C.L.A. 206.30-125.2088a.

30. Nanotechnology Advancement and New Opportunities Act, H.R. 820, available at: http://thomas.loc.gov/cgi-bin/bdquery/z?d111:HR00820:@@@L&summ2=m& (last visited June 26, 2010).

31. National Nanotechnology Initiative, *About the NNI-Home,* available at: http://www.nano.gov/html/about/home_about.html (last visited June 26, 2010).

32. Nanotechnology Safety Act of 2010, S. 2942, available at: http://thomas.loc.gov/cgi-bin/query/z?c111:S.2942: (last visited June 26, 2010).

33. Office of Science and Technology Policy Executive Office of the President (2010), "Independent Review Finds Federal Nanotechnology Initiative Highly Effective; Recommends Changes to Ensure Ongoing U.S. Dominance," Washington, D.C., available at: http://www.whitehouse.gov/sites/default/files/microsites/ostp/nano-release.pdf (last visited June 26, 2010).

34. President's Council of Advisors on Science and Technology (2010), *Report to the President and Congress on the Third Assessment of the National Nanotechnology Initiative*, Washington, D.C., at *x*.

35. Promote Nanotechnology in Schools Act, S. 3117, available at: http://thomas.loc.gov/cgi-bin/bdquery/D?d111:1:./temp/~bdZtWs: @@@L&summ2=m&|/home/LegislativeData.php| (last visited June 26, 2010).

36. Subcommittee on Nanoscale Science, Engineering, and Technology, Committee on Technology, National Science and Technology Counsel (2009), *National Nanotechnology Initiative, Research and Development Leading to a Revolution In Technology and Industry, Supplement to the President's FY 2010 Budget*, Executive Office of the President, Office of Science and Technology Policy, Washington, D.C., available at: http://nano.gov/NNI_2010_budget_supplement.pdf at 1 (last visited June 26, 2010) (hereinafter "NNI Budget").

37. Swiss Re (2004), *Nanotechnology, Small Matter Many Unknowns* available at: http://media.swissre.com/documents/nanotechnology_small_matter_many_unknowns_en.pdf (last visited June 26, 2010).

38. Swiss Re (2004), *Swiss Re investigates the opportunities and risks of nanotechnology from an insurance perspective*, available at: http://www.swissre.com/media/media_information/swiss_re_investigates_the_opportunities_and_risks_of_nanotechnology_from_an_insurance_perspective_.html (last visited June 26, 2010).

39. U.S. Department of Health and Human Services, Food and Drug Administration, *About the FDA: FDA Fundamentals*, available at: http://www.fda.gov/AboutFDA/Basics/ucm192695.htm (last visited June 26, 2010).

40. U.S. Department of Health and Human Services Food and Drug Administration, *Science and Research, Nanotechnology Task Force: About the Task Force* http://www.fda.gov/ScienceResearch/SpecialTopics/Nanotechnology/NanotechnologyTaskForce/default.htm (last visited June 26, 2010).

41. U.S. Department of Health and Human Services, Food and Drug Administration (2007), *Nanotechnology Task Force Report 2007*, Washington, D.C.

42. U.S. Department of Labor, Occupational Safety and Health Administration, *Health Effects and Workplace Assessments and Controls,* available at: http://www.osha.gov/dsg/nanotechnology/nanotech_healtheffects.html (last visited June 26, 2010).

43. U.S. Department of Labor, Occupational Safety and Health Administration, *Nanotechnology OSHA Standards,* available at: http://www.osha.gov/dsg/nanotechnology/nanotech_standards.html (last visited June 26, 2010).

44. U.S. Environmental Protection Agency, *Summary of the Toxic Substances Control Act,* available at: http://www.epa.gov/lawsregs/laws/tsca.html (last visited June 26, 2010).

45. U.S. Environmental Protection Agency, Office of Pollution Prevention and Toxics, *Nanoscale Materials Stewardship Program,* available at: http://www.epa.gov/oppt/nano/stewardship.html (last visited June 26, 2010).

46. U.S. Environmental Protection Agency Office of Pollution Prevention and Toxics (2009), *Nanoscale Materials Steward Program Interim Report,* Washington, D.C., (hereinafter *"Interim Report"*) available at: http://www.epa.gov/oppt/nano/nmsp-interim-report-final.pdf (last visited June 26, 2010), at 3.

47. U.S. Environmental Protection Agency Office of Pollution Prevention and Toxics, *Control of Nanoscale Materials under the Toxic Substances Control Act,* available at: http://www.epa.gov/oppt/nano/#existingmaterials (last visited June 26, 2010).

48. U.S. Food and Drug Administration, Science and Research, *Nanotechnology Frequently Asked Questions,* available at: http://www.fda.gov/ScienceResearch/SpecialTopics/Nanotechnology/FrequentlyAskedQuestions/default.htm (last visited June 26, 2010), at #13.

49. Wise, Amber R., Jackie Schwartz, and Tracey J. Woodruff (2010), A Nanotechnology Policy Framework for California: Policy recommendations for Addressing Potential Health Risks From Nanomaterials, University of California San Francisco Program on Reproductive Health and Environment.

50. 15 U.S.C. §2601 *et seq.*

51. 29 U.S.C. §654.

52. 29 C.F.R. §1910.1027.

53. 29 C.F.R. §1910.1026.

54. 73 Fed. Reg. 64946–64947 (Oct. 31, 2008).

55. 74 Fed. Reg. 57430–36 (Nov. 6, 2009).

56. 74 Okl. St. Ann. §5060.4(14).

57. A.C.A §15-4-2103(5).

58. America COMPETES Reauthorization Act of 2010, H.R.5116, available at: http://www.thomas.gov/cgi-bin/query/z?c111:H.R.5116.EH: (last visited June 26, 2010).

59. American Bar Association, Section of Environment, Energy, and Resources, *Section Nanotechnology Project*, available at: http://www.abanet.org/environ/nanotech/ (last visited June 26, 2010).

60. Baxter, David, *Nanotechnology: An Insurers' Perspective*, available at: http://www.safenano.org/Uploads/Features/SAFENANO_Nanotechnology Insurance.pdf (last visited June 26, 2010).

61. Berkeley Municipal Code Section 15.12.040(I); 15.12.050(7).

62. Center for Drug Evaluation and Research, Office of Pharmaceutical Science (2010), *Reporting Format for Nanotechnology-Related Information in CMC Review*, available at: http://www.fda.gov/downloads/AboutFDA/CentersOffices/CDER/ManualofPolicies Procedures/UCM214304.pdf (last visited June 26, 2010).

63. Centers for Disease Control and Prevention, National Institute for Occupational Safety and Health, *About NIOSH*, available at: http://www.cdc.gov/niosh/about.html (last visited June 26, 2010).

64. Centers for Disease Control and Prevention, National Institute for Occupational Safety and Health, *NIOSH Safety and Heath Topic: Nanotechnology*, available at: http://www.cdc.gov/niosh/topics/nanotech/ (last visited June 26, 2010).

65. Centers for Disease Control and Prevention, National Institute for Occupational Safety and Health, (2009) *Strategic Plan for NIOSH Nanotechnology Research and Guidance: Filling the Knowledge Gaps*, Washington, D.C.

66. Centers for Disease Control and Prevention, National Institute for Occupational Safety and Health, *NIOSH Safety and Heath Topic: Nanotechnology, Occupational Safety and Health Practitioners*, available at: http://www.cdc.gov/niosh/topics/nanotech/professionals.html (last visited June 26, 2010).

67. Centers for Disease Control and Prevention, National Institute for Occupational Safety and Health (2009), *Approaches to Safe Nanotechnology: Managing the Health and Safety Concerns Associated with Engineered Nanomaterials*, Washington, D.C.

68. City of Cambridge, Massachusetts (2008), *Recommendations for a Municipal Health & Safety Policy for Nanomaterials: A Report to the Cambridge City Manager*, Cambridge, MA.

69. Continental Western Group, *Notice to Policyholders,* available at: http://cwgins.com/mike/documents/PN61610708.pdf (last visited June 26, 2010).

70. Continental Western Group, *Nanotubes and Nanotechnology Exclusion,* available at: http://cwgins.com/mike/documents/CW33690608 NanotubesExclusion.pdf (last visited June 26, 2010).

71. *In re [Redacted],* Premanufacture Notice No. P-08-0177, available at: http://nanotech.law.asu.edu/Documents/2010/01/EPA-HQ-OPPT-2008-0252-0022%20MWCNT%20AOC_436_9286.pdf (last visited June 26, 2010).

72. Jordan, William (April 29, 2010), *Nanotechnology and Pesticides,* presented at the Pesticide Program Dialogue Committee, at page 5 (available at: http://www.nanotechproject.org/process/assets/files/8309/epa_newpolicy_nanomaterials.pdf) (last visited June 6, 2010).

73. Lexington Insurance Chartis, *LexNanoShield(SM),* available at: http://www.lexingtoninsurance.com/documents/lexHSLexNanoShield.pdf (last visited June 26, 2010).

74. The Lighthill Risk Network (2007), *Nanotechnology Seminar,* available at: http://www.lighthillrisknetwork.org/Nanotechnology/nanotechnology_seminar.html (last visited June 26, 2010).

75. The Lighthill Risk Network (2007), *Risks and Opportunities of Nanotechnology,* available at: http://www.lighthillrisknetwork.org/events/Risks%20and%20Opportunities%20of%20Nanotechnology.pdf (last visited June 26, 2010).

76. Lloyd's of London, *Nanotechnology: The Big Questions,* available at: http://www.lloyds.com/NR/rdonlyres/A3674547-BF46-42B2-ACF7-9DC17A574BD9/0/Nanotechnology_the_big_questions.pdf

77. Lloyd's of London (2007), *Nanotechnology-What are the risks?* Available at: http://www.lloyds.com/News-and-Insight/News-and-Features/Archive/2007/11/Nanotechnology_what_are_the_risks_23112007 (last visited June 26, 2010).

78. M.C.L.A. 206.30-125.2088a.

79. Nanotechnology Advancement and New Opportunities Act, H.R. 820, available at: http://thomas.loc.gov/cgi-bin/bdquery/z?d111: HR00820:@@@L&summ2=m& (last visited June 26, 2010).

80. National Nanotechnology Initiative, *About the NNI-Home,* available at: http://www.nano.gov/html/about/home_about.html (last visited June 26, 2010).

81. Nanotechnology Safety Act of 2010, S. 2942, available at: http://thomas.loc.gov/cgi-bin/query/z?c111:S.2942: (last visited June 26, 2010).

82. Office of Science and Technology Policy Executive Office of the President (2010), "Independent Review Finds Federal Nanotechnology Initiative Highly Effective; Recommends Changes to Ensure Ongoing U.S. Dominance," Washington, D.C., available at: http://www.whitehouse.gov/sites/default/files/microsites/ostp/nano-release.pdf (last visited June 26, 2010).

83. President's Council of Advisors on Science and Technology (2010), *Report to the President and Congress on the Third Assessment of the National Nanotechnology Initiative*, Washington, D.C., at x.

84. Promote Nanotechnology in Schools Act, S. 3117, available at: http://thomas.loc.gov/cgi-bin/bdquery/D?d111:1:./temp/~bdZtWs: @@@L&summ2=m&|/home/LegislativeData.php| (last visited June 26, 2010).

85. Subcommittee on Nanoscale Science, Engineering, and Technology, Committee on Technology, National Science and Technology Counsel (2009), *National Nanotechnology Initiative, Research and Development Leading to a Revolution In Technology and Industry, Supplement to the President's FY 2010 Budget*, Executive Office of the President, Office of Science and Technology Policy, Washington, D.C., available at: http://nano.gov/NNI_2010_budget_supplement.pdf at 1 (last visited June 26, 2010) (hereinafter "NNI Budget").

86. Swiss Re (2004), *Nanotechnology, Small Matter Many Unknowns* available at: http://media.swissre.com/documents/nanotechnology_small_matter_many_unknowns_en.pdf (last visited June 26, 2010).

87. Swiss Re (2004), *Swiss Re investigates the opportunities and risks of nanotechnology from an insurance perspective*, available at: http://www.swissre.com/media/media_information/swiss_re_investigates_the_opportunities_and_risks_of_nanotechnology_from_an_insurance_perspective_.html (last visited June 26, 2010).

88. U.S. Department of Health and Human Services, Food and Drug Administration, *About the FDA: FDA Fundamentals*, available at: http://www.fda.gov/AboutFDA/Basics/ucm192695.htm (last visited June 26, 2010).

89. U.S. Department of Health and Human Services Food and Drug Administration, *Science and Research, Nanotechnology Task Force: About the Task Force* http://www.fda.gov/ScienceResearch/SpecialTopics/

Nanotechnology/NanotechnologyTaskForce/default.htm (last visited June 26, 2010).

90. U.S. Department of Health and Human Services, Food and Drug Administration (2007), *Nanotechnology Task Force Report 2007*, Washington, D.C.

91. U.S. Department of Labor, Occupational Safety and Health Administration, *Health Effects and Workplace Assessments and Controls*, available at: http://www.osha.gov/dsg/nanotechnology/nanotech_healtheffects.html (last visited June 26, 2010).

92. U.S. Department of Labor, Occupational Safety and Health Administration, *Nanotechnology OSHA Standards*, available at: http://www.osha.gov/dsg/nanotechnology/nanotech_standards.html (last visited June 26, 2010).

93. U.S. Environmental Protection Agency, *Summary of the Toxic Substances Control Act*, available at: http://www.epa.gov/lawsregs/laws/tsca.html (last visited June 26, 2010).

94. U.S. Environmental Protection Agency, Office of Pollution Prevention and Toxics, *Nanoscale Materials Stewardship Program*, available at: http://www.epa.gov/oppt/nano/stewardship.html (last visited June 26, 2010).

95. U.S. Environmental Protection Agency Office of Pollution Prevention and Toxics (2009), *Nanoscale Materials Steward Program Interim Report*, Washington, D.C., (hereinafter "*Interim Report*") available at: http://www.epa.gov/oppt/nano/nmsp-interim-report-final.pdf (last visited June 26, 2010), at 3.

96. U.S. Environmental Protection Agency Office of Pollution Prevention and Toxics, *Control of Nanoscale Materials under the Toxic Substances Control Act*, available at: http://www.epa.gov/oppt/nano/#existingmaterials (last visited June 26, 2010).

97. U.S. Food and Drug Administration, Science and Research, *Nanotechnology Frequently Asked Questions*, available at: http://www.fda.gov/ScienceResearch/SpecialTopics/Nanotechnology/FrequentlyAskedQuestions/default.htm (last visited June 26, 2010), at #13.

98. Wise, Amber R., Jackie Schwartz, and Tracey J. Woodruff (2010), A Nanotechnology Policy Framework for California: Policy recommendations for Addressing Potential Health Risks From Nanomaterials, University of California San Francisco Program on Reproductive Health and Environment.

Index